W0111255

METHODS IN THE QUANTUM THEORY OF MAGNETISM

METHODS IN THE QUANTUM THEORY OF MAGNETISM

Sergei Vladimirovich Tyablikov
Chairman, Department of Statistical Mechanics
V. A. Steklov Mathematical Institute
Academy of Sciences of the USSR, Moscow

Translated from Russian by
Albin Tybulewicz
Editor, *Physics Abstracts*
and *Current Papers in Physics*, London

With a Foreword by
Daniel C. Mattis
Belfer Graduate School
Yeshiva University, New York

Springer Science+Business Media, LLC 1967

Sergei Vladimirovich Tyablikov was born in 1921 near Moscow. He studied at the physics department of the Moscow State University, where he continued his postgraduate studies. In 1946 he obtained a Candidate's degree, and in 1954 was awarded his doctorate. Currently, Dr. Tyablikov is a Professor of Theoretical and Mathematical Physics.

Professor Tyablikov's interests range over theoretical solid state physics, the theory of polarons, and the quantum theories of magnetism. He has published about 70 papers and was the author, with Professor V. L. Bonch-Bruevich, of "Green's Function Method in Statistical Mechanics" (Fizmatgiz, Moscow, 1961). Since 1947, Professor Tyablikov has worked at the V. A. Steklov Mathematical Institute of the Academy of Sciences of the USSR, and is now chairman of the department of statistical mechanics.

The original Russian text was published by Nauka Press in Moscow in 1965.

СЕРГЕЙ ВЛАДИМИРОВИЧ ТЯБЛИКОВ

МЕТОДЫ КВАНТОВОЙ ТЕОРИИ МАГНЕТИЗМА

METODY KVANTOVOI TEORII MAGNETIZMA

METHODS IN THE QUANTUM THEORY OF MAGNETISM

Library of Congress Catalog Card Number 65-27345

ISBN 978-1-4899-7091-6 *ISBN 978-1-4899-7182-1 (eBook)*
DOI 10.1007/978-1-4899-7182-1

© 1967 Springer Science+Business Media New York
Originally published by Plenum Press in 1967
Softcover reprint of the hardcover 1st edition 1967

All rights reserved

No part of this publication may be reproduced in any form without written permission from the publisher

Foreword

Only rarely does one find translations from Russian of scientific material that read as if they had originally been written in English. Mr. Tybulewicz has achieved this objective and is to be commended for it.

The present monograph will appeal to the students of magnetism and others who appreciated the Green function methods previously introduced by Bonch-Bruevich and Tyablikov,* but who felt that the sections on magnetism in the earlier book were too compact. Much the same material is discussed here, but in a much more leisurely manner than heretofore. Tyablikov might well have entitled the present monograph "Theories of the Heisenberg Model," for he treats this model from many points of view: (a) ground state properties (ferromagnetic, antiferromagnetic, helical, etc.), (b) dynamics and quantum mechanics, and (c) thermodynamical and statistical properties. Concepts of spin-wave theory, molecular-field theory, ferromagnetic resonance, and of Bogolyubov's "method of approximate second-quantization" are developed in the present context well before the Green function theory is even introduced. Some recent developments, such as H. B. Callen's improved decoupling scheme, are described, and, in general, the text covers many of the interesting applications of the formalism to cooperative magnetism.

It is only fair to point out that two important areas of research have been omitted entirely, namely, the origins of the magnetic interaction (e.g., superexchange, Hund's rules, etc.), and

---

*V. L. Bonch-Bruevich and S. V. Tyablikov "Green's Function Method in Statistical Mechanics" North-Holland Publ. Co., Amsterdam, 1962.

the band theory of magnetism (i.e., the magnetism of iron, nickel, and similar transition metals). English-speaking readers can satisfy their curiosity as to where the Heisenberg Hamiltonian finds its ultimate justification, or about where it is not valid, in several texts or reviews* where these subjects are adequately covered. But it is only here, in Tyablikov's monograph, that one finds an authoritative statement of recent Soviet work in the mathematical theory of the Heisenberg model of magnetism.

New York D.C. Mattis
September 1966

*G.T. Rado and H. Suhl, Eds. "Magnetism" Academic Press, New York; in several volumes, 1963 and following.
D. Mattis "The Theory of Magnetism" Harper & Row, New York, 1965.

Preface to the American Edition

The theory of magnetic phenomena in solids is being rapidly developed by many investigators in a number of countries. Obviously, it is well nigh impossible to write a book in which the results are right up to date. The present book has a more modest aim – to acquaint the reader with some of the fundamental methods in the quantum theory of magnetism; it does not pretend to give a complete account of all the methods or the results obtained by these methods. It is hoped that the book will be found useful by English-language readers.

S. V. Tyablikov

May 1966

Preface

Magnetic properties are exhibited to a greater or lesser extent by all substances. Among solids, we can distinguish a group of magnetic substances whose magnetic properties are particularly pronounced: ferromagnets, ferrimagnets, and antiferromagnets. The present treatment deals with some of the problems arising in the theory of the magnetic phenomena in such substances. We shall call the theory of phenomena in strongly magnetic substances the theory of strong magnetism or, simply, the theory of magnetism, because the weakly magnetic substances (paramagnets and diamagnets) will not be considered. The emphasis is on the methods giving general solutions and on the features characteristic of magnetic problems. The microscopic approach is used throughout.

Considerable progress has been made in recent years in the investigation of the many-body problem. The progress is, to a considerable extent, due to the application of quantum-field methods to statistical mechanics. Consequently, there has also been progress in the quantum theory of magnetism. However, the results of many interesting investigations have not yet been presented systematically. This creates certain difficulties for those who are new to the theory of magnetism, with its problems and methods of solution.

In this book, the author attempts to present, from a single standpoint, some of the methods used in the quantum theory of magnetism. The primary aim is to present systematically the theoretical apparatus in a form which would allow the reader to use it in investigations of still unsolved problems. The book also includes some additional topics in order to save the reader the trouble of referring to other sources, some of which may not be very accessible. The treatment does not pretend to be a comprehensive review of all the results obtained in the quantum theory of magnetism. It simply presents one of the directions of development pursued recently both in the Soviet Union and abroad.

The first part of the book (Chaps. I–III) is methodological in its approach. It begins with a brief introduction, in which the basic topics and definitions of the physics of magnetic phenomena and the elements of the phenomenological theory are presented. The second quantization method is described, the problem of spin Hamiltonians in the theory of strong magnetism is dealt with, and the essentials of statistical mechanics are presented briefly.

In the second part (Chaps. IV–VIII), the approximate second quantization (spin-wave) method (Chaps. IV–V), the molecular field method, elements of the perturbation theory at high temperatures (Chap. VI), and the method of Green's quantum functions (Chaps. VII–VIII) are presented.

This organization of the material makes it possible to consider systematically the calculation methods for low and high temperatures (including the vicinity of the Curie point), as well as the interpolation methods covering the whole range of temperatures. The general theories are presented first and the necessary mathematical apparatus is developed; this is followed by applications. The applications of a method are demonstrated by a more or less detailed analysis of one or two very simple problems in the theory of magnetism. Other applications are reviewed briefly at the ends of the chapters dealing with the applications of the methods. The applications referred to do not cover even a small part of the numerous results in the theory of magnetism, since this was not the purpose of the present monograph. In most cases, only those investigations are mentioned which are to some extent related to the problems discussed in the book. The exception is made for some theoretical problems which, in the opinion of the author, are of interest but which are not treated in the book because of lack of space.

The discussions always refer to single-crystal one-domain samples. To avoid repetition, this will not be mentioned again.

The system of units with $\hbar = 1$ is employed throughout the book.

The appendices contain information on the reciprocal lattice space, the formal Fourier transformations for a discrete medium, and problems associated with them.

The book is mainly intended for those who wish to become familiar with the elements of the quantum theory of magnetism and methods used in this theory, as well as postgraduate students and senior undergraduates. It is assumed that the reader is already familiar with the fundamentals of quantum and statistical mechanics as presented in the standard courses given by physics departments of universities and institutes of higher learning.

The book was read in manuscript by A.A. Gusev, A. G. Gurevich, and V.A. Moskalenko, to whom the author is grateful for their numerous comments and advice. He would also like to record his thanks to his colleagues for sending him preprints of their papers, which enabled him to become acquainted with a number of interesting results before their publication.

The author is grateful to Academician N.N. Bogolyubov for his valuable contribution to discussions of various problems in the quantum theory of magnetism and for reading the manuscript.

S. V. Tyablikov

Contents

List of Principal Symbols

$\mathcal{A}_1, \mathcal{A}_2, \ldots$	– one-particle, two-particle, etc., operators
$C, C(\ldots, n_f, \ldots)$	– wave function in the second quantization representation
$E_\nu, E(\nu)$	– energy of an elementary excitation
$\mathcal{E}, \mathcal{E}_n$	– energy of a system
F	– free energy
F_s	– distribution function of s particles
$G, G^{(j)}, \mathcal{G}$	– Green's quantum functions
H	– constant magnetic field
$H_\circ$	– demagnetizing field
H_c	– critical field
$\mathcal{H}$	– Hamiltonian of a system
I, I_1	– exchange integral for nearest neighbors
$I(f_1, f_2)$	– exchange integral for atoms at sites f_1 and f_2
$I(\omega)$	– spectral density or spectral function
$J(\nu)$	– Fourier transform of the exchange integral $I(f)$
K_i, K_i'	– i-th anisotropy constants
M	– magnetization of a sample
M_0	– saturation magnetization
$M_1, M(E)$	– mass operator for the first Green's function
$\mathcal{M}$	– magnetic moment operator of a system
N	– number of atoms in a lattice, number of particles in a system
$N_{\alpha\beta}$	– demagnetization factor tensor
$\bar{N}_\nu$	– average value of the number of particles (occupation number) in a state ν
$\mathcal{N}$	– operator for the total number of particles
$\mathcal{N}_\nu, n_\nu$	– operator for the number of particles in a state ν
P	– particle permutation operator, principal value symbol

$\mathscr{P}$ – projection operator
Q – partition function
S_f^α – α-component of the spin at a site f
S_f, S – value of the spin at a site f
S^α – operator for the total spin of a system
T – temperature
T_C – Curie temperature
T_N – Néel temperature
T_k – compensation temperature
V – volume of a system
$Z_p(x)$ – modified Riemann's zeta function
b_f^+, b_f – Pauli operators
f, g – numbers of lattice sites, numbers of one-particle states
$h(t)$ – alternating magnetic field
k – Boltzmann's constant
$\overline{m}$ – average magnetization per site
t – time
v – volume per site or per particle
z – number of nearest neighbors
Γ – damping of elementary excitations
γ^α – direction cosines of the magnetization vector
$\Delta(\nu_1 - \nu_2)$, δ_{ν_1, ν_2} – Kronecker's delta
δ – vector joining two nearest neighbors
$\delta(x)$ – Dirac's delta function
$\zeta(p)$ – Riemann's zeta function
ϑ – temperature in units of k
ϑ_C – Curie temperature
ϑ_N – Néel temperature
λ – chemical potential
μ, μ_f – magnetic moment of an atom; magnetic moment of an atom in units of S, S_f
μ_B – Bohr magneton
ν, $\varkappa$ – wave vector
ρ – density matrix
ρ_s – density matrix for a system of s particles
σ_s, σ_s^α – relative magnetization
τ – dimensionless temperature
τ_C – dimensionless Curie temperature
χ – susceptibility
Ω – thermodynamic potential

Chapter I
Introduction

In this chapter, we present general information and definitions relating to the theory of strong magnetism, and review briefly the theoretical ideas about the nature of strong magnetism. On this basis, we give a qualitative classification of the main types of magnetic substance.

Sec. 1. General Information and Definitions *

According to their magnetic properties, solids can be divided into weakly magnetic (diamagnetic and paramagnetic) and strongly magnetic (ferromagnetic, antiferromagnetic, and ferrimagnetic). Our treatment will deal only with the theoretical problems involving strongly magnetic substances.

* For detailed treatments of experimental data and theoretical problems relating to ferromagnetism, see the monographs of Akulov (1939), Vonsovskii and Shur (1948), Bogolyubov (1949), Belov (1951, 1959), Vonsovskii (1952), Bozorth (1956), Landau and Lifshits (1959), Gurevich (1960), and various collections: "Physics of Ferromagnetic Domains" (1951), "Ferromagnetic Resonance" (1952), "Magnetic Structure of Ferromagnets" (1959), "Ferromagnetic Resonance" (1961), "Theory of the Ferromagnetism of Metals and Alloys" (1963); for antiferromagnetism, see also the reviews of Nagamiya et al. (1955), Borovik–Romanov (1962), Belov et al. (1964), and "Antiferromagnetism" (1956); for weak ferromagnetism, see a review by Borovik–Romanov (1962) and a monograph by Turov (1963); for ferrimagnetism, see the reviews of Gorter (1955), Pakhomov and Smol'kov (1962), and a monograph by Smit and Wijn (1962).

Ferromagnets, antiferromagnets, and ferrimagnets are characterized by the existence, under certain conditions, of magnetic ordering and large macroscopic moments due to this ordering (such moments always appear in ferromagnets and ferrimagnets). The order of magnitude of these macroscopic moments is $N\mu_B$, where N is the number of atoms in a sample and μ_B is the Bohr magneton.

Transition metals (iron, cobalt, nickel) are typical ferromagnets; transition-metal oxides and other salts (FeO, CoO, CoF_2, $NiSO_4$, etc.), are typical antiferromagnets; typical ferrimagnets are compounds of transition elements in the form of complex salts ($MnO \cdot Fe_2O_3$, $3Y_2O_3 \cdot 5Fe_2O_3$, etc.) and some other compounds.

According to current ideas, the magnetism of solids is due to the electrons of partly filled inner shells of atoms. The strong magnetism appears only in those cases when the crystal lattice of a substance includes atoms with partly filled inner shells.

Partly filled inner shells are found in elements of the transition groups of iron (3d-shell), palladium (4d-shell), platinum (5d-shell), actinium (6d- and 5f-shells), as well as rare-earth elements (4f-shell). Table 1 gives the electron configurations of the partly filled shells, and of the shells next to them, for all these elements. The presence of atoms with partly filled inner shells is not a sufficient condition for the existence of strong magnetism.

Thus, in the iron group (3d-metals), pure Sc, Ti, and V are paramagnetic, Cr and Mn are antiferromagnetic, and Fe, Co, and Ni are ferromagnetic.

In the palladium group (4d-metals), Y, Zr, Nb, Mo, Tc, Ru, and Rh are all paramagnetic and the problem of the magnetic properties of Pd is not solved yet [according to Abragams (1963), this metal is not antiferromagnetic].

In the platinum group (5d-metals), La, Lu, Hf, Ta, W, Re, Os, and Ir are paramagnetic, while Pt is antiferromagnetic.

In the rare-earth group (4f-metals), some elements exist in two different magnetic phases at low temperatures. In the subgroup containing Ce, Pr, Nd, Pm, Sm, and Eu, antiferromagnet-

Table 1. Electron Configurations in Partly Filled Shells and in Shells which Follow Them, for Transition Elements in the Periodic System*

Iron group (electron configurations outside the Ar shell)

Sc	Ti	V	Cr	Mn	Fe	Co	Ni
$3d^14s^2$	$3d^24s^2$	$3d^34s^2$	$3d^54s$	$3d^54s^2$	$3d^64s^2$	$3d^74s^2$	$3d^84s^2$

Palladium group (electron configurations outside the Kr shell)

Y	Zr	Nb	Mo	Tc	Ru	Rh	Pd
$4d^15s^2$	$4d^25s^2$	$4d^45s$	$4d^55s$	$4d^55s^2$	$4d^75s$	$4d^85s$	$4d^{10}$

Platinum group (electron configurations outside the Xe shell for La and outside the Xe + $4f^{14}$ shell for remaining elements)

La							
$5d^16s^2$							
Lu	Hf	Ta	W	Re	Os	Ir	Pt
$5d^16s^2$	$5d^26s^2$	$5d^36s^2$	$5d^46s^2$	$5d^56s^2$	$5d^66s^2$	$5d^76s^2$	$5d^96s$

Rare-earth elements (electron configurations outside the Xe shell)

Ce	Pr	Nd	Pm	Sm	Eu	Gd
$4f^26s^2$	$4f^36s^2$	$4f^46s^2$	$4f^56s^2$	$4f^66s^2$	$4f^76s^2$	$4f^75d^16s^2$

Tb	Dy	Ho	Er	Tm	Yb
$4f^85d^16s^2$	$4f^{10}6s^2$	$4f^{11}6s^2$	$4f^{12}6s^2$	$4f^{13}6s^2$	$4f^{14}6s^2$

Actinium group (electron configurations outside the Rn shell)

Ac	Th	Pa	U	Transuranium elements
$6d'7s^2$	$6d^27s^2$	$5f^26d^17s^2$	$5f^36d^17s^2$	

* These configurations are taken from the monographs of Condon and Shortley (1949), and Landau and Lifshits (1963).

Table 2. Values of the Magnetomechanical Ratio for Some Magnetic Materials*

Material	g'
Fe	1.919 ± 0.002
Co	1.850 ± 0.004
Ni	1.835 ± 0.002
Supermalloy ($Ni_{0.79}Fe_{0.16}Mn_{0.05}$)	1.905 ± 0.002
Stoichiometric nickel ferrite ($NiO \cdot Fe_2O_3$)	1.849 ± 0.002
Heusler alloy (Cu_2MnAl)	1.993 ± 0.002
MnSb	1.978 ± 0.002
Pyrrhotite (Fe_7S_8)	1.9 ± 0.3

*According to Scott's data (1962).

ic properties are exhibited at very low temperatures by Ce, Nd, Sm, and Eu, while Pm is paramagnetic; all members of the subgroup comprising Gd, Tb, Dy, Ho, Er, and Tm are ferromagnetic in the "low-temperature phase" and antiferromagnetic in the "high-temperature phase" (the question of the existence of the antiferromagnetic phase in Gd still remains open); Yb is paramagnetic. For details of the magnetic properties of rare earths, the reader is directed to the review of K. P. Belov et al. (1964).

Actinides (6d- and 5f-metals)–Ac, Th, Pa, and U–are paramagnetic.

The magnetic properties of substances depend on the distribution of the electron density in their partly filled inner shells and the conduction-electron density in the crystal lattice. The present state of the theory does not allow us yet to formulate the necessary and sufficient conditions for the existence of strong magnetism in a given substance on the basis of information about the electron configurations of free atoms which compose the crystal lattice.

The magnetic moment of a sample is due to the intrinsic magnetic moments of atomic electrons as well as to the electron orbital moments. The magnetomechanical ratio g' (i.e., the ratio of the magnetic moment to the angular momentum), expressed in units of e/2mc, is equal to 2 for the intrinsic moments of electrons and 1 for their orbital moments. The first measurements of the magnetomechanical ratio were made using the Einstein–de Haas and Barnett effects. Recently, the magnetomechanical ratio has been measured with high accuracy in ferromagnetic resonance experiments. The results of these measurements give g' values close to g' for free electrons. Table 2 lists the results of the measurements of the magnetomechanical ratio for a number of magnetic materials.

From the available data, it is clear that the magnetic moments of strongly magnetic substances are principally due to the magnetic moments of the electrons in the partly filled shells and that the orbital moments of these electrons do not make an important contribution. We can also assume that the macroscopic magnetic moment appears due to the spin ordering of the electrons in the partly filled shells, which can occur only under certain conditions. The cause of the ordering is the interaction between these electrons.

This assumption, formulated first by Frenkel' (1928) and by Heisenberg (1928), is the foundation on which the modern quantum theory of strong magnetism is built.

Spin ordering is spontaneous when the temperature falls below a certain critical value. The magnetic moment per unit volume which appears below this temperature is known as the spontaneous magnetization. The value of this spontaneous magnetization depends on temperature but is practically independent of the external field. The highest theoretical value of the magnetization is known as the saturation magnetization.

The phenomenon of strong magnetization in substances is closely related to the existence of a crystal lattice because strong magnetism has not been found in liquids or gases. The influence of the crystal structure of a substance on its magnetic properties also manifests itself in the dependence of the magnetic properties on the direction along which they are measured. In other words, magnetic substances exhibit magnetocrystalline anisotropy.

The relative orientation of the electron spins of neighboring atoms is governed by the nature of the interaction between the electrons. The Coulomb interaction between the electrons may cause ordering in the relative positions of the spins, but it does not fix their general orientation. Spin–spin and spin–orbit interactions reduce considerably this directional degeneracy. Consequently, crystal lattices have only few directions of the spin orientation along which the thermodynamic potential is minimal. These directions are known as the axes of easy magnetization. Using the relationship between the magnetocrystalline anisotropy and magnetic interactions, Akulov (1939) developed a phenomenological theory of the anisotropy.

By way of example, we shall consider the three classical ferromagnetic materials: iron, nickel, and cobalt. Iron has the bcc lattice, and its axes of easy magnetization are directed along the edges of a unit cube. Nickel has the fcc lattice; its easy magnetization axes are along the principal diagonals of a cube. The low-temperature modification of cobalt has the hcp lattice and only one easy magnetization axis, which is the sixfold axis.

In a single-crystal sample, the spins may be oriented, in the absence of an external field, along any one of the easy magnetization axes. The situation favored by the energy considerations is obtained when a single crystal splits into a number of regions, in each of which the spins are oriented along a different axis. These regions are known as domains or spontaneous magnetization regions. The dimensions of domains, their shapes, and mutual positions are governed by the conditions for the existence of the minimum thermodynamic potential of a system. A quantitative theory of domain structures has been developed by Landau and Lifshits (1935a, 1959).

Under certain conditions, the behavior of the electrons in partly filled shells can be described by a system of spins located at lattice sites. The coefficient of proportionality which governs the intensity of the interaction between spins at different lattice sites (for example, at neighboring sites) is known as the exchange integral. It is assumed that the order of magnitude of the exchange integral I is equal to the exchange energy of the electrons at the corresponding sites.

Quantitative calculations, using even this simplified model of a real substance, are still quite complex. However, it is possible, under certain conditions and with a known degree of accuracy, to replace the electron spin operators with classical vectors. Then the magnetic properties of a substance are modeled by a system of dipoles located at lattice sites and interacting with one another with an energy equal to the exchange energy. This quasi-classical approach gives a sufficiently correct qualitative, and sometimes quantitative, description of strongly magnetic substances.

We shall use μ to denote the magnetic moment of an atom and N to denote the number of atoms in a lattice. Then the saturation magnetization is given by

$$M_0 = N\mu.$$

The measured values of the magnetization M are smaller than M_0. This is due to the demagnetizing influence of the thermal vibrations of the spin moments of atoms, the magnetocrystalline anisotropy, and the boundary effects of the sample. If a sample is placed in an external magnetic field H, its magnetization will increase with the magnetic field. The quantity

$$\chi(H) = \frac{\partial M}{\partial H}$$

is known as the magnetic susceptibility. As the field intensity is increased, the moments gradually rotate from the easy magnetization axes to the direction of H ("rotation" region) and at some values of H they all become, on the average, aligned along H. The further increase in M as H is increased is due to the suppression of the thermal vibrations of the spin moments by the applied external field ("saturation" region). In this region, the susceptibility decreases as H is increased. The absolute saturation mag-

Table 3. Curie (T_C) and Néel (T_N) Temperatures for Some Ferromagnets, Ferrimagnets, and Antiferromagnets.*

Material	T_C, °K	Material	T_N, °K
Ferromagnets		Antiferromagnets	
		$NiSO_4$	37
Fe	1043	$CoSO_4$	12
Co	1393	$FeSO_4$	21
Ni	631	$MnSO_4$	11.5
		NiO	520
Ferrimagnets		CoO	290
		FeO	188
$FeO \cdot Fe_2O_3$	863	MnO	118
$CoO \cdot Fe_2O_3$	793	NiF_2	73.2
$NiO \cdot Fe_2O_3$	863	CoF_2	37.2
$MnO \cdot Fe_2O_3$	593	FeF_2	78.3
$MgO \cdot Fe_2O_3$	643	MnF_2	66.5

* The values are taken from Bozorth's monograph (1956) and from the reviews published by Borovik–Romanov (1962) and Pakhomov and Smol'kov (1962).

netization is reached in the limiting case when $H \to \infty$; the susceptibility then tends to zero.

The quasi-classical treatment can be simplified even further by replacing the magnetic moments with an effective field, which is proportional to the exchange integral and the average magnetization of the substance. The effective field acting on the magnetic moments in a substance is known as the Weiss molecular field. This model of a ferromagnet was proposed by Rozing (1892, 1896, 1910) and by Weiss (1907) before the advent of quantum mechanics but it explains qualitatively the characteristic features of the behavior of ferromagnets when the temperature and the external field are varied.

The ordered distribution of spins is destroyed at a critical temperature known as the Curie temperature in the case of ferromagnets, and the Néel temperature in the case of antiferromagnets. It is evident that these are the temperatures at which the average thermal energy becomes of the same order as the exchange interaction energy. For typical ferromagnets, the Curie temperatures (Table 3) are of the order of 1000°K; $kT_C \approx 10^{-13}$ erg, where k is Boltzmann's constant. The exchange interaction energy I is of the order of e^2/a, where e is the electronic charge and a is the lattice constant. Assuming that $a \approx$ 2-3 Å, we obtain $I \approx 10^{-12}$-10^{-13} erg. It follows that at the Curie point these two energies are of the same order of magnitude.

The magnetic anisotropy energy is comparable with the energies of the magnetic interactions (spin–spin and spin–orbit) between electrons and its order of magnitude is $\mu^2/a^3 \approx 10^{-16}$-$10^{-17}$ erg (a is the lattice constant). If a sample of a strongly magnetic substance is placed in an external field H, this field should suppress the anisotropy if $H \gtrsim \mu/a^3$, i.e., if $H \gtrsim 10^3$-10^4 Oe, which is indeed observed experimentally. The magnitude of the magnetocrystalline anisotropy depends strongly on temperature.

The estimates given in the preceding paragraph show that the phenomenon of strong magnetism cannot be explained from the classical standpoint, by representing a substance as a system of magnetic moments located at lattice sites, with the usual dipole interaction between them. In fact, the order of magnitude of the dipole interaction energy is $\mu^2/a^3 \approx 10^{-16}$-$10^{-17}$ erg, and the corresponding Curie temperatures would have been of the order of 0.1-1°K. Nevertheless, dipole magnetism is observed, although rarely [Cooke et al. (1959)]. We shall not consider this type of magnetism, because it is outside the range of typical strongly magnetic substances. It should be noted, however, that the methods we shall present in later chapters can be applied, without any special changes, to dipole magnetism as well.

In compounds such as oxides or salts of transition metals, the magnetically active atoms are separated by magnetically neutral atoms. Since the direct exchange interaction decreases rapidly with distance, it must be very weak in such substances and cannot account for the orders of magnitude of the observed transition temperatures. It is assumed that in these cases we have the case of an indirect exchange, considered first by Kramers (1934), in which electrons of magnetic atoms interact through intermediate nonmagnetic atoms.* In the indirect exchange case, we can again

*The indirect exchange theory was developed further by Anderson (1950a, 1959), Van Vleck (1951), Shimizu (1952), Yamashita (1954), Vonsovskii and Seidov (1956), Keffer and Oguchi (1959), Fukuchi (1961a, b), and others.

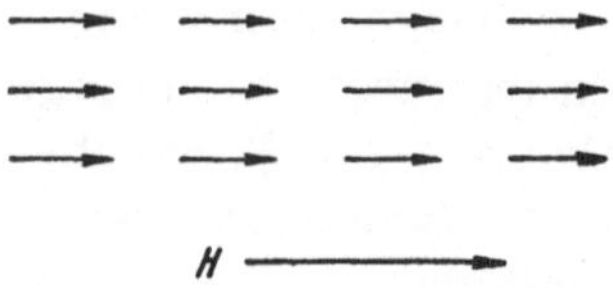

Fig. 1. Ferromagnetic ordering of spins.

represent the system of electrons responsible for magnetism by a system of spins located at lattice sites. The quantity I, analogous to the exchange integral of ferromagnets, is called the indirect exchange integral. Its values are, as a rule, one or two orders of magnitude smaller than the values of the exchange integral for typical ferromagnets (compare the values of the Curie and Néel temperatures in Table 3).

We shall now use the quasi-classical approximation in order to present graphically the models of ferromagnets, antiferromagnets, and ferrimagnets. For simplicity, we shall neglect the influence of the magnetic anisotropy by assuming it to be small.

Ferromagnets. Below the Curie temperature, all the spins are, on the average, oriented parallel to one another, giving rise to a large spontaneous magnetic moment. In an external field H the magnetic moments of the atoms and, consequently, the resultant moment, are oriented along the field (Fig. 1).

In the absence of an external field, the direction of the resultant magnetic moment **M** is indeterminate. However, since there is always some anisotropy, although it may be weak, the vector **M** is oriented along one of the easy magnetization axes.

The spontaneous magnetization decreases as the temperature rises and, in the absence of an external field, disappears at the Curie temperature T_C. A phase transition of the second kind (with a discontinuity in the second derivative of the thermodynamic potential) takes place at the Curie point.

Above the Curie point ($T \gg T_C$), a ferromagnet behaves like a classical paramagnetic substance and its paramagnetic susceptibility χ satisfies the Curie–Weiss law:

$$\chi = \frac{\text{const}}{T - T_C}. \tag{1.1}$$

In the vicinity of the Curie point at $T \leq T_C$, the temperature dependence of the spontaneous magnetization is given by the expression

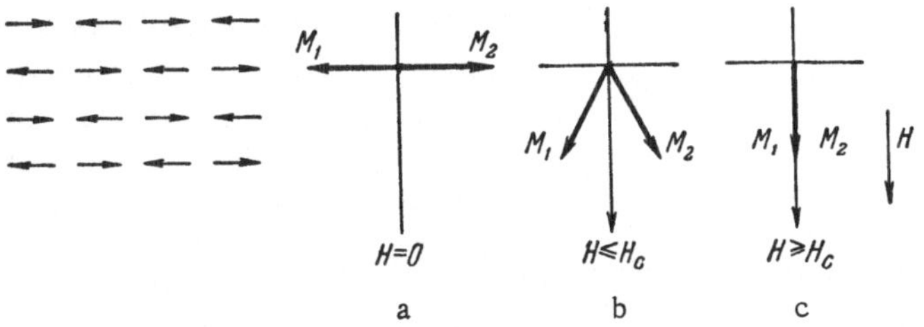

Fig. 2. Schematic representation of the spin configuration in an isotropic antiferromagnet: a) in the absence of an external field; b) in "weak" fields ($H \leq H_C$); c) in "strong" fields ($H \geq H_C$).

$$M(T) \simeq \text{const}\sqrt{1 - T/T_C}. \tag{1.2}$$

As $T \to 0$, the temperature dependence M(T) assumes the power form:

$$M(T) = M_0(1 - A_1T^{3/2} - A_2T^{5/2} - \ldots), \tag{1.3}$$

where A's are some constants, and M_0 is the saturation magnetization.

Antiferromagnets. According to Néel's hypothesis (1932, 1936), the antiferromagnetic spin configuration can be represented as an assembly of two or more superimposed ferromagnetic sublattices whose resultant moment is equal to zero. The spin configuration of an isotropic two-sublattice antiferromagnet is shown schematically in Fig. 2.

We shall represent the moments of the sublattices oriented to the "left" and "right" by the vectors $\mathbf{M}_1$ and $\mathbf{M}_2$ ($|\mathbf{M}_1| = |\mathbf{M}_2|$). If $H = 0$, the resultant magnetization is zero. If $H \leq H_C$, where H_C is some critical field, the magnetic moments become aligned as shown in Fig. 2b. In the range of fields $0 \leq H \leq H_C$, the resultant magnetization is aligned along the field and increases linearly with its intensity. If $H > H_C$, the spins of both sublattices become aligned along the field and the antiferromagnet behaves like a ferromagnet (Fig. 2c). The dependence of the resultant magnetization on the external field at 0°K is shown in Fig. 3.

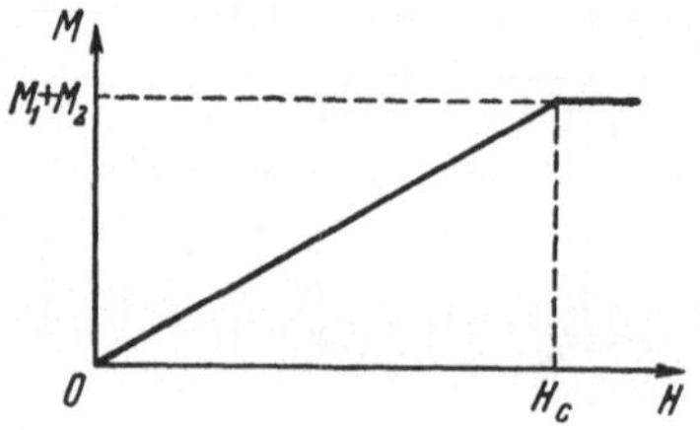

Fig. 3. Dependence of the magnetization of an isotropic antiferromagnet on the value of the external field at zero temperature.

The quantity μH_c is of the order of the exchange interaction energy and therefore the region of the curve in Fig. 3 at $H \gtrsim H_c$ may be reached in substances with a sufficiently low Néel temperature.

When the temperature rises, the sublattice magnetizations decrease and vanish at the Néel temperature T_N (at $H=0$). As in ferromagnets, a phase transition of the second kind takes place at the temperature $T = T_N$. Near this transition, the temperature dependence of the specific heat has the typical "lambda-like" form, and anomalies of other properties are also observed.

Above the Néel point the susceptibility varies in accordance with the Curie–Weiss law. Antiferromagnets are characterized by a susceptibility maximum at $T = T_N$ and by a strong temperature and field dependence of the susceptibility at $T < T_N$.

It should be mentioned that the anisotropy in antiferromagnets may basically alter their behavior, compared with that of isotropic antiferromagnets.

Weak Ferromagnets (Antiferromagnets with Weak Ferromagnetism). In these substances, the magnetic moments of the sublattices are not exactly antiparallel but are turned through a small angle (≈1°), due to the influence of the anisotropy. Consequently, in these antiferromagnets, the magnetic moments of the sublattices are not fully compensated (even at $H = 0$) and a nonzero spontaneous magnetization is observed. Examples of such substances are α-Fe_2O_3, $MnCO_3$, $CoCO_3$, $CuSO_4$, NiF_2, MnF_2, and several others.

The magnetization of weak ferromagnets can, under certain conditions, be described by the formula

$$M = m + \chi H, \tag{1.4}$$

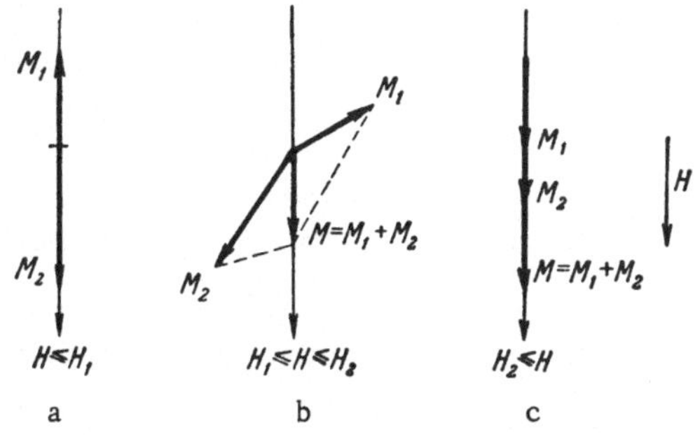

Fig. 4. Schematic representation of the spin configuration in an isotropic ferrimagnet: a) in weak fields ($H \leq H_1$); b) in strong fields ($H_1 \leq H \leq H_2$); c) in very strong fields ($H \geq H_2$).

where H is an external magnetic field, m and χ are constants which depend on temperature. Obviously, m is the resultant magnetization due to the noncollinear configuration of the sublattice moments. The temperature dependence of m is similar to that of the sublattice magnetization. The value of m is of the same order as the sublattice magnetization, reduced by the ratio of the energy of the magnetic interactions of the spins (spin–spin and spin–orbit) to the exchange interaction energy. Therefore, the spontaneous magnetization of weak ferromagnets is approximately 0.1% of the spontaneous magnetization of normal ferromagnets.

The phenomenon of weak ferromagnetism is closely related to the magnetic symmetry of a crystal; it can also be created artificially. One way of doing this is by the application of a magnetic field, which alters the magnetic structure so that an antiferromagnet becomes a weak ferromagnet. Another way of achieving the same result is to change the magnetic structure by deforming a crystal. For a detailed treatment of the phenomenon of weak ferromagnetism, see a review by Borovik–Romanov (1962) and Turov's monograph (1963).

Ferrimagnets. According to Néel (1948), ferrimagnets have several sublattices with a finite resultant moment, comparable with the spontaneous moments of the sublattices. The re-

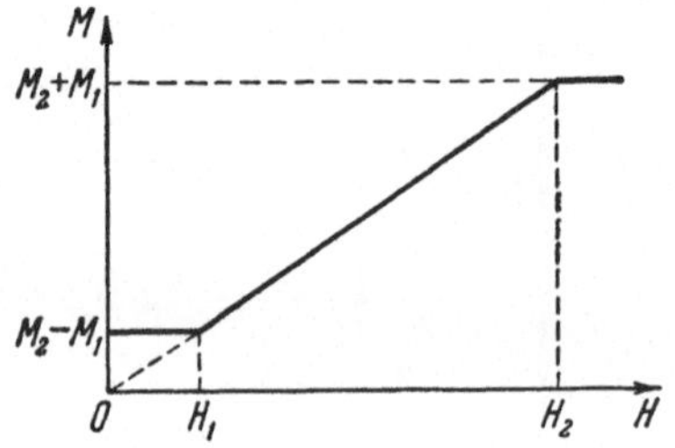

Fig. 5. Dependence of the magnetization of an isotropic two-sublattice ferrimagnet on the external field at zero temperature.

sultant moment may be due to a difference in the numbers of the "left" and "right" sites and spins at these sites, as well as to a noncollinear configuration of the sublattice moments.

We shall consider the behavior of a ferrimagnet in an external field. For simplicity we shall assume that there are only two sublattices with the resultant moments $\mathbf{M}_1$ and $\mathbf{M}_2$ ($|\mathbf{M}_1| < |\mathbf{M}_2|$) and we shall neglect the magnetic anisotropy. Then, depending on the magnitude of the external field, we have one of the possibilities shown in Fig. 4. In sufficiently weak fields, the resultant magnetic moment of a ferrimagnet is always directed along the field.

In weak fields $H \leq H_1$, where H_1 is the first critical field, the sublattice magnetization vectors are antiparallel; the resultant moment M is oriented along the field and equal to the difference $M_2 - M_1$. In strong fields, $H_1 \leq H \leq H_2$, where H_1 and H_2 are the first and second critical fields, the moments $\mathbf{M}_1$ and $\mathbf{M}_2$ do not coincide with the direction of H, but the total moment is oriented along H. In very strong fields, $H \geq H_2$, the sublattice moments are parallel to one another. It should be mentioned that, as in antiferromagnets, the values of μH_1 and μH_2 (μ is the magnetic moment of a crystal site) are of the order of the exchange interaction energy, i.e., H_1 and H_2 are of the order of 10^6-10^7 Oe (at $T < T_C$). Figure 5 shows the dependence of the magnetization of a ferrimagnet on the external field H at 0°K.

Rode and Vedyaev (1963) investigated the magnetization of the copper–cadmium ferrite $Cu_{0.8}Cd_{0.2} \cdot Fe_2O_3$ and of the gadolinium ferrite garnet $3Gd_2O_3 \cdot 5Fe_2O_3$ in strong pulse fields (up to 220 kOe). The copper–cadmium ferrite exhibited a linear rise of its magnetization with the field beginning at 110 kOe, and the same happened in the gadolinium ferrite garnet at 70 kOe, which was interpreted as a transition through the first critical field H_1.

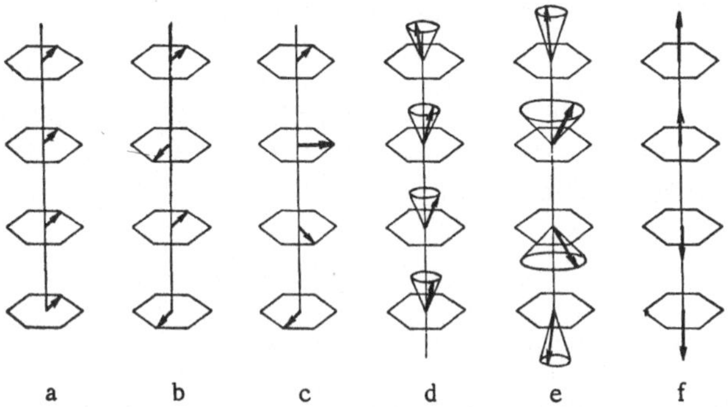

Fig. 6. Simplest types of helical structure: a, b) collinear structures (ferromagnetic and antiferromagnetic); c) simple helix; d) ferromagnetic helix; e) cycloidal structure; f) linear spin wave.

Table 4. Temperatures of Phase Transitions T_1 and T_2 and Values of Critical External Fields H_c for Rare-Earth Elements*

Rare-earth elements	T_1, °K	T_2, °K	H_c, Oe
Tb	219	230	200
Dy	85	179	11 000
Ho	20	133	18 000
Er	20	85	18 000
Tm	22	60	> 15 000

*Taken from the review of Belov et al. (1964).

In the case of ferrimagnets with more than two sublattices, there is an interesting possibility of the existence, under certain conditions, of a compensation temperature T_c ($T_c < T_C$), at which the resultant magnetization vanishes. The appearance of a compensation point is due to the fact that the magnetic moments of the sublattices compensate each other completely at $T = T_c$, due to the different temperature dependences of the sublattice magnetizations. Above this temperature, the compensation is no longer obtained, and the resultant moment does not disappear until the Curie point is reached.

Above the Curie point, ferrimagnets behave as paramagnetic materials, and the temperature dependence of their susceptibility is given by the Curie–Néel law:

$$\chi^{-1}=\chi_0^{-1}+\frac{T}{C}-\frac{\Delta}{T-T_C}, \tag{1.5}$$

where χ_0, Δ, and C are some constants.

Ferrimagnets include spinel ferrites, garnet ferrites, magnetoplumbites, and many other substances (some of the relevant literature is cited in the first footnote in the present chapter).

Magnetic Substances with Helical Structures. The possible types of magnetic structure are not limited to those mentioned so far. A number of substances (rare-earth metals, the compounds $MnAu_2$, MnO_2, $Mn_{2-x}Cr_xSb$, and the element Cr) have helical structures in which the spin vector components vary periodically along a certain special crystallographic direction. These substances usually have uniaxial symmetry. Figure 6 shows the simplest types of helical structure. The cases (a) and (b) of collinear ordering in ferromagnets or antiferromagnets are included in Fig. 6 for formal reasons, because they can be regarded as degenerate helical structures with the spins rotated by 0° or π, respectively, from one plane to the next. The ordering shown in Fig. 6c is called a simple or antiferromagnetic helix. In the case shown in Fig. 6d, the z-components of the spins have fixed values but the other two components, which lie in the basal plane, either vary periodically or remain unordered when they are translated along the principal axis. In the system represented in Fig. 6e, the z-component varies periodically and the two other components vary as in Fig. 6d. Finally, Fig. 6f depicts the case where the spins are aligned along the principal axis, but the z-component varies periodically.

Figure 6 shows that the helical structures can change gradually from one type to another, and that one cannot distinguish them very clearly. Clearly, a classification which takes into account the number of phase transitions is more satisfactory.

Table 5. Main Types of Magnetic Structure of Some Rare-Earth Elements (Columns represent the configurations shown in Fig. 6; *f* denotes the ferromagnetic, and *a* the antiferromagnetic phase)

Atomic number	Element	Type of structure					
		a	b	c	d	e	f
64	Gd	*f*	—	—	*a*?	—	—
65	Tb	*f*		*a*			
66	Dy	*f*		*a*			
67	Ho			*a*	*f*		
68	Er				*f*	*a*	
69	Tm					*a*	*f*

Substances with helical structures exhibit a number of features, which are exemplified by the behavior of rare-earth elements. According to neutron-diffraction data, Tb, Dy, Ho, Er, and Tm have two low-temperature phases: at lower temperatures they are ferromagnetic, while at higher temperatures they are antiferromagnetic. Consequently, these substances have two phase transition points: at T_1, they change from the ferromagnetic to the antiferromagnetic state, and at T_2, from the antiferromagnetic to the paramagnetic state. The helical configurations are destroyed by external fields H higher than some critical value H_c (Table 4).

The ferromagnetic phase (denoted by *f* in Table 5) can exist in the structures represented in Figs. 6a, 6d, 6e, and 6f, while the antiferromagnetic phase (denoted by *a* in Table 5) can exist in the structures shown in Figs. 6b, 6c, 6e, and 6f. Table 5 lists the types of magnetic ordering in Gd, Tb, Dy, Ho, Er, and Tm. In the ferromagnetic (*f*) phase of Tm, the spins are aligned along the principal axis: in four neighboring planes, they are directed one way, and in the next three planes, they are directed the opposite way (Fig. 6f).

The theory of magnetic substances with helical structures has been developed by Yoshimori (1959), Villain (1959), Yosida and Miwa (1961a, b), Kaplan (1961), and Elliott (1961) on the assumption that the exchange integrals between the nearest, second nearest, and third nearest neighbors differ in magnitude and sign; more detailed calculations allow for the uniaxial anisotropy and the anisotropy in the basal plane.* The conditions for the existence of such structures, their stability and behavior, when temperature and field are varied, were considered in these investigations.

It is assumed that the pitch of the helix in helical structures is incommensurable with the crystal lattice period [cf., for example, Naish (1963) and Dzyaloshinskii (1964a-c)].

The representation of strongly magnetic substances as systems of spins located at lattice sites has led to a number of important predictions which have been confirmed experimentally. However, this representation ignores such an important aspect as the interaction of the electrons in the partly filled shells with the conduction electrons. Consequently, this simplified model has been found to be more suitable for ferromagnetic dielectrics than for ferromagnetic metals. This model cannot explain a number of important experimental observations relating to ferromagnetic metals, for example, the anomalous (compared with nonferromagnetic metals) behavior of the electrical conductivity and specific heat, the anomalies present in the galvanomagnetic effects, and several other features. The simplified model describes more satisfactorily insulating or semiconducting antiferromagnets and ferrimagnets.

The problem of the fractional values of the magnetic moments in single crystals of the 3d-metals (Fe, Co, Ni), which is also not explained by the simple model, is of some interest.† If a ferromagnetic 3d-metal is regarded as a system of spin moments of partly filled shells, then, in the limiting case of low temperatures and sufficiently high external fields, the average magnetization per site should be identical with the magnetic moments of the free atoms or at least it should be close to these moments. In fact, for a number of substances considerable differences are observed between these quantities. Table 6 lists the magnetic moments of free atoms (μ_f) and the average values of the magnetic moment per atom (μ_a) for a number of crystals.

---

*See also the papers of Elliott (1962), Nagamiya et al. (1962a,b), Herpin (1962), Kaplan and Lyons (1963).

† For details, see Vonsovskii and Izyumov (1962a,b).

Table 6. Values of Magnetic Moments of Atoms of Some Transition Elements* (μ_f – in the free state; μ_a – in crystals)

Element	μ_f, in μ_B	μ_a, in μ_B
3d-metals		
Cr	5	0.2—0.4
Mn	5	1.5
Fe	4	2.218
Co	3	1.715
Ni	2	0.604
4f-metals		
Gd	7	7.12
Er	9	8—9

*Taken from Vonsovskii and Izyumov's review (1962a).

Table 7. Values of the Magnetic Moments of Some Spinel and Garnet Ferrites* (Values of magnetic moments per one "molecule": μ_f – in the free state; μ_a – in crystals)

Compound	μ_f, in μ_B	μ_a, in μ_B
$MnO \cdot Fe_2O_3$	5	4.40—5.00
$FeO \cdot Fe_2O_3$	4	4.03—4.08
$CoO \cdot Fe_2O_3$	3	3.30—3.94
$NiO \cdot Fe_2O_3$	2	2.22—2.40
$CuO \cdot Fe_2O_3$	1	1.30—1.70
$MgO \cdot Fe_2O_3$	0	0.86—2.20
$3Y_2O_3 \cdot 5Fe_2O_3$	10	9.44
$3Gd_2O_3 \cdot 5Fe_2O_3$	32	30.3
$3Tb_2O_3 \cdot 5Fe_2O_3$	26	31.4
$3Dy_2O_3 \cdot 5Fe_2O_3$	20	32.5

*Taken from the reviews of Gorter (1954) and Pakhomov and Smol'kov (1962).

Table 6 shows that metals with partly filled 3d-shells have values of μ_f and μ_a which differ considerably from one another, while for metals with partly filled 4f -shells these quantities are quite similar. Moreover, the moments μ_a differ considerably from integral values (in units of μ_B) for 3d-metals but are closer

Table 8. Electronic Specific Heats of Transition and Other Metals* (Experimental values of C_{el}, measured at 1°K)

Element	Li	Cu	Cd	Fe	Co	Ni	Mn	Pd	Pt
$C_{el} \cdot 10^{-3}$, $J \cdot mole^{-1} \cdot deg^{-2}$	1.75	0.686	0.71	5.0	5.0	7.3	15.8	13.0	6.6

* Taken from Vonsovskii and Izyumov's review (1962a).

to integral values for 4f-metals. In those spinel and garnet ferrites which are semiconductors or dielectrics, the average values μ_a are close to the values calculated from the moments of free atoms (cf., Table 7).

The considerable differences between the values of the atomic and average magnetic moments and their fractional values in 3d-metals may be due, first to the quite strong "collectivization" of the electrons in the partly filled shells, which makes them somewhat similar to the collective states of electrons in the conduction bands, and, secondly, to the interaction between the electrons in the partly filled and valence shells. Of course, real substances depart considerably from the assumption that magnetic substances can be represented in the form of systems of spins located at crystal lattice sites. Definite confirmation of this explanation is provided by the anomalously high values of the electronic specific heat of transition metals compared with those of other metals. The values of the electronic specific heat of transition and other metals are compared in Table 8. It is evident that the values for transition metals from Fe to Pt are much higher than for Li, Cu, and Cd.

These and other experimental observations have served as the basis for the development of a different approach to the theory of strong magnetism, known as the band theory of magnetism, in which the electrons in the partly filled shells are regarded, like the conduction electrons, as a system which is wholly in the "collective" state.* The band theory gives a simple explanation of the fractional values of the magnetic moment, of the anomalously high values of the electronic specific heat, and of the paramagnetic susceptibility of 3d-metals. On the other hand, the present versions of this theory deal very unsatisfactorily with the spatial inhomogeneity of the electron density, cannot explain the measured form factors, and do not provide a quantitative description of the processes of neutron and electron scattering by spin waves. The band theory exaggerates the "collective" effect of electrons and is therefore unable to explain some of the purely magnetic properties.

Finally, we must mention one more variant of the theory, namely, the hybrid s–d exchange model of Vonsovskii (1946). According to this model, the interaction between the electrons in the partly filled and valence shells is considered as a small perturbation. Accordingly, it is assumed approximately that the d-electrons are mainly responsible for the magnetic effects, and the s-electrons for the effects associated with currents.†

According to this model, the interaction between the s- and d-electrons is assumed to be small and therefore it can hardly explain such observations as the fractional values of the average atomic moments of the 3d-metals and the anomalies of the electronic specific heat of transition metals.‡

*The first attempts to establish a band theory of magnetism were made by Frenkel' (1928) and Bloch (1929). The band theory was developed further by Slater (1936), Stoner (1936, 1938, 1939, 1948), Mott (1935), Herring and Kittel (1951, 1952a, b), Bohm and Pines (1951, 1953), Pines (1953), Pines and Bohm (1952), Matsubara (1954), Shimizu (1960), Vonsovskii and Kobelev (1961), and others.

†In exactly the same way, we can develop a similar model for the 4f-metals; by analogy it should be called the s–f exchange model. When we subsequently refer to the s–d exchange model, we shall imply also the s–f exchange model, unless otherwise stated.

‡Further improvements of this theory and investigations of its consequences can be found in the papers of Vonsovskii and Turov (1953), Abrahams (1954), Turov (1953, 1954, 1955a, b, 1957, 1958), Turov and Vonsovskii (1953), Turov and Shavrov (1958),

We shall be interested mainly in the magnetic phenomena, and, therefore, we shall restrict ourselves to the investigation of the Heisenberg model of magnetic substances.

Sec. 2. Elements of the Phenomenological Theory*

From now on, we shall deal mainly with the following two groups of problems in the theory of magnetism: first, the problems arising in the determination of the temperature and field dependences of the static characteristics of magnetic materials (for example, the magnetization); secondly, the problems met with in the theory of ferromagnetic resonance. Therefore, we shall give here some assumptions and formulations of the classical theory of magnetism which we shall later translate to the terms of the quantum-mechanical theory.

For simplicity, we shall consider in the present section only ferromagnets, because the extension of the results to antiferromagnets and ferrimagnets should not, in principle, present any difficulties.

First, we shall derive an expression for the free-energy density in a ferromagnet, since the main characteristics can be determined easily from this density.

The free energy has several components, of which the exchange interaction is one.

We shall represent a substance by its spin density, i.e., by the spin moment per unit volume. If we restrict ourselves to oscillations of wavelength much longer than the distance between neighboring spins, the spin density may be regarded as a continuous function of points in space. This transition to a continuous

---

Berdyshev and Vonsovskii (1954), Irkhin and Turov (1957), Yosida (1957), Mitchell (1957), Kasuya (1956, 1958), Vonsovskii and Izyumov (1960), Potapkov and Tyablikov (1960), Tahir-Kheli and ter Haar (1963), Bar'yakhtar and Peletminskii (1960), et al.

*For a more detailed presentation of the problems of the phenomenological theory, see monographs of Akulov (1939), Vonsovskii and Shur (1948), Vonsovskii (1952), Belov (1951, 1959), Landau and Lifshits (1959).

distribution of spins is similar to the transition, in the limit, from a discrete distribution of masses to a continuous one in the theory of crystal-lattice vibrations, and it represents the Debye approximation. As an example, we shall consider the expression for the exchange energy in the quasi-classical model. * After the transition, in the limit, from a discrete to a continuous distribution of spins, we obtain

$$F_{exc} = -\frac{1}{2} NIz \left(\frac{M}{M_0}\right)^2 + \frac{NI}{2a} \sum_{\alpha} (\nabla\gamma^{\alpha})^2, \tag{2.1}$$

where M is the magnetization at temperature T, M_0 is the saturation magnetization, a is the lattice constant, N is the number of atoms in the lattice, I is the value of the exchange integral for the nearest neighbors, z is the number of such neighbors, and γ is a unit vector in the direction of the vector **M**.

Next, we must allow for the contribution of external fields to the energy of the ferromagnet. This contribution is

$$F_e = -(H, M). \tag{2.2}$$

In a finite sample, magnetic "charges" appear on the surface (and in the interior, if the magnetization is inhomogeneous) and these "charges" produce fields which are proportional to the magnetization and directed opposite to it. These are known as the demagnetizing fields $H_{\circ}$ and they also contribute to the free energy of a ferromagnet; the magnitude of this contribution depends strongly on the geometry of the sample.

In general, the relationship between $H_{\circ}$ and M is a tensor:

$$H^{\alpha}_{\circ} = -\sum_{\beta} N_{\alpha\beta} M^{\alpha}; \tag{2.3}$$

the quantities $N_{\alpha\beta}$ are known as the demagnetization factors. For a triaxial ellipsoid, a disk, and a long cylinder, only the diagonal elements of the tensor $N_{\alpha\beta}$ are not equal to zero. The density of

*See, for example, Turov's paper (1963).

the demagnetizing field energy is equal to

$$F_{\text{demag}} = -\frac{1}{2}(H_{\circ}, M) = \frac{1}{2}\sum_{\alpha,\beta} N_{\alpha\beta} M^{\alpha} M^{\beta}. \tag{2.4}$$

Magnetic materials have, as a rule, anisotropic magnetic properties. This anisotropy is associated with the dependence of the free energy of a ferromagnet on the orientation of the magnetization vector with respect to the crystal-lattice axes. Therefore, the free energy of a crystal should include a term which depends on the direction cosines γ^{α} of the magnetization vector; moreover, this term may depend also on the components of the deformation tensor. The presence of such a term in the free energy makes it possible to explain the phenomenon of magnetostriction, i.e., the change in the shape and dimensions of a sample when one varies the orientation of the magnetization vector with respect to the crystallographic axes. This term will be called the magneto-elastic part of the free energy:

$$F_{\text{m.e.}} = F_{\text{m.e.}}(\gamma, A). \tag{2.5}$$

We shall expand Eq. (2.5) as a series in powers of the deformations:

$$F_{\text{m.e.}} = F_{\text{an}}(\gamma) + \sum F^{ik}(\gamma) A_{ik} + \sum F^{ik,\,lm}(\gamma) A_{ik} A_{lm} + \ldots \tag{2.6}$$

The first term in Eq. (2.6) depends only on the direction cosines of the magnetization vector and is called the free energy of the magnetocrystalline anisotropy. We shall now represent $F_{\text{an}}(\gamma)$ in the form of a power function of γ. The values of γ may be present in $F_{\text{an}}(\gamma)$ only in the form of invariant combinations, permitted by the lattice symmetry.

For crystals of the cubic system, the free energy of the anisotropy can be written as follows*:

$$F_{\text{an}} = K_1(\gamma_1^2\gamma_2^2 + \gamma_2^2\gamma_3^2 + \gamma_3^2\gamma_1^2) + K_2\gamma_1^2\gamma_2^2\gamma_3^2 + K_3(\gamma_1^2\gamma_2^2 + \gamma_2^2\gamma_3^2 + \gamma_3^2\gamma_1^2)^2 + \ldots \tag{2.7}$$

*Noting that $\gamma_1^2 + \gamma_2^2 + \gamma_3^2 = 1$, the expression (2.7) may be written in the equivalent form

$$F_{\text{an}} = \frac{1}{2}K_1 - \frac{1}{2}K_1(\gamma_1^4 + \gamma_2^4 + \gamma_3^4) + \ldots$$

For hexagonal crystals

$$F_{an} = K_1'\gamma_3^2 + K_2'\gamma_3^4 + \dots \qquad (2.8)$$

(here γ_3 is the cosine of the angle between the magnetization vector and the hexagonal axis of the crystal).

The constants $K_1, K_2, \dots$ and $K_1', K_2', \dots$ are known as the first, second, etc., anisotropy constants. The absolute values of the magnetocrystalline anisotropy constants usually decrease as one goes up the sequence.

The second term in Eq. (2.6) correlates the changes in the magnetic and elastic quantities and is known as the free energy of magnetostriction:

$$F_{m.s.} = \sum F^{ik}(\gamma) A_{ik}. \qquad (2.9)$$

Usually, the quantities $F^{ik}(\gamma)$ are also represented in the form of series in powers of γ, and only the first terms of these series are used.

If the dependence of the coefficients in the series on the direction cosines is ignored in the last term of Eq. (2.6), this term transforms into the usual expression for the elastic deformation energy:

$$F_{el} = \sum F^{ik,lm}(0) A_{ik} A_{lm}. \qquad (2.10)$$

A ferromagnet placed in an alternating magnetic field may absorb energy from this field. Under certain conditions, such absorption is of the resonance kind: it increases sharply when the alternating field frequency approaches some frequency which is characteristic of a given sample. * Usually, a constant magnetizing field H_0 is applied in addition to an alternating field h(t) in order to observe ferromagnetic resonance; this is known as resonance in an external field. However, if a sample has magnetocrystalline anisotropy, the resonance absorption may be observed also in $H_0 = 0$. This is known as intrinsic resonance.

* The classical theory of ferromagnetic resonance is presented in detail in a paper by Skrotskii and Kurbatov ["Ferromagnetic Resonance" (1961)] and in Gurevich's book (1960).

The alternating field displaces the spin moments from their equilibrium positions and these moments begin to precess about the direction of the constant magnetizing field (or the anisotropy field in the case of intrinsic resonance) at the frequency of the alternating field. A sharp increase in the absorption is observed at the alternating field frequencies close to the free precession frequency or the resonance frequency.

The resultant effect is the appearance, in the magnetization vector, of a component δM, which is proportional to the alternating field amplitude:

$$\delta M(t) = \chi h(t). \tag{2.11}$$

Such a dependence is observed only in weak fields. In strong fields, δM depends nonlinearly on h. The quantity χ is a tensor and is known as the complex magnetic susceptibility tensor. In general, we must assume that the total field acting on M is the sum of the external and internal fields: $H_{tot} = -\partial F/\partial M$, where F is the free energy of a sample. The nature of the phenomenon of ferromagnetic resonance depends strongly on the nature of the internal field, which is related, in particular, to the demagnetization factors and the anisotropy.

Using the phenomenological approach, we can establish only the general form of the expression for the free energy. In those cases when the concrete form of any dependence is required, for example, the temperature dependence of the anisotropy constants, it is necessary to use definite models. Later, we shall calculate the thermodynamic properties of a system from its Hamiltonian.

Chapter II

Spin Hamiltonians

We shall now discuss the auxiliary mathematical apparatus which is needed later (the second quantization method of diagonalization of the forms which are quadratic with respect to the second quantization operators). We shall also discuss the conditions under which the exact Hamiltonian of a system can be reduced to a spin Hamiltonian of the Heisenberg type, and we shall consider typical spin Hamiltonians.

Sec. 3. Second Quantization*

The second quantization representations of operators and wave functions are very effective in investigations of many-body problems in quantum mechanics. We shall present here the main ideas of the second quantization method, which we shall use extensively hereafter. The general description of the method can be found in any sufficiently full course on quantum mechanics, for example, that of Blokhintsev (1963) or Landau and Lifshits (1963).

We shall consider a system of N identical particles. The system of coordinates of the i-th particle, for example, three Cartesian coordinates and a spin variable, will be denoted by x_i ($i = 1, 2, \ldots, N$). We shall use $\mathscr{H}$ to denote the Hamiltonian of the system. For clarity, we shall assume that $\mathscr{H}$ has the following form:

$$\mathscr{H} = \sum_i T(i) + \frac{1}{2} \sum_{i \neq j} \Phi(i, j). \tag{3.1}$$

*See Fock (1932, 1957), Bogolyubov (1949).

where T(i) is the Hamiltonian of the i-th "free" particle, which is the sum of the kinetic energy operators and the energy operators for the particle in external fields (if such are present); $\Phi(i, j)$ is the energy of interaction between the i-th and j-th particles.

The problem of describing the behavior of a system is assumed to be solved, in principle, if we find the solution of Schrödinger's equation:

$$i\frac{\partial\psi}{\partial t} = \mathscr{H}\psi. \tag{3.2}$$

To simplify the writing down of the solution, we shall use the identity of the particles. We shall introduce a complete orthonormalized system of one-particle functions $\{\varphi_f(x)\}$; for example, we can use the eigenfunctions of the Hamiltonian T (x) of a free particle; then f may be interpreted as the number of a one-particle state. We shall expand an arbitrary wave function ψ as a Fourier series in terms of the functions φ_f:

$$\psi(x_1, \ldots, x_N) = \sum_{f_1, \ldots, f_N} \psi(f_1, \ldots, f_N)\, \varphi_{f_1}(x_1) \cdots \varphi_{f_N}(x_N). \tag{3.3}$$

The wave function $\psi(x_1, \ldots, x_N)$ should be symmetrical with respect to the permutation of the particles if the particles obey the Bose statistics, and antisymmetrical if the particles obey the Fermi statistics. Therefore, the Fourier transform $\psi(f_1, \ldots, f_N)$ of the wave function can be either a symmetrical or antisymmetrical function with respect to the permutation of the indices f.

Let us consider the Bose statistics case. We shall use P to denote the particle permutation operator. We shall multiply both sides of Eq. (3.3) by P and sum for all possible permutations of P:

$$\sum_P P\psi(x_1, \ldots, x_N) = \sum_{f_1, \ldots, f_N} \psi(f_1, \ldots, f_N) \sum_P P\varphi_{f_1}(x_1) \cdots \varphi_{f_N}(x_N).$$

Since $\psi(x_1, \ldots, x_N)$ is symmetrical with respect to the particle permutation, $P\psi(x_1, \ldots, x_N) = \psi(x_1, \ldots, x_N)$ and, consequently, we obtain

$$\psi(x_1, \ldots, x_N) = \sum_{f_1, \ldots, f_N} \frac{\psi(f_1, \ldots, f_N)}{\sum_P 1} \varphi^{(s)}_{f_1 \ldots f_N}(x_1, \ldots, x_N), \tag{3.4}$$

where $\sum_P 1 = N!$ is the number of all the permutations of N particles, including the trivial case, and $\varphi^{(s)}_{f_1 \ldots f_N}$ is the symmetrized product of one-particle functions:

$$\varphi^{(s)}_{f_1 \ldots f_N}(x_1, \ldots, x_N) = \sum_P P\varphi_{f_1}(x_1) \cdots \varphi_{f_N}(x_N). \tag{3.5}$$

Thus, any symmetrical wave function of N variables can be expanded as a Fourier series of symmetrized products of one-particle functions of the type given in Eq. (3.5). In this sense, the system of functions given by Eq. (3.5) is complete.

Since the particles are identical, it is not important which particular particles are in the state f_i. All that is important is the total number of particles n_{f_i} in that state. Therefore, the functions in Eq. (3.5) can be numbered by means of the numbers n_f for each one-particle state. Moreover, we can show that the functions (3.5) are orthogonal and that we can construct from them an orthonormalized system of functions:

$$\psi^{(s)}_{\ldots n_f \ldots}(x_1, \ldots, x_N) = \sqrt{\frac{\prod_f (n_f)!}{N!}} \varphi^{(s)}_{f_1 \ldots f_N}(x_1, \ldots, x_N), \tag{3.6}$$

$$\int \psi^{*(s)}_{\ldots n_f \ldots}(x_1, \ldots, x_N) \psi^{(s)}_{\ldots n'_f \ldots}(x_1, \ldots, x_s) \, dx_1 \ldots dx_N = \prod_f \Delta(n_f - n'_f),$$

where $\Delta(n - n') = 1$ if $n = n'$, and $\Delta(n - n') = 0$ if $n \neq n'$. The system (3.6) is also complete in the space of symmetrical wave functions. The function $\psi^{(s)}_{\ldots n_f \ldots}$ of Eq. (3.6) describes the state of a system of N free particles in which n_{f_1} particles are in the state f_1, n_{f_2} particles are in the state f_2, etc. The numbers n_f are called the occupation numbers, and they can assume any integral positive value:

$$n_f = 0,\ 1,\ 2,\ \ldots, \tag{3.7}$$

with the additional condition that the total number of occupied one-particle states is equal to the number of particles in the system:

$$\sum_f n_f = N. \tag{3.8}$$

The number of the wave function in Eq. (3.6) is the set of occupation numbers $\{\ldots n_f \ldots\}$. Therefore, the expansion of an arbitrary wave function as a Fourier series of functions such as in Eq. (3.6) has the form

$$\psi(x_1, \ldots, x_N) = \sum_{\ldots n_f \ldots} C(\ldots, n_f, \ldots)\varphi^{(s)}_{\ldots n_f \ldots}(x_1, \ldots, x_N). \tag{3.9}$$

The representation of wave functions in the form of Eq. (3.9) is known as the second quantization representation, and the functions $C(\ldots, n_f, \ldots)$ of the occupation numbers (Fourier coefficients) are called wave functions in the second quantization representation.

In the Fermi statistics case, we multiply Eq. (3.3) by $(-1)^P P$ and sum over all P [by definition, $(-1)^P$ is equal to 1 if the permutation of particles represented by P is even, and −1 if the permutation P is odd). Since $\psi(x_1, \ldots, x_N)$ is antisymmetrical, therefore

$$(-1)^P P\psi(x_1, \ldots, x_N) = \psi(x_1, \ldots, x_N)$$

and, consequently, we obtain an expression of the type

$$\psi(x_1, \ldots, x_N) = \sum_{f_1, \ldots, f_N} \frac{\psi(f_1, \ldots, f_N)}{\sum_P 1} \varphi^{(a)}_{f_1 \ldots f_N}(x_1, \ldots, x_N), \tag{3.10}$$

where $\varphi^{(a)}_{f_1 \ldots f_N}$ is the antisymmetrized product of one-particle functions:

$$\varphi^{(a)}_{f_1 \ldots f_N}(x_1, \ldots, x_N) = \sum_P (-1)^P P\varphi_{f_1}(x_1) \ldots \varphi_{f_N}(x_N) = \mathrm{Det}\,\|\varphi_{f_r}(x_s)\|; \tag{3.11}$$

the order of indices f in Eq. (3.10) is assumed to be fixed. The functions (3.11) are orthogonal; from them we can construct, exactly as before, a system of normalized functions:

$$\psi^{(a)}_{\ldots n_f \ldots}(x_1, \ldots, x_N) = \sqrt{\frac{\prod_f (n_f!)}{N!}}\, \varphi^{(a)}_{f_1 \ldots f_N}(x_1, \ldots, x_N). \tag{3.12}$$

The occupation numbers n_f can, in this case, assume only two values:

$$n_f = 0,\ 1, \tag{3.13}$$

since, even if two particles are in the same state, the function (3.11) [or (3.12)] is identically zero [two rows in the determinant (3.11) are then identical]; the occupation numbers satisfy the obvious additional condition

$$\sum_f n_f = N. \tag{3.14}$$

The second quantization representation for wave functions in the Fermi statistics case has the form

$$\psi(x_1, \ldots, x_N) = \sum_{\ldots, n_f, \ldots} C(\ldots, n_f, \ldots)\, \psi^{(a)}_{\ldots n_f \ldots}(x_1, \ldots, x_N), \tag{3.15}$$

where $C(\ldots, n_f, \ldots)$ is a wave function in the second quantization representation for the Fermi statistics case; its arguments n_f assume the values 0 and 1.

Let us assume that the functions $\psi(x_1, \ldots x_N)$ are normalized; then, from the orthogonality and normalization of the functions $\psi^{(s)}_{\ldots n_f \ldots}$ given in Eq. (3.6) or the functions $\psi^{(a)}_{\ldots n_f \ldots}$ given in Eq. (3.12), it follows that in both cases:

$$\int |\psi(x_1, \ldots, x_N)|^2 dx_1 \ldots dx_N = \sum_{\ldots, n_f, \ldots} |C(\ldots, n_f, \ldots)|^2. \tag{3.16}$$

Consequently, the quantity $|C(\ldots,n_f,\ldots)|^2$ may be interpreted as the probability that a system will remain in the state characterized by a given set of the occupation numbers $\{\ldots,n_f,\ldots\}$ of one-particle states.

We shall now establish the rules by means of which the operators in the coordinate space are associated with the corresponding operators in the second quantization representation.

To define an operator completely, it is also necessary to specify the space of functions in which it applies. Therefore, when we speak of operators, we understand that they apply in the space of symmetrical wave functions if we are dealing with a system of Bose particles, and in the space of antisymmetrical functions if we are dealing with a system of Fermi particles.

We shall consider one-particle, two-particle, etc., operators (i.e., operators which depend on the coordinates of single particles, pairs of particles, etc.):

$$\mathcal{A}_1 = \sum_{1\leq i\leq N} \mathcal{A}_1(i), \quad \mathcal{A}_2 = \sum_{1\leq i<j\leq N} \mathcal{A}_2(i, j), \ \ldots, \tag{3.17}$$

where $\mathcal{A}(i)$, $\mathcal{A}(i, j), \ldots$ are the operators which depend on the coordinates of the i-th particle, i-th and j-th particles, etc. In the coordinate representation, they are described by the matrices

$$\begin{aligned} \mathcal{A}_1(i) &= \mathcal{A}_1(x_i \mid x_i') \prod_{r\neq i} \delta(x_r - x_r'), \\ \mathcal{A}_2(i, j) &= \mathcal{A}_2(x_i, x_j \mid x_j' x_i') \prod_{r\neq i, j} \delta(x_r - x_r'), \end{aligned} \tag{3.18}$$

where $\delta(x - x')$ is the Dirac δ-function. We shall confine ourselves to the consideration of one-particle and two-particle operators; the generalization to more complex cases is self-evident.

For brevity, we shall use x to denote the system of the coordinates of all the particles, and n to denote the set of occupation numbers which satisfy the functions (3.6) or (3.12):

$$x = \{x_1, \ \ldots, \ x_N\},$$

$$n = \{\ldots, \ n_f, \ \ldots\}.$$

Since the system of functions $\{\varphi_f(x)\}$ is complete, the formulas (3.17)-(3.18) may be rewritten in the following form*:

$$\mathcal{A}_1 = \mathcal{A}_1(x, x') = \\ = \sum_{f, f'} A_1(f \mid f') \sum_i \varphi_f(x_i) \varphi^*_{f'}(x'_i) \prod_{r \neq i} \delta(x_r - x'_r),$$

$$\mathcal{A}_2 = \mathcal{A}_2(x, x') = \\ = \frac{1}{2} \sum_{f_1, f_2, f'_2, f'_1} A_2(f_1, f_2 \mid f'_2 f'_1) \sum_{i, j} \varphi_{f_1}(x_i) \varphi_{f_2}(x_j) \times \\ \times \varphi^*_{f'_2}(x_j) \varphi^*_{f'_1}(x_i) \prod_{r \neq i, j} \delta(x_r - x_r). \tag{3.19}$$

where

$$A_1(f \mid f') = \int \varphi^*_f(x) A_1(x \mid x') \varphi_{f'}(x')\, dx\, dx',$$

$$A_2(f_1, f_2 \mid f'_2, f'_1) = \\ = \int \varphi^*_{f_1}(x_1) \varphi^*_{f_2}(x_2) A_2(x_1, x_2 \mid x'_2, x'_1) \times \\ \times \varphi_{f'_2}(x'_2) \varphi_{f'_1}(x'_1)\, dx_1 \ldots dx'_1. \tag{3.20}$$

We shall now employ standard formulas to transform the coordinate representation of the operators $\mathcal{A}_1$ and $\mathcal{A}_2$ into the second quantization representation.

We shall now consider some operator $\mathcal{A}$ (for example, $\mathcal{A}_1$ or $\mathcal{A}_2$) in the coordinate representation. By definition, the application of the operator $\mathcal{A}$ (x, x') to some function ψ (x) yields a new function ψ '(x), where

$$\int \mathcal{A}(x, x') \psi(x')\, dx' = \psi'(x). \tag{3.21}$$

We shall expand the functions ψ, ψ ' as a complete orthonormalized system (3.6) or (3.12), and substitute them into Eq. (3.21):

---

* The equivalence of the formulas (3.17)-(3.18) and (3.19)-(3.20) can be easily checked using the relationship

$$\sum_f \varphi^*_f(x) \varphi_f(x') = \delta(x - x').$$

$$\sum_{n'} \int \mathcal{A}(x, x') \psi_{n'}(x') \, dx' \, C(n') = \sum_{n''} \psi_{n''}(x) C'(n'').$$

We shall multiply the above equation on the left by $\psi_n^*(x)$ and integrate with respect to x. Since the functions $\psi_n(x)$ are orthonormalized, we shall obtain

$$\sum_{n'} \mathcal{A}(n, n') C(n') = C'(n), \tag{3.22}$$

where

$$\mathcal{A}(n, n') = \int \psi_n^*(x) \mathcal{A}(x, x') \psi_{n'}(x') \, dx \, dx'. \tag{3.23}$$

The matrix $\mathcal{A}(n, n')$ transforms a wave function in the second quantization representation, C(n), into some other function C'(n). Therefore, the expression (3.23) can be considered as an operator $\mathcal{A}$ in the second quantization representation.

In accordance with Eq. (3.23), the expressions for the operators $\mathcal{A}_1$ and $\mathcal{A}_2$ have the following form in the second quantization representation:

$$\begin{aligned} \mathcal{A}_1(n, n') &= \sum_{f, f'} A(f, f') a_{f'f}(n, n'), \\ \mathcal{A}_2(n, n') &= \frac{1}{2} \sum_{f_1, f_2, f_2', f_1'} A(f_1, f_2 | f_2', f_1') a_{f_1' f_2' f_2 f_1}(n, n'), \end{aligned} \tag{3.24}$$

where the quantities

$$\begin{aligned} a_{f'f}(n, n') = \\ = N \int \psi^*_{\ldots n_f \ldots}(x_1, x_2, \ldots, x_N) \varphi_f(x_1) \varphi^*_{f'}(x_1') \times \\ \times \psi_{\ldots n_f' \ldots}(x_1', x_2, \ldots, x_N) \, dx_1' \, dx_1 \ldots dx_N, \end{aligned} \tag{3.25}$$

$$\begin{aligned} a_{f_1' f_2' f_2 f_1}(n, n') = \\ = N(N-1) \int \psi^*_{\ldots n_f \ldots}(x_1, x_2, x_3, \ldots, x_N) \times \\ \times \varphi_{f_1}(x_1) \varphi_{f_2}(x_2) \varphi^*_{f_2'}(x_2') \varphi^*_{f_1'}(x_1') \times \\ \times \psi_{\ldots n_f' \ldots}(x_1', x_2', x_3, \ldots, x_N) \, dx_1' \, dx_2' \, dx_1 \ldots dx_N \end{aligned}$$

are considered to be some new operators in the second quantization representation.

Direct calculations show [cf. Bogolyubov (1949)] that the quantities (3.25) are expressed in terms of matrix elements of the products of operators, which, when applied to functions of the occupation numbers, alter the arguments of these functions by unity.

For a system of Bose particles we obtain the following expressions for the quantities given in Eq. (3.25):

$$a_{f'f}(n, n') = \{\beta_f^+ \beta_{f'}\}_{nn'},$$
$$a_{f_1' f_2' f_2 f_1}(n, n') = \left\{\beta_{f_1}^+ \beta_{f_2}^+ \beta_{f_2'} \beta_{f_1'}\right\}_{nn'}; \tag{3.26}$$

the operators β, β^+ satisfy the commutation relationships

$$\beta_f \beta_{f'}^+ - \beta_{f'}^+ \beta_f = \Delta(f - f'); \quad \beta_f \beta_{f'} - \beta_{f'} \beta_f = 0,$$
$$n_f = \beta_f^+ \beta_f = 0, \ 1, \ 2, \ \ldots \tag{3.27}$$

The operators β_f and β_f^+ are adjoint, and the operator n_f is self-adjoint. The matrix elements of the operators β, β^+ have the form

$$\{\beta_f\}_{\ldots n_f \ldots; \ldots n_f' \ldots} = \{\beta_f^+\}_{\ldots n_f' \ldots; \ldots n_f \ldots} =$$
$$= \sqrt{n_f}\, \Delta(n_f' - n_f - 1) \prod_{g \neq f} \Delta(n_f - n_g). \tag{3.28}$$

Relationships of the (3.27) type are known as the Bose commutation relationships, and the operators which satisfy them are called the Bose operators.

To prove this, we shall expand the function (3.6) as a Fourier series in one-particle functions $\varphi_g(x_1)$. The quantity (3.5), considered as a function of the variable x_1, is a linear combination of the functions

$$\varphi_{f_1}(x_1), \ \varphi_{f_2}(x_1), \ldots, \ \varphi_{f_N}(x_1).$$

Therefore, the coefficients of the expansion will be symmetrized products of the type:

$$\varphi_{\bar{f}_2}(x_2)\, \varphi_{\bar{f}_3}(x_3) \ldots \varphi_{\bar{f}_N}(x_N),$$

where $(\bar{f}_2, \ldots, \bar{f}_N)$ is found from the set of indices $(f_1, f_2, \ldots, f_N)$ by deleting one of them which is equal to g. Therefore, the function (3.5) can be written in the form:

$$\varphi^{(s)}_{f_1 \ldots f_N}(x_1, \ldots, x_N) = \sum_g \varphi_g(x_1)\, \varphi^{(s)}_{\bar{f}_2 \ldots \bar{f}_N}(x_2, \ldots, x_N).$$

Using this expression, we shall write Eq. (3.6) as follows:

$$\psi^{(s)}_{\ldots n_f \ldots}(x_1, \ldots, x_N) = \sum_g \sqrt{\frac{\bar{n}_g \prod_f (\bar{n}_f)!}{N\,(N-1)!}}\, \varphi_g(x_1)\, \varphi^{(s)}_{\bar{f}_2, \ldots \bar{f}_N}(x_2, \ldots, x_N),$$

where $(\ldots, \bar{n}_f, \ldots)$ is obtained from the set of occupation numbers $(\ldots, n_f, \ldots)$ by replacing n_g with $n_g - 1$. By definition:

$$\sqrt{\frac{\prod_f (\bar{n}_f)!}{(N-1)!}}\, \varphi^{(s)}_{\bar{f}_2 \ldots \bar{f}_N}(x_2, \ldots, x_N) = \psi^{(s)}_{\ldots \bar{n}_f \ldots}(x_2, \ldots, x_N).$$

Consequently, the required expansion has the form:

$$\psi^{(s)}_{\ldots n_f \ldots}(x_1, \ldots, x_N) = \sum_g \sqrt{\frac{n_g}{N}}\, \varphi_g(x_1)\, \psi^{(s)}_{\ldots \bar{n}_f \ldots}(x_2, \ldots, x_N).$$

We can easily see that the integrals with respect to x_1 and x_1' in Eq. (3.25) are the Fourier coefficients for the functions $\psi^*{}_{\ldots n_f \ldots}$ and $\psi_{\ldots n_f \ldots}$. Consequently:

$$a_{f'f}(n, n') =$$

$$= \sqrt{n_f n_{f'}} \int \overset{*}{\psi}{}^{(s)}_{\ldots \bar{n}_f \ldots}(x_2, \ldots, x_N)\, \psi^{(s)}_{\ldots \bar{n}_{f'}' \ldots}(x_2, \ldots, x_N)\, dx_2 \ldots dx_N =$$

$$= \sqrt{n_f n_{f'}} \prod_{\bar{f}} \Delta\left(n_{\bar{f}} - n_{\bar{f}}' - \Delta(\bar{f} - f) + \Delta(\bar{f} - f')\right).$$

We shall take into account that, for f = f':

$$a_{ff}(n, n') = n_f \prod_{\bar{f}} \Delta(n_{\bar{f}} - n_{\bar{f}}') = n_f \{1\}_{n, n'}.$$

and for f ≠ f':

$$a_{f'f}(n, n') = \sqrt{n_f n_f'}\, \Delta(n_f - n_f' - 1)\, \Delta(n_{f'} - n_{f'}' + 1) \prod_{\bar{f} \neq f, f'} \Delta(n_{\bar{f}} - n_{\bar{f}}') =$$

$$= \sqrt{n_f}\,\sqrt{n_{f'}+1}\,\Delta(n_f - n_f' + 1)\,\Delta(n_{f'} + n_{f'}' + 1)\prod_{\bar{f}\neq f, f'}\Delta(n_{\bar{f}} - n_{\bar{f}}').$$

We shall now introduce the operators β_f, $\overset{+}{\beta}_f$, which have the following form in the matrix representation:

$$\{\beta_f\}_{n,n'} = \sqrt{1+n_f}\,\Delta(n_f - n_f' + 1)\prod_{\bar{f}\neq f}\Delta(n_{\bar{f}} - n_{\bar{f}}');$$

$$\{\overset{+}{\beta}_f\}_{n,n'} = \sqrt{n_f}\,\Delta(n_f - n_f' - 1)\prod_{\bar{f}\neq 1}\Delta(n_{\bar{f}} - n_{\bar{f}}').$$

They can also be introduced as operators which act in the following manner on functions of occupation numbers:

$$\beta_f F(n_f) = \sqrt{1+n_f}\,F(n_f+1);$$

$$\overset{+}{\beta}_f F(n_f) = \sqrt{n_f}\,F(n_f-1).$$

Consequently, the formula for $a_{f'f}(n, n')$ transforms into:

$$a_{f'f}(n,n') = \sqrt{n_f}\,\Delta(n_f - n_f' - 1)\prod_{\bar{f}\neq f f'}\Delta(n_{\bar{f}} - n_f')\times$$

$$\times\sqrt{1+n_{f'}}\,\Delta(n_{f'} - n_{f'}' + 1)\prod_{\bar{f}\neq f f'}\Delta(n_{\bar{f}} - n_f') = \left\{\overset{+}{\beta}_f\,\beta_{f'}\right\}_{nn}.$$

This completes the derivation of the first of the two relationships in Eq. (3.26). To derive the second relationship, we shall consider the product of two arbitrary dynamic quantities of the additive type:

$$\sum_{1\leq i\leq N}\mathcal{A}(i)\cdot\sum_{1\leq j\leq N}\mathcal{B}(j) \equiv$$

$$\equiv\sum_{1\leq i\leq N}\mathcal{A}(i)\,\mathcal{B}(i) + \sum_{1\leq i<j\leq N}\{\mathcal{A}(i)\,\mathcal{B}(j) + \mathcal{A}(j)\,\mathcal{B}(i)\}.$$

We shall now go over to the second quantization representation in the above identity. Using the expressions in Eq. (3.24), we obtain:

$$\sum_{f_1 f_1'} A(f_1, f_1')\,a_{f_1', f_1}\sum_{f_2 f_2'} B(f_2, f_2')\,a_{f_2' f_2} =$$

$$= \sum_{f_1 f_2' f_1'} A(f_1, f_1') B(f_1', f_2') a_{f_2' f_1}$$

$$+ \frac{1}{2} \sum_{f_1 f_2 f_2' f_1'} \left\{ A(f_1 f_1') B(f_2 f_2') + A(f_2 f_2') B(f_1 f_1') \right\} a_{f_1' f_2' f_2 f_1} .$$

We shall take into account the fact that the following symmetry condition follows from the definition of the operators $a_{f_1' f_2' f_2 f_1}$ in Eq. (3.25):

$$a_{f_1 f_2 f_2' f_1'} = a_{f_2 f_1 f_1' f_2'},$$

and we shall rewrite the preceding equation in the form:

$$\sum A(f_1, f_1') B(f_2, f_2') \left\{ a_{f_1' f_1} a_{f_2' f_2} - \Delta(f_1' - f_2) a_{f_2' f_1} - a_{f_1' f_2' f_2 f_1} \right\} = 0$$

Because the quantities $\mathscr{A}$, $\mathscr{B}$ are arbitrary, we obtain:

$$a_{f_1' f_2' f_2 f_1} = a_{f_1' f_1} a_{f_2' f_2} - \Delta(f_1' - f_2) a_{f_2' f_1} .$$

Substituting into the above equations the expressions for $a_{f'f}$ in the terms of $\beta_f^+ \beta_f$, we finally have:

$$a_{f_1' f_2' f_2 f_1}(n, n') = \left\{ \overset{+}{\beta}_{f_1} \beta_{f_1'} \overset{+}{\beta}_{f_2} \beta_{f_2'} - \Delta(f_1' - f_2) \overset{+}{\beta}_{f_1} \beta_{f_2'} \right\}_{nn'} =$$

$$= \left\{ \overset{+}{\beta}_{f_1} \overset{+}{\beta}_{f_2} \beta_{f_2'} \beta_{f_1'} \right\}_{nn'} .$$

Similarly, we can obtain the representation of the quantities $a_{f_1' \ldots f_s' f_s \ldots f_1}(n, n')$ in terms of the second quantization operators (3.27)-(3.28).

For a system of Fermi particles, we correspondingly have

$$a_{f'f}(n, n') = \{\alpha_f^+ \alpha_{f'}\}_{nn'},$$
$$a_{f_1' f_2' f_2 f_1}(n, n') = \left\{ \alpha_{f_1}^+ \alpha_{f_2}^+ \alpha_{f_2'} \alpha_{f_1'} \right\}_{nn'} \tag{3.29}$$

and the commutation relationships for the operators α, α^+ are

$$\alpha_f \alpha_{f'}^+ + \alpha_{f'}^+ \alpha_f = 0, \quad \alpha_f \alpha_{f'} + \alpha_{f'} \alpha_f = 0,$$
$$n_f = \alpha_f^+ \alpha_f = 0, 1; \tag{3.30}$$

the operators α_f and α_f^+ are also adjoint and n_f is self-adjoint. Relationships of the type of Eq. (3.30) are known as the Fermi commutation relationships, and the operators α, α^+ which satisfy these relationships are called the Fermi operators.

Formulas (3.29)-(3.30) can be derived in the same way as in the case of Bose particles.

From the formulas just given, it is evident that the operator β_f (or α_f), acting on a function of the occupation numbers, reduces its argument n_f by unity and the operator β_f^+ (or α_f^+) increases it by unity. Therefore, they are called the particle annihilation and creation operators, respectively. The application of the operator β_f (or α_f) to a function with the argument $n_f = 0$ yields zero; this means that we cannot reduce the number of particles in a state f if there are no particles in that state. The application of the Fermi operators α_f^+ to a function with the argument $n_f = 1$ also gives zero because not more than one Fermi particle can be in any one state.

The operators given in Eq. (3.24) act on functions of the occupation numbers; they can also be considered as the matrix elements of the operators $\mathcal{A}_1$, $\mathcal{A}_2$ in terms of functions of the occupation numbers. Consequently, in the operator form they can be written in the following way:

$$\begin{aligned} \mathcal{A}_1 &= \sum_{f,f'} A(f\,|\,f')\,\Delta_f^+\Delta_{f'}, \\ \mathcal{A}_2 &= \frac{1}{2}\sum_{f_1,f_2,f_2',f_1'} A(f_1,f_2\,|\,f_2',f_1')\,\Delta_{f_1}^+\Delta_{f_2}^+\Delta_{f_2'}\Delta_{f_1'}, \end{aligned} \tag{3.31}$$

where $\Delta_f = \beta_f$ of Eq. (3.27) in the Bose-system case, and $\Delta_f = \alpha_f$ of Eq. (3.30) in the Fermi-system case.

It should be mentioned that, if initially $\mathcal{A}_1$, $\mathcal{A}_2$ are defined for the space of symmetrical functions, when they are transformed to the second quantization representation they are defined in the space of the functions of the occupation numbers $n_f = 0, 1, 2, \ldots$; if initially $\mathcal{A}_1$, $\mathcal{A}_2$ are defined in the space of antisymmetrical functions, in the second quantization representation they are defined in the space of the functions of the occupation numbers $n_f = 0, 1$.

In the second quantization representation, the matrix form of

Schrödinger's equation (3.2) is

$$i\frac{\partial}{\partial t}C(n)=\sum_{n'}\mathscr{H}(n,n')C(n'), \tag{3.32}$$

where $\mathscr{H}$ (n,n') is a Hamiltonian in the second quantization representation. According to Eqs. (3.23) and (3.26) or (3.29), $\mathscr{H}$ has the form

$$\mathscr{H}(n,n')=\sum T(f\,|\,f')\{\Delta_f^+\Delta_{f'}\}_{nn'}+ \\ +\frac{1}{2}\sum\Phi(f_1,f_2\,|\,f_2',f_1')\left\{\Delta_{f_1}^+\Delta_{f_2}^+\Delta_{f_2'}\Delta_{f_1'}\right\}_{nn'}, \tag{3.33}$$

where

$$T(f\,|\,f')=\int\varphi_f^*(x)\,T(x)\,\varphi_{f'}(x)\,dx, \\ \Phi(f_1,f_2\,|\,f_2',f_1')= \\ =\int\varphi_{f_1}^*(x_1)\,\varphi_{f_2}^*(x_2)\,\Phi(x_1,x_2)\,\varphi_{f_2'}(x_2)\,\varphi_{f_1'}(x_1)\,dx_1\,dx_2. \tag{3.34}$$

In the operator form, Schrödinger's equation is given by

$$i\frac{\partial}{\partial t}C(n)=\mathscr{H}C(n), \tag{3.35}$$

where

$$\mathscr{H}=\sum T(f\,|\,f')\,\Delta_f^+\Delta_{f'}+ \\ +\frac{1}{2}\sum\Phi(f_1,f_2\,|\,f_2',f_1')\,\Delta_{f_1}^+\Delta_{f_2}^+\Delta_{f_2'}\Delta_{f_1'}. \tag{3.36}$$

To make our discussion definite, we shall consider the Bose statistics case. We shall define the operator function

$$\psi(x)=\sum_f\beta_f\varphi_f(x);\quad \psi^+(x)=\sum_f\beta_f^+\varphi_f^*(x). \tag{3.37}$$

It is easy to check that, by virtue of the commutation relationships for the operators β, β^+ given by Eq. (3.27) and the condition of completeness of $\{\varphi_f(x)\}$, the functions ψ (x) and ψ^+(x) satisfy the commutation relationships

$$\psi(x)\,\psi^+(x')-\psi^+(x')\,\psi(x)=\delta(x-x'), \\ \psi(x)\,\psi(x')-\psi(x')\,\psi(x)=0. \tag{3.38}$$

In the Bose statistics case, the corresponding operator functions

$$\psi(x)=\sum_f a_f \varphi_f(x), \quad \psi^+(x)=\sum_f a_f^+ \varphi_f^*(x) \tag{3.39}$$

satisfy the commutation relationships

$$\begin{gathered}\psi(x)\psi^+(x')+\psi^+(x')\psi(x)=\delta(x-x'),\\ \psi(x)\psi(x')+\psi(x')\psi(x)=0.\end{gathered} \tag{3.40}$$

The functions (3.37), (3.39) are known as the second quantization wave functions.

Using the function (3.37) or (3.39) – depending on the type of statistics – the operators $\mathcal{A}_1$, $\mathcal{A}_2$ of Eq. (3.31) can be written in the second quantization representation:

$$\mathcal{A}_1=\int \psi^+(x_1)\mathcal{A}(1)\psi(x_1)\,dx_1, \tag{3.41}$$

$$\mathcal{A}_2=\frac{1}{2}\int \psi^+(x_1)\psi^+(x_2)\mathcal{A}(1,\ 2)\psi(x_2)\psi(x_1)\,dx_1\,dx_2. \tag{3.42}$$

The Hamiltonian of the system (3.36) can also be written in terms of the second quantization wave functions:

$$\begin{aligned}\mathcal{H}=&\int \psi^+(x_1)T(1)\psi(x_1)\,dx_1+\\ &+\frac{1}{2}\int \psi^+(x_1)\psi^+(x_2)\Phi(1,\ 2)\psi(x_2)\psi(x_1)\,dx_1\,dx_2.\end{aligned} \tag{3.43}$$

A second quantization function can be considered to be a dynamic variable, whose evolution with time is governed by the equations of motion in the Heisenberg form:

$$i\frac{\partial}{\partial t}\psi(x)=\psi(x)\mathcal{H}-\mathcal{H}\psi(x). \tag{3.44}$$

The commutator on the right-hand side of Eq. (3.44) should be calculated taking into account the commutation relationships (3.38) or (3.40). Thus, we obtain

$$i\frac{\partial}{\partial t}\psi(x_1)=T(1)\psi(1)+\int \Phi(1,\ 2)\psi^+(x_2)\psi(x_2)\,dx_2\cdot\psi(x_1). \tag{3.45}$$

The form of the second term on the right-hand side resembles the potential for the interaction of particles in the self-consistent field equations. However, in contrast to these equations, Eq. (3.45) is an exact equation. This is because the quantity $\psi^{+}(x_2)\psi(x_2)dx_2$ is the exact value of the number of particles in the volume dx_2.

Sec. 4. Operator Form of the Perturbation Theory *

Without specifying a particular model, we shall consider the equation

$$(\mathscr{H}_0+\varepsilon\mathscr{H}_1+\varepsilon^2\mathscr{H}_2+\ldots-\mathscr{E})C=0, \tag{4.1}$$

where $\mathscr{H}=\mathscr{H}_0+\varepsilon\mathscr{H}_1+\varepsilon^2\mathscr{H}_2+\ldots$ is a self-adjoint operator, $\mathscr{E}$ is an eigenvalue, C is the corresponding eigenfunction, and ε is a small parameter ($\varepsilon \ll 1$). We shall present the operator form of the perturbation theory, assuming $\varepsilon\mathscr{H}_1, \varepsilon^2\mathscr{H}_2, \ldots$ to be small quantities of increasing order of smallness.

We shall use $\mathscr{E}_0$ and C_0 to denote an eigenvalue and the corresponding eigenfunction of the operator $\mathscr{H}_0$:

$$(\mathscr{H}_0-\mathscr{E}_0)C_0=0. \tag{4.2}$$

We shall assume that the level $\mathscr{E}_0$ is degenerate. Then the set of functions C_0, which are associated with this level, forms a linear space (any linear combination of these functions is also an eigenfunction of the operator $\mathscr{H}_0$, which is also associated with the level $\mathscr{E}_0$). We shall denote this space by $\mathscr{L}$. We shall also assume that all the remaining levels $\mathscr{E}_n$ (the "excited" states) of the operator $\mathscr{H}_0$ are separated by a gap from the level $\mathscr{E}_0$:

$$\mathscr{E}_n-\mathscr{E}_0\geqslant\Delta>0, \tag{4.3}$$

where Δ is some constant.

If Q is the degree of degeneracy of the level $\mathscr{E}_0$, we can use the functions C_0 to obtain Q linearly independent orthonormalized solutions $C_{0,q}$ ($q = 1, \ldots, Q$).

---

*See Bogolyubov (1949); and Bogolyubov and Tyablikov (1949a).

We shall use $\mathscr{P}$ to denote the operator which projects an arbitrary wave function C into the space $\mathscr{L}$:†

$$\mathscr{P}C = \sum_{q=1}^{Q} C_{0,q}(C^*_{0,q}, C). \tag{4.4}$$

We note that, by definition,

$$\mathscr{P}^2 = \mathscr{P}, \quad \mathscr{P}C = C_0. \tag{4.5}$$

We shall now substitute into Eq. (4.1) the expression

$$C = \mathscr{P}C + (1 - \mathscr{P})C = \mathscr{P}C + C_1. \tag{4.6}$$

The above two-term expression for C represents the separation of the part $\mathscr{P}$C which belongs to the space $\mathscr{L}$ of unperturbed functions of Eq. (4.2), from the orthogonal supplement $C_1 = (1 - \mathscr{P})C$ [according to Eq. (4.5), $\mathscr{P}C_1 = \mathscr{P}(1 - \mathscr{P})C \equiv 0$], which is due to the perturbation $\varepsilon\mathscr{H}_1 + \varepsilon^2\mathscr{H}_2 + \ldots$ As $\varepsilon \to 0$, Eq. (4.1) should assume the form of Eq. (4.2); consequently, C_1 should vanish at the same time as ε. We shall henceforth assume that $C_1 \sim O(\varepsilon)$.

The substitution gives

$$(\mathscr{E} - \mathscr{H}_0 - \varepsilon\mathscr{H}_1 - \varepsilon^2\mathscr{H}_2 - \ldots)\mathscr{P}C + \\ + (\mathscr{E} - \mathscr{H}_0 - \varepsilon\mathscr{H}_1 - \varepsilon^2\mathscr{H}_2 - \ldots)C_1 = 0. \tag{4.7}$$

We shall multiply the equation on the left by $\mathscr{P}$, bearing in mind that

$$\mathscr{P}\mathscr{H}_0 = \mathscr{H}_0\mathscr{P},$$
$$\mathscr{P}(\mathscr{E} - \mathscr{H}_0)C_1 = (\mathscr{E} - \mathscr{H}_0)\mathscr{P}(1 - \mathscr{P})C = 0.$$

This gives the following equation:

† (C_1^*, C_2) is used to denote the scalar product of the functions C_1, C_2, where $(C_1^*, C_2) = \int C_1^*(x)\, C_2(x)\, dx$ if these functions have continuous arguments, and $(C_1^*, C_2) = \sum_n C_1^*(n)\, C_2(n)$, if the functions have discrete arguments. The functions C_1 and C_2 are orthogonal if their product is equal to zero.

$$(\mathcal{E} - \mathcal{H}_0 - \varepsilon\mathcal{P}\mathcal{H}_1\mathcal{P} - \varepsilon^2\mathcal{P}\mathcal{H}_2\mathcal{P} - \ldots)\mathcal{P}C - \\ - \varepsilon\mathcal{P}\mathcal{H}_1C_1 - \varepsilon^2\mathcal{P}\mathcal{H}_2C_1 - \ldots = 0. \qquad (4.8)$$

We shall subtract Eq. (4.8) from Eq. (4.7):

$$(\mathcal{E} - \mathcal{H}_0 - \varepsilon\mathcal{H}_1 - \varepsilon^2\mathcal{H}_2 - \ldots + \varepsilon\mathcal{P}\mathcal{H}_1 + \\ + \varepsilon^2\mathcal{P}\mathcal{H}_2 + \ldots)C_1 + \varepsilon(\mathcal{P}\mathcal{H}_1\mathcal{P} - \mathcal{H}_1)\mathcal{P}C + \\ + \varepsilon^2(\mathcal{P}\mathcal{H}_2\mathcal{P} - \mathcal{H}_2)\mathcal{P}C + \ldots = 0. \qquad (4.9)$$

Since $\mathcal{E} - \mathcal{E}_0$ and C_1 are quantities of the first order of smallness, we shall assume

$$\mathcal{E} - \mathcal{E}_0 = \varepsilon\mathcal{E}_0^{(1)} + \varepsilon^2\mathcal{E}_0^{(2)} + \ldots, \\ C_1 = \varepsilon K_1 + \varepsilon^2 K_2 + \ldots \qquad (4.10)$$

Substituting these expansions into (4.9) and collecting terms of the same order of smallness, we obtain

$$(\mathcal{E}_0 - \mathcal{H}_0)K_1 = (\mathcal{H}_1 - \mathcal{P}\mathcal{H}_1\mathcal{P})\mathcal{P}C, \\ (\mathcal{E}_0 - \mathcal{H}_0)K_2 = (\mathcal{H}_2 - \mathcal{P}\mathcal{H}_2\mathcal{P})\mathcal{P}C + \\ + (\mathcal{H}_1 - \mathcal{P}\mathcal{H}_1 - \mathcal{E}_0^{(1)})K_1, \qquad (4.11) \\ \ldots\ldots\ldots\ldots\ldots\ldots$$

Hence, we find the expression for K_1:

$$K_1 = (\mathcal{H}_0 - \mathcal{E}_0)^{-1}(\mathcal{P}\mathcal{H}_1\mathcal{P} - \mathcal{H}_1)\mathcal{P}C. \qquad (4.12)$$

We note that the function $C' = (\mathcal{H}_1 - \mathcal{P}\mathcal{H}_1\mathcal{P})\mathcal{P}C$ is orthogonal to all functions in the space $\mathcal{L}$. Therefore, the result of applying the operator $(\mathcal{H}_0 - \mathcal{E}_0)^{-1}$ to this function will be finite by virtue of the condition (4.3), which defines the finite width of the gap between the levels $\mathcal{E}_0$ and $\mathcal{E}_n$ of the operator $\mathcal{H}_0$.

We shall denote the excited-state eigenfunctions of the operator $\mathcal{H}_0$ by C_n ($n \neq 0$) and assume, as usual, that the system of functions $\{C_0, C_n\}$ is complete. The expansion of the functions C' in terms of eigenfunctions of the operator $\mathcal{H}_0$ has the form

$$C' = \sum_{n \neq 0} C_n(C_n^*, C')$$

Applying the operator $(\mathscr{H}_0 - \mathscr{E}_0)^{-1}$ to both parts of the above equation, we obtain

$$(\mathscr{H}_0 - \mathscr{E}_0)^{-1} C' = \sum_{n \neq 0} C_n \frac{(C_n^*, C')}{\mathscr{E}_n - \mathscr{E}_0}.$$

Since the denominators of the terms on the right-hand side are finite by virtue of Eq. (4.3), the result of the application of the operator $(\mathscr{H}_0 - \mathscr{E}_0)^{-1}$ to C' will also be finite.

The substitution of Eq. (4.12) into the second of the two equations denoted by (4.11) gives us

$$\begin{aligned} K_2 = {} & (\mathscr{H}_0 - \mathscr{E}_0)^{-1} (\mathscr{P}\mathscr{H}_2\mathscr{P} - \mathscr{H}_2)\mathscr{P}C + \\ & + (\mathscr{H}_0 - \mathscr{E}_0)^{-1} (\mathscr{E}_0^{(1)} + \mathscr{P}\mathscr{H}_1 - \mathscr{H}_1) K_1 = \\ & = (\mathscr{H}_0 - \mathscr{E}_0)^{-1} (\mathscr{P}\mathscr{H}_2\mathscr{P} - \mathscr{H}_2)\mathscr{P}C + \\ & + (\mathscr{H}_0 - \mathscr{E}_0)^{-1} (\mathscr{E}_0^{(1)} + \mathscr{P}\mathscr{H}_1 - \mathscr{H}_1)(\mathscr{H}_0 - \mathscr{E}_0)^{-1} \times \\ & \times (\mathscr{P}\mathscr{H}_1\mathscr{P} - \mathscr{H}_1)\mathscr{P}C. \end{aligned} \tag{4.13}$$

We shall now substitute the expansions of Eq. (4.10) into Eq. (4.8):

$$\begin{aligned} & (\mathscr{E} - \mathscr{H}_0 - \varepsilon\mathscr{P}\mathscr{H}_1\mathscr{P} - \varepsilon^2\mathscr{P}\mathscr{H}_2\mathscr{P} - \ldots)\mathscr{P}C - \\ & - \varepsilon^2\mathscr{P}\mathscr{H}_1 K_1 - \varepsilon^3\mathscr{P}\mathscr{H}_1 K_2 - \varepsilon^3\mathscr{P}\mathscr{H}_2 K_1 - \ldots = 0. \end{aligned} \tag{4.14}$$

Since

$$\mathscr{P}K_1 = 0, \quad \mathscr{P}K_2 = 0,$$

we can write identically:

$$\begin{aligned} \mathscr{P}\mathscr{H}_1 K_1 &= \mathscr{P}(\mathscr{H}_1 - \mathscr{P}\mathscr{H}_1\mathscr{P}) K_1, \\ \mathscr{P}\mathscr{H}_1 K_2 &= \mathscr{P}(\mathscr{H}_1 - \mathscr{P}\mathscr{H}_1\mathscr{P}) K_2, \\ \mathscr{P}\mathscr{H}_2 K_1 &= \mathscr{P}(\mathscr{H}_2 - \mathscr{P}\mathscr{H}_2\mathscr{P}) K_1. \end{aligned} \tag{4.15}$$

Substituting into Eq. (4.14) the expressions (4.12) and (4.13) for K_1, K_2, ... and using Eqs. (4.15) and (4.5), we finally obtain an equation for the determination of the function C_0:

$$\begin{aligned} (\mathscr{E} - \mathscr{E}_0) C_0 = {} & \mathscr{P} \{\varepsilon\mathscr{H}_1 + \varepsilon^2\mathscr{H}_2 + \varepsilon^3\mathscr{H}_3 + \ldots - \\ & - \varepsilon^2 (\mathscr{H}_1 - \mathscr{P}\mathscr{H}_1\mathscr{P})(\mathscr{H}_0 - \mathscr{E}_0)^{-1}(\mathscr{H}_1 - \mathscr{P}\mathscr{H}_1\mathscr{P}) - \\ & - \varepsilon^3 (\mathscr{H}_2 - \mathscr{P}\mathscr{H}_2\mathscr{P})(\mathscr{H}_0 - \mathscr{E}_0)^{-1}(\mathscr{H}_1 - \mathscr{P}\mathscr{H}_1\mathscr{P}) + \end{aligned}$$

$$+\varepsilon^3(\mathcal{H}_1-\mathcal{P}\mathcal{H}_1\mathcal{P})(\mathcal{H}_0-\mathcal{E}_0)^{-1}(\mathcal{H}_1-\mathcal{P}\mathcal{H}_1-\mathcal{E}_0^{(1)})\times$$
$$\times(\mathcal{H}_0-\mathcal{E}_0)^{-1}(\mathcal{H}_1-\mathcal{P}\mathcal{H}_1\mathcal{P})-\varepsilon^3(\mathcal{H}_1-\mathcal{P}\mathcal{H}_1\mathcal{P})\times$$
$$\times(\mathcal{H}_0-\mathcal{E}_0)^{-1}(\mathcal{H}_2-\mathcal{P}\mathcal{H}_2\mathcal{P})+\varepsilon^4\ldots\}\mathcal{P}C_0. \quad (4.16)$$

The quantity $\mathcal{E}_0^{(1)}$ has the coefficient ε^3 in Eq. (4.16). Using Eq. (4.10), we shall write the expression for this quantity in the form

$$\varepsilon^3\mathcal{E}_0^{(1)}=\varepsilon^2(\mathcal{E}-\mathcal{E}_0)+\varepsilon^4\ldots$$

Therefore, to quantities of the order of ε^4, we have

$$\varepsilon^3\mathcal{E}_0^{(1)}=\varepsilon^2(\mathcal{E}^{(1)}-\mathcal{E}_0), \quad (4.17)$$

where $\mathcal{E}^{(1)}$ has the value of $\mathcal{E}$, calculated using Eq. (4.16), accurate to within quantities of the order of ε, inclusive.

In all these approximations, the equation for the wave function C_0 can be written in the form

$$(\mathcal{E}-\mathcal{E}_0)C_0=\widetilde{\mathcal{H}}C_0, \quad (4.18)$$

where the "equivalent" Hamiltonian $\widetilde{\mathcal{H}}$ is a self-adjoint operator which transforms the space $\mathcal{L}$ into itself.

We shall now consider the normalization condition for the function C_0.

The complete wave function C is normalized, as usual, to unity:

$$(C^*, C)=1. \quad (4.19)$$

Using Eq. (4.6), we obtain

$$(C_0^*, C_0)+(C_1^*, C_1)=1. \quad (4.20)$$

Hence, in accordance with Eqs. (4.10) and (4.12), we obtain

$$(C_0^*, C_0)=1-\varepsilon^2(K_1^*, K_1)-\varepsilon^3\ldots=$$
$$=1-\varepsilon^2(C_0^*, \mathcal{P}(\mathcal{P}\mathcal{H}_1\mathcal{P}-\mathcal{H}_1)(\mathcal{H}_0-\mathcal{E}_0)^{-2}\times$$
$$\times(\mathcal{P}\mathcal{H}_1\mathcal{P}-\mathcal{H}_1)C_0)-\varepsilon^3\ldots \quad (4.21)$$

This solves the problem of determining the operator form of the perturbation theory for a degenerate level.

The operator form of the perturbation theory is identical with the standard form of the perturbation theory for a degenerate level [cf., for example, textbooks on quantum mechanics by Blokhintsev (1963) or Landau and Lifshits (1963)]. The operator form is more convenient for higher degrees of degeneracy.

Sec. 5. Transformation Formulas for Spin Operators

Later, we shall use Hamiltonians written in terms of spin variables. In solving actual problems, it is sometimes more convenient to transform spin operators into new "particle creation and annihilation" operators. We shall therefore consider the formulas for the transformation into the Pauli operators, for the Holstein–Primakoff–Izyumov transformation, and for the transformation into Dyson's ideal spin-wave operators.

The components of a vector spin operator satisfy the following commutation relationships:

$$S_f^x S_{f'}^y - S_{f'}^y S_f^x = iS_f^z \Delta(f - f'),\quad S_f^y S_{f'}^z - S_{f'}^z S_f^y = iS_f^x \Delta(f - f'),$$
$$S_f^z S_{f'}^x - S_{f'}^x S_f^z = iS_f^y \Delta(f - f'), \tag{5.1}$$

where f is the number of a lattice site.

We shall allow for the fact that the square of a spin operator is $S(S+1)$, and that its z-component can only assume $2S+1$ values (in the representation in which the z-component is diagonal). This gives the following equations:

$$(S_f,\ S_f) = S(S+1), \tag{5.2}$$

$$\prod_{r=-S}^{S} (S_f^z - r) = 0. \tag{5.3}$$

In place of the operators S_f^α ($\alpha = x, y, z$), we shall introduce the operators

$$S_f^{\pm} = S_f^x \pm iS_f^y \quad \left[(S_f^+)^+ = S_f^-\right] \quad S_f^z. \tag{5.4}$$

It follows from Eq. (5.1) that these operators satisfy the following commutation relationships:

$$S_f^+ S_{f'}^- - S_{f'}^- S_f^+ = 2S_f^z \Delta(f - f'), \tag{5.5}$$

$$S_f^\pm S_{f'}^z - S_{f'}^z S_f^\pm = \mp S_f^\pm \Delta(f - f'), \tag{5.6}$$

and from Eqs. (5.2), (5.4), and (5.5) it follows that

$$\begin{aligned} S_f^- S_f^+ &= S(S+1) - S_f^z - (S_f^z)^2, \\ S_f^+ S_f^- &= S(S+1) + S_f^z - (S_f^z)^2. \end{aligned} \tag{5.7}$$

In some applications, it is more convenient to transform from the representation in which the quantization axis is the z-axis to the characteristic representation in which the quantization axis is the direction of a unit vector γ.

We shall consider the operators

$$S_f'^\alpha = \gamma_f^\alpha S_f^z + A_f^\alpha S_f^+ + A_f^{*\alpha} S_f^-, \quad \alpha = x, y, z, \tag{5.8}$$

where $S_f^\pm$, S_f^z are the spin operators of Eq. (5.4); the transformation coefficients (vectors γ_f, A_f) satisfy the conditions

$$\begin{gathered} (\gamma_f, \gamma_f) = 1, \ \gamma_f^\alpha = \gamma_f^{*\alpha}, \\ (A_f^*, A_f) = \frac{1}{2}, \quad (A_f, A_f) = 0, \quad (A_f, \gamma_f) = 0, \\ |A_f^\alpha|^2 = \frac{1}{4}(1 - (\gamma_f^\alpha)^2), \quad iA_f^\alpha = [\gamma_f \times A_f]_\alpha. \end{gathered} \tag{5.9}$$

In particular, we can use the following expressions for the vectors A_f:

$$\begin{aligned} A_f^x &= -e^{i\varphi_f}\frac{1-\gamma_f^z}{4} + e^{-i\varphi_f}\frac{1-\gamma_f^z}{4}, \\ A_f^y &= i\left(e^{i\varphi_f}\frac{1-\gamma_f^z}{4} + e^{-i\varphi_f}\frac{1-\gamma_f^z}{4}\right), \\ A_f^z &= \frac{1}{2}\sqrt{1-(\gamma_f^z)^2}, \ \tan\varphi_f = \frac{\gamma_f^y}{\gamma_f^x}. \end{aligned} \tag{5.10}$$

It can be easily checked that the operators S_f' of Eq. (5.8) satisfy the communtation relationships for the spin operators of Eq. (5.1) if the transformation coefficients γ_f, A_f satisfy the conditions of Eq. (5.9).

The quantization axis of the new spin operators S_f' coincides with the direction of the vector γ_f, because the projection of the vector S'_f onto γ_f is equal to S^z_f and is diagonal by definition, while the projection of S'_f onto any direction a, perpendicular to γ_f, is equal to zero:

$$(S'_f, \gamma_f) = S^z_f, \quad (S'_f, [a \times \gamma_f]) = 0. \tag{5.11}$$

In the special case where $\gamma^z_f = 1$, $\gamma^x_f = \gamma^y_f = 0$, $\varphi_f = \pi$, the quantization axis is the z-axis:

$$S'^x_f = \frac{1}{2}(S^-_f + S^+_f), \quad S'^y_f = \frac{i}{2}(S^-_f - S^+_f), \quad S'^z_f = S^z_f. \tag{5.12}$$

Representation of Spin Operators in Terms of Pauli Operators ($S = 1/2$). The Pauli operators obey the following commutation relationships:

$$\begin{gathered} b_f b^+_{f'} - b^+_{f'} b_f = (1 - 2n_f)\Delta(f - f'), \\ b_f b_{f'} - b_{f'} b_f = 0, \quad b^2_f = b^{+2}_f = 0, \\ n_f = b^+_f b_f = 0,\ 1. \end{gathered} \tag{5.13}$$

For spin $S = 1/2$, the operators b_f, b^+_f are identical with the operators S^+_f, S^-_f. The general transformation has the form

$$S^a_f = \gamma^a_f\left(\frac{1}{2} - n_f\right) + A^a_f b_f + A^{*a}_f b^+_f; \tag{5.14}$$

the transformation coefficients are given by Eqs. (5.9)-(5.10). In the special case given by Eq. (5.12), we have

$$S^x_f = \frac{1}{2}(b^+_f + b_f), \quad S^y_f = \frac{i}{2}(b^+_f - b_f), \quad S^z_f = \frac{1}{2} - n_f. \tag{5.15}$$

Holstein – Primakoff – Izyumov Representation ($S \geq 1/2$). Holstein and Primakoff (1940) proposed the following representation for spin operators in terms of the second quantization operators:

$$\begin{gathered} S^+_f = \sqrt{2S}\,\varphi(n_f)\,a_f, \quad S^-_f = \sqrt{2S}\,a^+_f\varphi(n_f), \\ S^z_f = S - n_f, \end{gathered} \tag{5.16}$$

where

$$\varphi(n_f) = \left(1 - \frac{n_f}{2S}\right)^{1/2}, \quad n_f = a^+_f a_f \tag{5.17}$$

and where the operators a, a^+ satisfy the Bose commutation relationships:

$$a_f a_{f'}^+ - a_{f'}^+ a_f = \Delta(f - f'), \quad a_f a_{f'} - a_{f'} a_f = 0, \\ n_f = a_f^+ a_f = 0, 1, \ldots \tag{5.18}$$

The operators (5.16)-(5.17) are defined in the space of functions of the occupation numbers $n_f = 0, 1, 2, \ldots$. They remain invariant (each of them transforms into itself) in the subspace of the functions of the occupation numbers $n_f = 0, 1, \ldots, 2S$ and $n_f \geq 2S + 1$ [cf., Holstein and Primakoff (1940)]. The subspace of functions of the occupation numbers $n_f \geq 2S + 1$ is called the nonphysical space; it corresponds to spin values higher than S. Only the physical states ($n_f = 0, 1, \ldots, 2S$) can be used in calculations, since the nonphysical states ($n_f \geq 2S + 1$) can give uncontrolled errors. It is difficult to allow for this additional condition directly. We can satisfy this condition by assuming [cf., Izyumov (1959)]

$$a_f a_{f'}^+ - a_{f'}^+ a_f = \left(1 - \frac{2S+1}{(2S)!} a_f^{+2S} a_f^{2S}\right) \Delta(f - f'), \\ a_f a_{f'} - a_{f'} a_f = 0, \quad a_f^{+2S+1} = a_f^{2S+1} = 0. \tag{5.19}$$

The operator on the right-hand side of the first formula in Eq. (5.19) can be written as follows:

$$a_f^{+2S+1} a_f^{2S+1} = \prod_{p=0}^{2S-1} (a_f^+ a_f - p). \tag{5.20}$$

Since

$$n_f C(\ldots, n_f^0, \ldots) = n_f^0 C(\ldots, n_f^0, \ldots),$$

therefore,

$$\prod_{p=0}^{2S-1} (n_f - p) C(\ldots, n_f^0, \ldots) = \prod_{p=0}^{2S-1} (n_f^0 - p) C(\ldots, n_f^0, \ldots).$$

Hence, we find that if $n_f^0 \leq 2S - 1$, one of the factors in the product on the right-hand side vanishes. Consequently, when the operator of Eq. (5.20) acts on a function of the occupation numbers, we obtain a nonzero result if the argument of the wave function is $n_f^0 \geq 2S$. Thus, the difference between the operators of Eq. (5.19) and the Bose operators is of importance only when strongly degenerate states are included (states with high values of n_f^0). At low temperatures, the probability of such states is not very high ($\langle n_f \rangle \ll 1$), and the operators of Eq. (5.19) may be regarded as the Bose operators. This approximation improves at higher values of S.

For S = ½, the operators of Eq. (5.19) are identical with the Pauli operators of Eq. (5.13), while for S ≥ 1, they can be regarded as generalizations of the Pauli operators.

The approximate formulas for the transformation of the spin operators into the operators a_f^+, a_f are obtained by expanding the operators of Eq. (5.17) as a series:

$$\varphi(n_f) = 1 - \frac{n_f}{4S} - \frac{n_f^2}{32S^2} - \dots \qquad (5.21)$$

Consequently, we obtain

$$\begin{aligned} S_f^+ &= \sqrt{2S}\left(a_f - \frac{1}{4S} n_f a_f - \frac{1}{32S^2} n_f^2 a_f - \dots\right), \\ S_f^- &= \sqrt{2S}\left(a_f^+ - \frac{1}{4S} a_f^+ n_f - \frac{1}{32S^2} a_f^+ n_f^2 - \dots\right). \end{aligned} \qquad (5.22)$$

We note that the use of the approximate formulas of Eq. (5.22) is justified for S >> 1; for S = ½, they are inaccurate.

Dyson's Ideal Spin-Wave Operators (S ≥ ½). According to Dyson (1956a, b), the operators for a real spin system may be associated, in some hypothetical space, with "ideal spin-wave operators," which possess Bose properties. Nearly independent excitations are meaningful only at low temperatures when the probabilities of the processes which are calculated by means of ideal spin waves are equal to the probabilities of the processes in a real system. Using these considerations, we can obtain [Maleev (1957)] the following representation for the spin operators:

$$\begin{aligned} S_f^+ &= \sqrt{2S}\left(1 - \frac{1}{2S}\alpha_f^+\alpha_f\right)\alpha_f, \quad S_f^- = \sqrt{2S}\,\alpha_f^+, \\ S_f^z &= S - \alpha_f^+\alpha_f; \end{aligned} \qquad (5.23)$$

here, α_f^+, α_f are the creation and annihilation operators for Dyson's ideal spin waves which obey the Bose commutation relationships. We note that the operators S_f^+ and S_f^- are not adjoint. Consequently, the transformation of Eq. (5.23) yields a non-Hermitian Hamiltonian.

Oguchi (1961) has shown that Dyson's (5.23) and Holstein–Primakoff's (5.16) representations are equivalent.

Let the Hamiltonian of a system $\mathscr{H}$ be written in terms of Holstein–Primakoff variables of Eq. (5.16). We shall transform $\mathscr{H}$ by means of a nonunitary matrix T:

$$\mathscr{H}' = T^{-1}\mathscr{H}T \quad (TT^{-1} = 1), \tag{5.24}$$

where $\mathscr{H}'$ is the Hamiltonian $\mathscr{H}$, in which the operators a^+, a are replaced with

$$a^{+\prime} = T^{-1}a^+T, \quad a' = T^{-1}aT. \tag{5.25}$$

We can easily see that the primed operators obey the same commutation relationships as the unprimed operators. The operators $a^{+\prime}$, a' are no longer adjoint in the Hermitian sense, but the eigenvalues of $\mathscr{H}'$ are the same as for $\mathscr{H}$.

The operators of Eq. (5.16) assume the form

$$\begin{aligned} S_f^+ &= \sqrt{2S}\left(1 - \frac{a_f^{+\prime} a_f'}{2S}\right)^{\frac{1}{2}} a_f', \\ S_f^- &= \sqrt{2S}\, a_f^{+\prime}\left(1 - \frac{a_f^{+\prime} a_f'}{2S}\right)^{\frac{1}{2}}, \\ S_f^z &= S - a_f^{+\prime} a_f'. \end{aligned} \tag{5.26}$$

We shall introduce the operators α^+, α using the formulas

$$a_f^{+\prime} = \alpha_f\left(1 - \frac{\alpha_f^+\alpha_f}{2S}\right)^{-\frac{1}{2}}, \quad a_f' = \left(1 - \frac{\alpha_f^+\alpha_f}{2S}\right)^{\frac{1}{2}} \alpha_f. \tag{5.27}$$

We can easily see that $\alpha_f^+\alpha_f = a_f^{+\prime}a_f'$, and that the new operators α^+, α satisfy the same commutation relationships as $a^{+\prime}$, a'. Substituting Eq. (5.27) into Eq. (5.26), we obtain a representation of the type of Eq. (5.23) for the operators $S_f^{\pm}$, S_f^z. This proves the equivalence of the Dyson and Holstein–Primakoff representations.

The problem of deriving the creation and annihilation operators in Dyson's theory was considered also by Praveczki (1961), who obtained formulas equivalent to Eqs. (5.16) and (5.19). A critical analysis of the problems associated with the introduction of the ideal spin-wave operators was carried out by Dembinski (1964). In particular, Dembinski showed that, strictly speaking, these operators can be used only in the physical space.

Sec. 6. Heisenberg Model

To describe the behavior of a many-body system (in our case, a strongly magnetic substance), it is necessary to know the explicit form of the Hamiltonian of the system. Usually, no attempt is made to include all the features of the system in the Hamiltonian and only the most important characteristics are allowed for.

The following statement of the problem may be considered to be sufficiently general in the quantum theory of strong magnetism.

It is assumed that: 1) we know the type of atoms of which the crystal is composed, as well as the electron configurations of the partly filled (inner) atomic shells and the valence (outer) shells in the free state; 2) we know the external forces acting on the system; 3) we know the distribution of atoms over the crystal lattice sites; 4) the lattice is ordered.

From the Hamiltonian of such a system, we have to determine, at least approximately, the energy spectrum of the system, the lifetimes of the excited states, and such macroscopic quantities as the magnetization, specific heat, anisotropy constants, electrical conductivity, line width of (ferro) magnetic reasonance, dependence of these quantities on temperature, external fields, etc., allowing for the crystal lattice symmetry.

The problem presented in this form is far too complex for solution and even qualitative results can hardly be obtained. Therefore, in the theory of magnetism, it is normal to consider much simpler systems, which model only the most important (or those considered most important) characteristics of real magnetic substances. We shall follow the same procedure and consider only several simple models which are widely used in the quantum theory of magnetism.

A strongly magnetic substance is frequently considered as a system of spins located at lattice sites and coupled by the exchange interaction. We shall deduce the Hamiltonian for such a model.

We shall assume that a crystal consists of N identical atoms located at sites of a simple lattice. We shall assume that the interaction between the electrons in the partly filled inner shells and the electrons in the valence shells (or the conduction electrons) is weak. Then, the electrons in the partly filled shells and the conduction electrons can be regarded approximately as two independent subsystems. Since we are interested in the magnetic properties of the system, we shall consider only the electrons in the partly filled inner shells (d- and f -shells).

We shall assume that each atom has one d-electron. We shall neglect the orbital moment of the d-electron and the interaction of the electron magnetic moments with the orbital moments, as well as the interaction between the electrons themselves. In other words, we shall consider the d-electron as if it were an s-electron. For the time being, we shall assume that there is no external magnetic field.

Under these conditions, a real ferromagnet (antiferromagnet or ferrimagnet) is modeled by a ferromagnetic (antiferromagnetic or ferrimagnetic) dielectric, whose lattice consists of atoms of one kind, each having one valence s-electron in the normal state, and these s-electrons are regarded as responsible for the magnetism. We shall assume that the lowest energy level $\mathcal{E}_0$ of the electron system is characterized, in the zeroth approximation, by the occupation numbers N_f of the electrons at lattice sites being equal to unity:

$$N_f = n_{f,-\frac{1}{2}} + n_{f,\frac{1}{2}} = 1, \tag{6.1}$$

where $n_{f\sigma}$ ($\sigma = \pm\, ^1/_2$) is the number of electrons at a site f with the spin directed to the "left" ($\sigma = -\frac{1}{2}$) or to the "right" ($\sigma = \frac{1}{2}$), and that the level $\mathcal{E}_0$ is separated by a gap from the remaining (excited) states of the system. The interaction between electrons will be regarded as a perturbation.

We shall show that the level $\mathscr{E}_0$ is degenerate, and we shall investigate the splitting of the level $\mathscr{E}_0$ under the action of a perturbation [Bogolyubov (1949), Bogolyubov and Tyablikov (1949a)].

We shall use $\mathscr{H}_0$ to denote the zeroth-approximation Hamiltonian and C_0 to denote the eigenfunction of the ground state:

$$(\mathscr{H}_0 - \mathscr{E}_0) C_0 = 0. \tag{6.2}$$

The level $\mathscr{E}_0$ will be spin-degenerate, since it is defined only by specifying the value (unity) of the occupation numbers of electrons at the lattice sites, while the value of the spin at the site remains indeterminate ($\sigma = -\frac{1}{2}$ or $\frac{1}{2}$).

The functions C_0 are defined by the sets of the occupation numbers $n_{f\sigma}$:

$$C_0 = C_0(\ldots, n_{f\sigma}, \ldots). \tag{6.3}$$

Since any linear combination of these functions is also an eigenfunction of the operator $\mathscr{H}_0$, which is associated with the level $\mathscr{E}_0$, they form a linear space. We shall denote this space by $\mathscr{L}$.

We shall now calculate the splitting of the level $\mathscr{E}_0$ by a perturbation. Using the operator form of the perturbation theory of Sec. 4, we can show that in any given order of the perturbation theory, the splitting of the level $\mathscr{E}_0$ can be found from the equation

$$(\mathscr{E} - \mathscr{E}_0) C_0 = \tilde{\mathscr{H}} C_0, \tag{6.4}$$

where C_0 is a function in the $\mathscr{L}$ space, and $\tilde{\mathscr{H}}$ is some self-adjoint operator which transforms any function in the $\mathscr{L}$ space into another function in the same space.

Since there is one electron at each lattice site, and the direction of the spin of this electron is not specified ("left" or "right"), the level $\mathscr{E}_0$ may be determined by specifying the z-components of the electron spins at each site. Then, C_0 can be regarded as a function of the z-components of the spin operators:

$$C_0 = C_0(\ldots, S_f^z, \ldots). \tag{6.5}$$

Consequently, the equivalent Hamiltonian $\tilde{\mathscr{H}}$ of Eq. (6.4) can also be considered to be a function of the spin operators. Representing

$\tilde{\mathscr{H}}$ in the form of a series in powers of the spin operators S_f^α (α= x, y, z), we find

$$\tilde{\mathscr{H}} = G_0 + \sum G_\alpha(f) S_f^\alpha + \sum G_{\alpha_1\alpha_2}(f_1, f_2) S_{f_1}^{\alpha_1} S_{f_2}^{\alpha_2} + \\ + \sum G_{\alpha_1\alpha_2\alpha_3}(f_1, f_2, f_3) S_{f_1}^{\alpha_1} S_{f_2}^{\alpha_2} S_{f_3}^{\alpha_3} + \ldots, \qquad (6.6)$$

where the coefficients G are the usual functions of the numbers (coordinates) of the lattice sites. Summation is carried out over all values of f and α which occur in the above expressions, including only the combinations of unequal indices f. If this is not done, then, in accordance with the commutation relationships for the spin operators in the case S = $^1/_2$,

$$2S_f^x S_f^y = iS_f^z, \quad 2S_f^y S_f^z = iS_f^x, \quad 2S_f^z S_f^x = iS_f^y,$$
$$(S_f^x)^2 = (S_f^y)^2 = (S_f^z)^2 = \frac{1}{4},$$

these operators can be reduced to terms of lower order, in which all the indices f are different. Under these conditions, the summations of the operators S_f^α under the summation sign in Eq. (6.6) always commute. Since the operator $\tilde{\mathscr{H}}$ is self-adjoint, the coefficients G in the expansion of Eq. (6.6) are real functions.

We shall consider first only the electrostatic forces, ignoring the magnetic forces.

Since the electrostatic forces are independent of the spin orientation, $\tilde{\mathscr{H}}$ should be invariant under the spin-rotation transformation. Therefore, all the terms in Eq. (6.6) should be scalar functions of the spin vectors.

We shall carry out a canonical transformation of the spin operators*:

$$S_f^\alpha \to -S_f^\alpha, \quad i \to -i. \qquad (6.7)$$

The operator $\tilde{\mathscr{H}}$ remains invariant under this transformation, because the initial Hamiltonian is independent of the spin direction and is a function of operators with real coefficients.

*The transformations are called canonical if they conserve the commutation relationships.

Consequently, the operator $\tilde{\mathcal{H}}$ should be scalar, consisting of an even number of spin operators. Therefore, Eq. (6.6) may be rewritten in the form

$$\tilde{\mathcal{H}} = G_0 + \sum G(f_1, f_2)(S_{f_1}, S_{f_2}) + O\,(S_{f_1}^{\alpha_1} S_{f_2}^{\alpha_2} S_{f_3}^{\alpha_3} S_{f_4}^{\alpha_4}).$$

Introducing the symmetrized notation for the coefficients, we can also write

$$\tilde{\mathcal{H}} = G_0 - \frac{1}{2} \sum I(f_1, f_2)(S_{f_1}, S_{f_2}), \tag{6.8}$$

where I is the exchange integral:

$$I(f_1, f_2) = -G(f_1, f_2) - G(f_2, f_1),$$
$$I(f_1, f_2) = I(f_2, f_1). \tag{6.9}$$

Higher terms in Eq. (6.8) are terms of the fourth order in S_f^α.

Electrostatic interaction is unaffected by the reversal of the axes in the coordinate space:

$$f^\alpha \to -f^\alpha \quad (\alpha = x, y, z); \tag{6.10}$$

therefore,

$$I(f_1, f_2) = I(-f_1, -f_2). \tag{6.11}$$

Finally, we shall take into account the fact that atoms are located at crystal lattice sites, and crystal lattices are invariant under translations by a distance f_0, which is a multiple of the lattice constant:

$$f^\alpha \to f^\alpha + f_0^\alpha \quad (\alpha = x, y, z). \tag{6.12}$$

The quantities I should not be affected by this transformation and, consequently, they should be functions of relative distances only. If all lattice sites are equivalent, then

$$I(f_1, f_2) = I(|f_1 - f_2|). \tag{6.13}$$

Let us assume now that an external magnetic field H acts on the system. We shall use μ to denote the magnetic moment of an atom. Then, the Hamiltonian of an isotropic substance has the form

$$\tilde{\mathcal{H}} = \mathcal{C}_0 - \mu \sum (H, S_f) - \frac{1}{2} \sum I(f_1, f_2)(S_{f_1}, S_{f_2}). \tag{6.14}$$

Thus, using only the general features of the given model (cf. the beginning of the present section), we have established that the Hamiltonian for this model is written in terms of the spin operators of the electrons in the "unfilled" (partly filled) shells. This allows us to describe magnetic materials as assemblies of spins located at crystal lattice sites and interacting (in pairs) with one another with an energy equal to I. The explicit form of the integrals I is not defined in terms of atomic quantities, which are simply phenomenological parameters. Ferromagnetic ordering of the spins corresponds to a positive value of I, while antiferromagnetic ordering has negative I.

This model of a strongly magnetic substance is usually called the Heisenberg model and the corresponding Hamiltonian of Eq. (6.8) or Eq. (6.14) is known as the Heisenberg spin Hamiltonian. This model was discussed first by Frenkel' (1928) and Heisenberg (1928), who founded the modern theory of magnetism.

The possibility of reducing the Coulomb interaction to the effective spin Hamiltonian of Eq. (6.14) was first established by Dirac (1929,1960). Using a similar treatment, Arai (1928) has shown in a very general form that the Hamiltonian of Eq. (6.14) is the first approximation to the exact Hamiltonian, and that further corrections can be neglected if the overlap of the corresponding atomic wave functions can be regarded as small.

The Heisenberg Hamiltonian is frequently derived by means of the Heitler–London approximation, which is used in the theory of covalent bonding in molecules [Heitler and London (1927)]. Free atoms are regarded as the zeroth approximation, and the overlap between atomic wave functions of neighboring sites is treated as a small perturbation [Heisenberg (1928)]. In this way, we can obtain explicit expressions for the exchange integrals in terms of both the atomic wave functions and the energy of interaction between electrons [Slater (1953), Löwdin (1962), Bogolyubov (1949), Bogolyubov and Tyablikov (1949b)].

However, numerical calculations of I in terms of the atomic functions, carried out for metals in the iron group, gave contradictory results both in respect of the value of the integral I and its sign. None of these results was in agreement with expectations. Thus, Wohlfarth (1949) obtained a negative value for I. Stuart and Marshall (1960) obtained a positive value, but about two orders of magnitude smaller than that which would be expected from the Curie temperature. Freeman and Watson (1960) found that in the $3d^1$-configuration case, the exchange integral was large and negative, but in the $3d^9$-configuration it was small and negative. Fuller calculations by Freeman et al. (1962) also yielded negative values for the exchange integral.

In view of these results, suggestions were made [for example, Zener (1951), Goodenough (1960)] that the direct exchange could not be the cause of spin ordering. Allowance for the indirect exchange through the conduction electrons gives more reasonable values for the exchange integral of some rare-earth metals.

In spite of this, the spin Hamiltonians describe sufficiently well the magnetic properties of strongly magnetic substances. Therefore, we shall consider the exchange integrals as phenomenological quantities in our theory.

Sec. 7. Generalizations of the Heisenberg Model

In this section, we shall consider some simple generalizations of the Heisenberg model which can be obtained by taking into account interactions other than the electrostatic interaction between electrons in partly filled shells.

The Introduction of Magnetic Anisotropy. We shall consider first of all the influence of the magnetic interaction of electrons on the properties of magnetic substances. Without quoting in full, we shall simply indicate those changes which must be made in the expression for $\mathscr{H}$ [Bogolyubov (1949), Bogolyubov and Tyablikov (1949a)]. For simplicity, we shall assume that there is no external magnetic field.

The magnetic interaction between electrons [cf., for example, Bethe and Salpeter (1960)] represents the mutual interaction of the electron spins (the spin–spin interaction) and the interaction of the spin and orbital moments of electrons (the spin–orbit interaction). We shall consider now the derivation of these quantities because they are important in our treatment.

The contribution of the spin–spin interaction to the total Hamiltonian of a many-electron system has the form

$$\mathcal{H}'_{ss}=\frac{1}{2}\sum_{f_1,f_2} D(f_1, f_2)\{(S_{f_1}, S_{f_2})-3(e_{f_1f_2}, S_{f_1})(e_{f_1f_2}, S_{f_2})\}-$$
$$-\frac{1}{2}\sum_{f_1,f_2} 4\mu_B^2\frac{8\pi}{3}(S_{f_1}, S_{f_2})\delta(r_{f_1f_2}). \tag{7.1}$$

where

$$D(f_1, f_2)=\frac{4\mu_B^2}{|r_{f_1f_2}|^3}, \quad e_{f_1f_2}=\frac{r_{f_1f_2}}{|r_{f_1f_2}|}, \tag{7.2}$$

$\mu_B = e\hbar/2mc$ is the magnetic moment of an electron; $r_{f_1f_2} = r_{f_1} - r_{f_2}$ is the distance between the electrons at sites f_1 and f_2.

The contribution of the spin–orbit interaction has the form

$$\mathcal{H}'_{s.o}=\frac{\mu_B}{mc}\sum_{f}([\nabla_{r_f}U\times p_f], S_f)+$$
$$+\frac{\mu_B}{mc}\sum_{f_1,f_2}\frac{2e}{r^3_{f_1f_2}}([r_{f_1f_2}\times p_{f_2}], S_f). \tag{7.3}$$

where $\nabla_r U$ is the Coulomb field of all charges (atomic cores and electrons) acting on a given electron, and $p_f = -i(\partial/\partial r_f)$ is the momentum operator of the f-th electron. Sometimes, the treatment is restricted to the interaction of the electron spin with the orbital moment for a given atom. In that case, the expression (7.3) may be simplified to

$$\mathcal{H}''_{s.o}=\frac{\mu_B}{mc}\sum_{f}\frac{1}{r_f}\frac{\partial U}{\partial r_f}(L_f, S_f), \tag{7.4}$$

where L_f is the orbital moment operator of the electron of the f-th atom, and U is the potential energy of the electron in the field of the core of that atom.

It is important to note that the spin–spin interaction energy of Eq. (7.1) is a quadratic form in terms of the spin operators S_f, while the energy of the spin–orbit interaction of Eq. (7.3) or (7.4) is a bilinear form in terms of the spin operators S_f and the momentum operators p_f with real coefficients. In this case, the total Hamiltonian is not invariant under the spin-rotation transformation, but is invariant under the transformation of Eq. (6.7).

We shall now consider again the equivalent Hamiltonian in the general form of Eq. (6.6). We shall assume that in its derivation we included the magnetic interactions between electrons of the type given by Eqs. (7.1) and (7.3) or (7.4). Since the equivalent Hamiltonian $\tilde{\mathcal{H}}$ can undergo the same transformations as the initial Hamiltonian, it follows that $\tilde{\mathcal{H}}$ will not be invariant under the spin-rotation transformation, but will be invariant under the transformation of Eq. (6.7).

We shall bear in mind that the coefficients in the expansion in powers of the spin operators are real functions (because $\tilde{\mathcal{H}}$ is self-adjoint). Consequently, the equivalent Hamiltonian is a paired scalar with respect to the spin operators.

We shall write $\tilde{\mathcal{H}}$ in the following form:

$$\tilde{\mathcal{H}} = G_0 - \frac{1}{2}\sum I_{\alpha_1\alpha_2}(f_1, f_2)\, S_{f_1}^{\alpha_1} S_{f_2}^{\alpha_2} - \\ - \frac{1}{4!}\sum I_{\alpha_1\alpha_2\alpha_3\alpha_4}(f_1, f_2, f_3, f_4)\, S_{f_1}^{\alpha_1} S_{f_2}^{\alpha_2} S_{f_3}^{\alpha_3} S_{f_4}^{\alpha_4}. \tag{7.5}$$

We note that the quantities $I_{\alpha_1\alpha_2}$, $I_{\alpha_1\alpha_2\alpha_3\alpha_4}$ transform like tensors under the permissible transformations of spin and coordinate variables.

The quantity $I_{\alpha_1\alpha_2}$ is called the exchange interaction tensor. It satisfies the condition

$$I_{\alpha_1\alpha_2}(f_1, f_2) = I_{\alpha_2\alpha_1}(f_2, f_1) = I_{\alpha_2\alpha_1}(f_1, f_2). \tag{7.6}$$

Since the interactions considered here are not affected by the reversal of direction in the coordinate space, we have, in accordance with Eq. (6.10),

$$\begin{gathered} I_{\alpha_1\alpha_2}(f_1, f_2) = I_{\alpha_1\alpha_2}(-f_1, -f_2), \\ I_{\alpha_1\alpha_2\alpha_3\alpha_4}(f_1, f_2, f_3, f_4) = I_{\alpha_1\alpha_2\alpha_3\alpha_4}(-f_1, -f_2, -f_3, -f_4). \end{gathered} \tag{7.7}$$

Finally, from the invariance of the Hamiltonian of the system under translation by distances which are multiples of the lattice period [Eq. (6.12)] and from Eq. (7.7), it follows that the values of I for a simple lattice should be functions of relative distances:

$$I_{\alpha_1\alpha_2}(f_1, f_2)=I_{\alpha_1,\alpha_2}(|f_1-f_2|),$$

$$I_{\alpha_1\alpha_2\alpha_3\alpha_4}(f_1, f_2, f_3, f_4)=I_{\alpha_1\alpha_2\alpha_3\alpha_4}(|f_1-f_2|, |f_2-f_3|, \dots). \tag{7.8}$$

As in the preceding section, we can now include an external magnetic field by adding the Zeeman energy operator to the expression obtained.

In the spin Hamiltonian, we shall distinguish the isotropic and anisotropic parts of the spin interaction energy. The anisotropic part will be called the magnetocrystalline anisotropy energy operator. *

The isotropic part of the interactions is usually written in the form of Eq. (6.8). However, generally speaking, it may also include scalar functions of higher orders in powers of the spin operators:

$$-\tfrac{1}{2}\sum I(f_1, f_2)(S_{f_1}, S_{f_2})-$$
$$-\tfrac{1}{4!}\sum I(f_1, f_2, f_3, f_4)([S_{f_1}\times S_{f_2}], [S_{f_3}\times S_{f_4}])-\dots$$

---

*The exact form of the magnetocrystalline anisotropy energy operator is determined by the symmetry group of the magnetic lattice.

Belov, Neporova, and Smirnova (1956) and Tavger and Zaitsev (1956) tabulated the magnetic space and point groups. Indenbom (1960) showed that these groups were isomorphous with the usual space and point groups. Villain (1959), Yoshimori (1959), Kaplan (1959), Kaplan and Lyons (1960), Bertaut (1960, 1961a, b) et al. considered the distribution of magnetic moments in a lattice, making certain assumptions about the interaction energy of these moments (assumptions about the exchange integrals). Usually, only the translation symmetry of the lattice was allowed for.

Alexander (1962) considered the relationship between the magnetic structures and the crystal lattice symmetry. Dimmock and Wheeler (1962) investigated the symmetry properties of wave functions of magnetic crystals in the spin-wave approximation.

In calculations, the contribution of the fourth-power and higher forms is regarded as negligibly small and is ignored.* We shall ignore it in our treatment.

In a uniaxial ferromagnet, the lowest-order anisotropy energy has the following form in powers of the spin operators:

$$\mathcal{H}_A = -\frac{1}{2}\sum I_a(f_1, f_2) S^z_{f_1} S^z_{f_2}. \tag{7.9}$$

For a ferromagnet with cubic symmetry, the lowest-order anisotropy energy has the form

$$\mathcal{H}_A = -\frac{1}{4!}\sum_{(\alpha_1 \neq \alpha_2)} I_a(f_1, f_2, f_3, f_4) S^{\alpha_1}_{f_1} S^{\alpha_1}_{f_2} S^{\alpha_2}_{f_3} S^{\alpha_2}_{f_4}. \tag{7.10}$$

The quantities $I_a(f_1, f_2)$, $I_a(f_1, f_2, f_3, f_4)$ will be called the magnetocrystalline anisotropy integrals. From previous considerations, it is clear that these integrals are of the order of the spin–spin and spin–orbit interaction energies.

If the spin at a site is $S \geq 1$, then the expression for the Hamiltonian in terms of the spin operators may include products of the operators for a given site. In this case, the anisotropy energy operator will contain one term for each site. This part of the anisotropy is known as the one-ion anisotropy. For uniaxial and cubic crystals, this anisotropy is given by the expressions

$$\mathcal{H}'_A = -\frac{1}{2}\sum I_a(f)(S^z_f)^2, \tag{7.11}$$

$$\mathcal{H}'_A = -\frac{1}{4!}\sum I_a(f)\{(S^x_f)^4 + (S^y_f)^4 + (S^z_f)^4\}. \tag{7.12}$$

Expressions of this type are used, for example, in the theory of paramagnetic resonance [cf., Al'tshuler and Kozyrev (1961)].

Sometimes, the uniaxial anisotropy is introduced as an effective anisotropy field similar to an external magnetic field. We shall consider the expression (7.9) for the anisotropy energy. If the quantity $I_a(f_1, f_2)$ decreases sufficiently slowly when the distance $|f_1 - f_2|$ is increased, and if the lattice is simple, then the

*See also Arai (1962).

sum

$$\sum_{f'} I_a(f-f')\,S_{f'}^z$$

can be replaced approximately by a constant. We shall write this constant as μH_a^z, and then the expression (7.9) can be rewritten as follows:

$$\mathscr{H}_A \approx -\mu \sum H_a^z S_f^z. \tag{7.13}$$

Approximations of this type are used, for example, in the investigation of uniaxial antiferromagnets. However, in view of the very approximate nature of the expression (7.13), even the qualitative nature of the effects associated with the anisotropy is not described quite correctly (cf., Sec. 19).

The phenomenon of magnetic anisotropy in single crystals is associated, as already mentioned, with the existence of magnetic interactions between the electrons in the partly filled inner shells. The microscopic theory of magnetic anisotropy, based on the allowance for the magnetic interactions between electrons, has been considered by Heisenberg (1930), Powell (1930), Bloch and Gentile (1931), Van Vleck (1937, 1947), Vonsovskii (1940a), Potapkov (1962a), and others.

Hamiltonian with Dipole Interaction. Such a Hamiltonian was proposed by Holstein and Primakoff (1940). It can also be considered within the framework of our scheme.

From the quadratic form in Eq. (7.5) we shall take out the isotropic part, associated with the intrinsic exchange energy, and we shall assume that the anisotropic part is entirely due to the dipole interaction between the magnetic moments of electrons at different sites. The spin–orbit interaction will be ignored. Then, the Hamiltonian of the system becomes

$$\begin{aligned}\tilde{\mathscr{H}} = &-\sum (H, S_f) - \frac{1}{2}\sum I(f_1, f_2)(S_{f_1}, S_{f_2}) - \\ &-\frac{1}{2}\sum D(f_1, f_2)\{(S_{f_1}, S_{f_2}) - 3(e_{f_1f_2}, S_{f_1})(e_{f_1f_2}, S_{f_2})\},\end{aligned} \tag{7.14}$$

where $D(f_1, f_2)$ and $e^{\alpha}_{f_1f_2}$ are the coefficients of the dipole interaction:

$$D(f_1, f_2) = \frac{\mu^2}{|f_1 - f_2|^3}, \quad e^{\alpha}_{f_1f_2} = \frac{f_1^{\alpha} - f_2^{\alpha}}{|f_1 - f_2|}. \tag{7.15}$$

The dipole interaction is usually two to three orders of magnitude weaker than the exchange interaction, and it can therefore be considered to be a small perturbation. However, because of its long-range nature, the dipole interaction is, in some cases, very important. In particular, it leads to the appearance of the demagnetization factors.

Demagnetization Factors. In the case of an ellipsoid, these factors can be introduced without making an explicit allowance for the dipole interaction, using a simpler but less accurate method [Oguchi (1957)]. For this purpose, we add to the Hamiltonian of the system a demagnetization energy operator which we shall write, by analogy with the classical expression of Eq. (2.4), as follows:

$$\mathcal{H}_{\text{demag}} = \frac{1}{2N} \sum N_{\alpha\beta} \mathcal{M}^{\alpha} \mathcal{M}^{\beta}, \tag{7.16}$$

where $N_{\alpha\beta}$ are the classical demagnetization factors, and $\mathcal{M}^{\alpha}$ are the components of the operator of the total magnetic moment:

$$\mathcal{M}^{\alpha} = \mu \sum S_f^{\alpha}. \tag{7.17}$$

We note that in our previous considerations we made no assumptions about the nature of the spin ordering and, therefore, the results apply equally to ferromagnets, antiferromagnets, and ferrimagnets. Moreover, the results may be generalized to the case of lattices containing several types of "magnetic" ion.

Hamiltonian of the s – d Exchange Model.* In this model, we shall consider two groups of levels in the electron system: the d- (or f-) levels of the electrons in the inner partly filled shells and the s-levels of the valence electrons. The interaction between the electrons at the d- and s-levels is regarded as a small perturbation.

Using these assumptions [Vonsovskii and Turov (1953)], the Hamiltonian for this model may be written as follows:

$$\tilde{\mathcal{H}} = \mathcal{H}_{dd} + \mathcal{H}_{ss} + \mathcal{H}_{sd}, \tag{7.18}$$

*Second footnote, p. 21.

where $\mathscr{H}_{dd}$ represents the operator for the energy of interaction of the d-electrons among themselves, $\mathscr{H}_{ss}$ is the corresponding operator for the s-electrons, and $\mathscr{H}_{sd}$ is the operator for the energy of interaction between the d- and s-electrons:

$$\mathscr{H}_{dd}=-\frac{1}{2}\sum I(f_1, f_2)(S_{f_1}, S_{f_2}). \tag{7.19}$$

$$\mathscr{H}_{ss}=\sum E_{\nu\sigma}a^+_{\nu\sigma}a_{\nu\sigma}, \tag{7.20}$$

$$\begin{aligned}\mathscr{H}_{sd}=-\frac{1}{2N}\sum B(\nu_1-\nu_2)\,e^{-i(\nu_1-\nu_2,f)}\times\\ \times\Big\{S^z_f\Big(a^+_{\nu_1,-\frac{1}{2}}a_{\nu_2,-\frac{1}{2}}-a^+_{\nu_1,\frac{1}{2}}a_{\nu_2,\frac{1}{2}}\Big)+\\ +S^x_f\Big(a^+_{\nu_1,-\frac{1}{2}}a_{\nu_2,\frac{1}{2}}+a^+_{\nu_1,\frac{1}{2}}a_{\nu_2,-\frac{1}{2}}\Big)+\\ +iS^y_f\Big(a^+_{\nu_1,\frac{1}{2}}a_{\nu_2,-\frac{1}{2}}-a^+_{\nu_1,-\frac{1}{2}}a_{\nu_2,\frac{1}{2}}\Big)\Big\},\end{aligned} \tag{7.21}$$

where S^α_f is the spin operator of an electron at a site f; $a^+_{\nu\sigma}$, $a_{\nu\sigma}$ ($\sigma=\pm\,{}^1/_2$) are the Fermi operators representing the creation and annihilation of an s-electron in a state with a wave vector ν and a spin σ; $E_{\nu\sigma}$ is the energy of an electron in the state $(\nu\,,\sigma)$; I is the d–d exchange integral; B is the s–d exchange integral; N is the number of sites in the lattice.

We shall consider the physical meaning of the operator $\mathscr{H}_{sd}$. We shall use the notation

$$a_{g\sigma}=\frac{1}{\sqrt{N}}\sum_{\nu}a_{\nu\sigma}e^{i(\nu,g)},\quad I(f)=\frac{1}{N}\sum_{\nu}B(\nu)\,e^{i(f,\nu)}. \tag{7.22}$$

It is evident that the operators $a^+_{g\sigma}$, $a_{g\sigma}$ may be interpreted as the operators for the creation and annihilation of an s-electron with a spin σ at a site g. Using this notation, Eq. (7.21) can be rewritten in the form

$$\begin{aligned}\mathscr{H}_{sd}=-\frac{1}{2}\sum I(f-g)\Big\{S^z_f\Big(a^+_{g,-\frac{1}{2}}a_{g,-\frac{1}{2}}-a^+_{g,\frac{1}{2}}a_{g,\frac{1}{2}}\Big)+\\ +S^x_f\Big(a^+_{g,-\frac{1}{2}}a_{g,\frac{1}{2}}+a^+_{g,\frac{1}{2}}a_{g,-\frac{1}{2}}\Big)+\\ +iS^y_f\Big(a^+_{g,\frac{1}{2}}a_{g,-\frac{1}{2}}-a^+_{g,-\frac{1}{2}}a_{g,\frac{1}{2}}\Big)\Big\}.\end{aligned} \tag{7.23}$$

We shall define the following operators [Bogolyubov (1949)]:

$$\begin{aligned} \sigma_g^x &= \frac{1}{2}\left(a^+_{g,-\frac{1}{2}}a_{g,\frac{1}{2}} + a^+_{g,\frac{1}{2}}a_{g,-\frac{1}{2}}\right), \\ \sigma_g^y &= \frac{i}{2}\left(a^+_{g,\frac{1}{2}}a_{g,-\frac{1}{2}} - a^+_{g,-\frac{1}{2}}a_{g,\frac{1}{2}}\right), \\ \sigma_g^z &= \frac{1}{2}\left(a^+_{g,-\frac{1}{2}}a_{g,-\frac{1}{2}} - a^+_{g,\frac{1}{2}}a_{g,\frac{1}{2}}\right); \end{aligned} \tag{7.24}$$

we shall assume that these operators are defined in the space of functions of the occupation numbers satisfying the condition

$$n_{f,-\frac{1}{2}} + n_{f,\frac{1}{2}} = 1, \quad n_{f\sigma} = a^+_{f\sigma}a_{f\sigma} \quad \left(\sigma = \pm\frac{1}{2}\right). \tag{7.25}$$

It can easily be verified that the operators given in Eq. (7.24) will satisfy the commutation relationships of Eq. (5.1) for the spin operators (in the case of S = ½). The condition of Eq. (7.25) does not apply to the s-electrons, but if this can be ignored, then the operator for the s–d interaction assumes the form of the usual exchange interaction

$$\mathcal{H}_{sd} = -\sum I(f-g)(S_f, \sigma_g). \tag{7.26}$$

Thus, the expression for the interaction of the s- and d-electrons, assumed in the s–d model theory, can indeed be interpreted as the exchange interaction:

Spin – Phonon Hamiltonian. So far, we have assumed the lattice sites to be fixed. However, at nonzero temperatures, atoms execute thermal vibrations. These vibrations alter the energy of the exchange interaction and give rise to an interaction between spins and lattice vibrations (phonons).

For simplicity, we shall consider an isotropic ferromagnet. We shall use δf to denote the value of a displacement of an atom from its equilibrium position, and we shall expand the exchange integral in terms of relative displacements, using only terms linear in δf. Moreover, we shall represent the displacements in the form of a sum of normal modes, adding to the initial operator an operator for the energy of the natural vibrations of the lattice, and we shall go over to the second quantization representation.

Consequently, we shall obtain the following expression for the total Hamiltonian of the system:

$$\tilde{\mathcal{H}} = \mathcal{H}_{dd} + \mathcal{H}_{ph} + \mathcal{H}_{d\text{-}ph}, \tag{7.27}$$

where $\mathcal{H}_{dd}$ and $\mathcal{H}_{ph}$ are the operators for the internal energies of the spin and phonon subsystems:

$$\begin{aligned} \mathcal{H}_{dd} &= -\mu \sum (H, S_f) - \frac{1}{2} \sum I(f_1 - f_2)(S_{f_1}, S_{f_2}), \\ \mathcal{H}_{ph} &= \sum \omega_k \beta_k^{+} \beta_k, \end{aligned} \tag{7.28}$$

where β_k^{+}, β_k are the operators for the creation and annihilation of a phonon with energy ω_k (the subscript k represents both the wave vector of the phonon and the number of the corresponding mode of the phonon spectrum), which satisfy the Bose commutation relationships; $\mathcal{H}_{d\text{-}ph}$ is the operator for the spin–phonon interaction energy:

$$\begin{aligned} \mathcal{H}_{d\text{-}ph} &= \sum A_k(f_1 - f_2)(S_{f_1}, S_{f_2})(b_k - b_{-k}^{+})\left(e^{i(f_1, \nu)} - e^{i(f_2, \nu)}\right), \\ A_k(f) &= i\sqrt{\frac{1}{2V\rho\omega_k}}(e_k, \nabla_f I(f)), \end{aligned} \tag{7.29}$$

where V is the volume of the system, ρ is the density of the substance, e_k is the unit polarization vector of the phonon k.

Weak Ferromagnetism. This phenomenon is due to a slight deviation of the magnetic moments in the sublattices of an antiferromagnet from the strictly antiparallel alignment (cf. also, Sec. 1). This deviation is explained [Dzyaloshinskii (1957, 1958)] by the existence of a magnetocrystalline anisotropy of the type

$$\sum_{f_1, f_2} (D(f_1, f_2), [S_{f_1} \times S_{f_2}]), \tag{7.30}$$

where D (f_1, f_2) is some function of the lattice site coordinates; the order of magnitude of this function of the same as that of the spin – spin or spin – orbit interaction energy. Moria (1960) and Seidov (1963) have shown that the indirect exchange may also lead to an expression of the type represented by Eq. (7.30). The problems of the phenomenological theory of weak ferromagnetism are discussed in detail in Turov's monograph (1963).

Chapter III

Elements of Statistical Mechanics

The main problems in quantum statistical mechanics are the determination of thermodynamic properties of a system from a given Hamiltonian and the calculation of the average values of the dynamic variables, which are the observable quantities.

In this chapter, we shall summarize the assumptions and the results, which we shall use later in our treatment. We shall consider separately the problem of the degeneracy of states in the case of statistical equilibrium, and the problem of quasi-averages.

Sec. 8. Density Matrix and Thermodynamic Functions*

We shall take a system of N particles enclosed in a volume V; the Hamiltonian of the system will be denoted by $\mathscr{H}$. We shall consider only volume properties which are conserved in the transition to the limit $N \to \infty$, $V \to \infty$, $N/V = v^{-1} = \text{const}$.

The most complete description of a system in quantum statistical mechanics is given by the density matrix ρ, introduced first by von Neumann (1927a, b) for arbitrary temperatures, and by Landau (1927) for zero temperature. In the "coordinate" representation, the density matrix† is written in the form

*See the monographs by Gibbs (1946), von Neumann (1932), and Bogolyubov (1949), as well as the papers by von Neumann (1927a, b) and Landau (1927).

† The density matrix is also known as the statistical operator.

$$\rho(\xi, \xi', t) = \sum_n W_n \psi_n(\xi, t)\psi_n^*(\xi', t),$$
$$\sum_n W_n = 1 \qquad (W_n \geqslant 0), \tag{8.1}$$

where $\psi_n(\xi, t)$ is the normalized wave function of the n-th state of the system, ξ is a set of independent variables ("coordinates"), t is the time, and W_n is the probability that the system is in the n-th state. For example, for a system of N particles, ξ is understood to be the set of all the coordinates x and spins σ of the particles of the system, $\xi = (x_1\sigma_1; x_2\sigma_2; \ldots; x_N\sigma_N)$, or the set of their momenta and spins, $\xi = (p_1\sigma_1; p_2\sigma_2; \ldots; p_N\sigma_N)$, etc.

Since the development in time of the n-th state of the system is given by the equation

$$i\frac{\partial}{\partial t}\psi_n(\xi, t) = \mathscr{H}\psi_n(\xi, t),$$

where $\mathscr{H}$ is the operator of the total energy of the system, the equation of motion for the operator ρ has the form

$$i\frac{\partial\rho}{\partial t} = \mathscr{H}\rho - \rho\mathscr{H}. \tag{8.2}$$

According to the definition in Eq. (8.1), the operator ρ is normalized to unity:

$$\operatorname*{Sp}_{(\xi)} \rho = 1; \tag{8.3}$$

the sign under the spur (trace) symbol denotes the variables over which the spur is taken.

In quantum mechanics, each physical quantity A is associated with an operator $\mathscr{A}$. The experimentally observed values of a quantity in a state characterized by the eigenfunction ψ_n are the quantum-mechanical averages $A = (\psi_n^*, \mathscr{A}\psi_n)$. If ψ_n is the eigenfunction of an operator $\mathscr{A}$, the observed value of the quantity A will be an eigenvalue of the operator $\mathscr{A}$.

In quantum statistical mechanics, the observed values are the average statistical values or the averages over an ensemble, which are defined as follows:

$$A = \langle\mathscr{A}\rangle = \mathrm{Sp}(\mathscr{A}\rho). \tag{8.4}$$

At statistical equilibrium, the expression (8.1) may be rewritten in the form

$$\rho(\xi, \xi') = \sum_n W_n \psi_n(\xi) \psi_n^*(\xi'), \qquad (8.5)$$

where $\psi_n(\xi)$ are the eigenfunctions of the operator for the total energy of the system:

$$\mathcal{H} \psi_n(\xi) = \mathcal{E}_n \psi_n(\xi).$$

For a system of identical particles, the wave functions ψ_n should satisfy the symmetry or antisymmetry conditions, depending on whether the particles obey Bose–Einstein or Fermi–Dirac statistics. Accordingly, the summation in Eq. (8.5) applies only to those states which have the required symmetry or antisymmetry with respect to the particle permutations.

In the stationary case, Eq. (8.2) assumes the form

$$\mathcal{H}\rho - \rho\mathcal{H} = 0, \qquad (8.6)$$

and then the expression (8.5) may be considered to be the solution of Eq. (8.6).

The distribution of probabilities W_n at statistical equilibrium is selected in the form

$$W_n = Q^{-1} e^{-\frac{\mathcal{E}_n}{\vartheta}}, \qquad (8.7)$$

where Q is a normalizing factor, ϑ = kT is the modulus of the canonical distribution (T is absolute temperature and k is Boltzmann's constant), and $\mathcal{E}_n$ are the eigenvalues of the Hamiltonian of the system.

The quantity Q is known as the sum-over-states or the partition function and is found from the condition of normalization of W_n in Eq. (8.7) to unity:

$$Q = \sum_n e^{-\frac{\mathcal{E}_n}{\vartheta}}. \qquad (8.8)$$

The partition function Q is related to the free energy F of a system by the expression

$$F = -\vartheta \ln Q. \tag{8.9}$$

Using Eqs. (8.7) and (8.9), we shall rewrite the expression (8.5) for ρ in the following form:

$$\rho(\xi, \xi') = \sum_n e^{\frac{F - \mathcal{E}_n}{\vartheta}} \psi_n(\xi) \psi_n^*(\xi'). \tag{8.10}$$

The quantity ρ can be regarded as an operator which is defined only in the subspace of functions satisfying the symmetry or antisymmetry conditions, and we can rewrite the expression for this operator in the form

$$\rho = \exp \frac{1}{\vartheta}(F - \mathcal{H}). \tag{8.11}$$

The formulas (8.8) and (8.9) for the partition function and the free energy then assume the form

$$Q = \mathrm{Sp}\left(e^{-\frac{\mathcal{H}}{\vartheta}}\right) = \sum_n e^{-\frac{\mathcal{E}_n}{\vartheta}}, \tag{8.12}$$

$$F = -\vartheta \ln \mathrm{Sp}\left(e^{-\frac{\mathcal{H}}{\vartheta}}\right). \tag{8.13}$$

The expression (8.11) is known as the canonical distribution, and can easily be shown to be the generalization of the well-known Gibbs distribution in classical statistics to the case of quantum statistics. We note that the distribution (8.11) can be regarded as the solution of the operator equation (8.6).

If, in addition to the total energy, there are other integrals of motion $I_1, \ldots, I_q$, for example, the components of the total momentum of the system, of the total angular momentum, etc., then the distribution (8.11) is replaced by the following expression:

$$\rho = \exp \frac{1}{\vartheta}\left(F - \mathcal{H} + \sum_{1 \leq \alpha \leq q} \lambda_\alpha I_\alpha\right), \tag{8.14}$$

where the quantities λ_α are some numbers.

If the distribution (8.11) is used in systems with a constant number of particles N, we must bear in mind that the spurs (traces) in the formulas for the free energy and for the average values of the dynamic variables should be taken only over states with given N. Usually, this complicates the calculations. In such cases, it is more convenient to use the grand canonical distribution of Gibbs.

We shall introduce the operator $\mathcal{N}$ for the total number of particles in the system. In the second-quantization representation, we have

$$\mathcal{N} = \sum_r n_r, \tag{8.15}$$

where n_r is the particle-number operator in the r-th single-particle state. In Bose–Einstein statistics,

$$n_r = \beta_r^* \beta_r \qquad (= 0, 1, 2, \ldots),$$

and in Fermi–Dirac statistics,

$$n_r = \alpha_r^+ \alpha_r \qquad (= 0, 1).$$

Since the number of particles N in the system is given, the operator for the total number of particles, $\mathcal{N}$, is an integral of motion and the distribution should have the form of Eq. (8.14).

We shall replace this distribution with the grand canonical distribution of Gibbs*:

$$\rho = \exp \frac{1}{\vartheta} (\Omega - \mathcal{H} + \lambda \mathcal{N}) \tag{8.16}$$

(states with every value of N are included); Ω is the thermodynamic potential of the system in terms of the variables λ, ϑ, V, which is determined from the condition for the normalization of

---

*The possibility of such a replacement is based on the thermodynamic equivalence of ensembles. See, for example, Hill's book (1960).

the distribution (8.16) to unity:

$$\Omega = -\vartheta \ln \mathrm{Sp} \exp \frac{1}{\vartheta}(\lambda \mathcal{N} - \mathcal{H}), \tag{8.17}$$

where λ is the chemical potential, defined by the condition of equality of the average value of $\mathcal{N}$ and the number of particles in the system:

$$N = \langle \mathcal{N} \rangle = \mathrm{Sp}\, \langle \mathcal{N} \rho \rangle. \tag{8.18}$$

In accordance with the definition given in Eq. (8.16), the spurs (traces) are calculated for all states without any restriction on the number of particles in the system.

For systems in which the number of particles is indeterminate, $\lambda = 0$. In such a case, the Hamiltonian of the system is always of a form such that $\mathcal{N}$ is not an integral of the motion.

In the case of ideal gases (bosons or fermions) having a varying number of particles, an apparent contradiction may appear because the total energy and the total-number-of-particles operators commute for such gases, and it would seem that the chemical potential should have a nonzero value. This contradiction is removed by the fact that, to establish statistical equilibrium, the Hamiltonian must include also a very small but finite interaction which does not conserve the number of particles in the system.

We shall now establish some thermodynamic relationships for the canonical [(8.11)] and grand canonical [(8.16)] Gibbs distributions.

We shall consider a set of external parameters $a_1, \ldots, a_r$, representing a macroscopic state of statistical equilibrium. Such parameters may be, for example, the volume V of the system, the intensity H of an external field, etc. Let the parameters a vary so slowly that, in time intervals which are of the order necessary to establish statistical equilibrium, these parameters may be taken to be constant. Then we may assume that at any given moment the system is in the state of statistical equilibrium.

We shall now deal with the canonical Gibbs distribution given by Eq. (8.11).

We shall consider the quantities a to be the generalized coordinates. Then a change in these coordinates causes the appearance of the corresponding generalized forces:

$$\mathcal{A}_i = -\frac{\partial \mathcal{H}}{\partial a_i}.$$

The observed values of these forces are defined, in accordance with the usual rules, as follows:

$$\langle \mathcal{A}_i \rangle = -\left\langle \frac{\partial \mathcal{H}}{\partial a_i} \right\rangle, \tag{8.19}$$

where averaging is carried out with the weighting ρ given by Eq. (8.11). On the other hand, differentiating Eq. (8.13) with respect to a, we obtain

$$\left\langle \frac{\partial \mathcal{H}}{\partial a_i} \right\rangle = \frac{\partial F}{\partial a_i}. \tag{8.20}$$

Comparison of the last two expressions shows that

$$\langle \mathcal{A}_i \rangle = -\frac{\partial F}{\partial a_i}. \tag{8.21}$$

If, for example, H is an external magnetic field, then the average magnetization is given by

$$M^\alpha = -\frac{\partial F}{\partial H^\alpha} \qquad (\alpha = x,\ y,\ z). \tag{8.22}$$

The average energy U of the system is

$$U = \langle \mathcal{H} \rangle. \tag{8.23}$$

We shall differentiate, with respect to ϑ, the equation

$$e^{-\frac{F}{\vartheta}} = \mathrm{Sp}\left(e^{-\frac{\mathcal{H}}{\vartheta}}\right),$$

which is the definition of the free energy. This gives

$$\left(\frac{F}{\vartheta^2} - \frac{1}{\vartheta}\frac{\partial F}{\partial \vartheta}\right) e^{-\frac{F}{\vartheta}} = \frac{1}{\vartheta^2}\,\mathrm{Sp}\left(\mathcal{H} e^{-\frac{\mathcal{H}}{\vartheta}}\right).$$

Using the average-energy definition of Eq. (8.23), we obtain the Gibbs–Helmholtz equation:

$$F = \langle \mathcal{H} \rangle + \vartheta \frac{\partial F}{\partial \vartheta}. \tag{8.24}$$

At zero temperature, the free and average energies are identical. Therefore, Eq. (8.24) can be integrated with the initial condition:

$$F_{\vartheta=0} = \langle \mathcal{H} \rangle_{\vartheta=0}$$

As a result, we obtain the following expression for the free energy in terms of the average energy:

$$F = \langle \mathcal{H} \rangle_{\vartheta=0} - \vartheta \int_0^{\vartheta} \frac{d\vartheta'}{\vartheta'^2} \left\{ \langle \mathcal{H} \rangle_{\vartheta'} - \langle \mathcal{H} \rangle_{\vartheta=0} \right\} . \qquad (8.24a)$$

The comparison of Eq. (8.24) with the thermodynamic formula for the free energy,

$$F = U - \vartheta S,$$

where S is the entropy of the system, gives

$$S = -\frac{\partial F}{\partial \vartheta} = -\frac{1}{\vartheta}(F - \langle \mathcal{H} \rangle). \qquad (8.25)$$

The entropy of the system can also be written in the following form [von Neumann (1927b)]:

$$S = -\,\mathrm{Sp}\,(\rho \ln \rho). \qquad (8.26)$$

The identity of the formulas (8.25) and (8.26) is evident from the fact that, according to Eq. (8.11),

$$\ln \rho = \frac{1}{\vartheta}(F - \mathcal{H}).$$

The expression for the magnetization may be written in a somewhat different form. To obtain the latter, we shall bear in mind that for any Hamiltonian $\mathcal{H}_0$, which describes the spin system, the operator for the energy of interaction with an external field [cf., for example, Eq. (6.14) or (7.14) has the form

$$\mathcal{H}_1 = -\sum_f \mu_f (H, S_f),$$

where S_f is the spin operator at a site f; μ_f is the corresponding magnetic moment. The total Hamiltonian of the system is then given by

$$\mathcal{H} = \mathcal{H}_0 + \mathcal{H}_1,$$

where $\mathcal{H}_0$ is independent of H. Using Eqs. (8.9) and (8.12), we find that

$$\frac{\partial F}{\partial H^\alpha} = -\vartheta \frac{1}{Q} \frac{\partial Q}{\partial H^\alpha} = -\frac{1}{Q} \operatorname{Sp}\left\{\sum_f \mu_f S_f^\alpha e^{-\frac{1}{\vartheta}(\mathcal{H}_0 + \mathcal{H}_1)}\right\} = -\sum_f \mu_f \langle S_f^\alpha \rangle.$$

Since $\mathcal{M}^\alpha = \sum \mu_f S_f^\alpha$ is the operator for total magnetic moment of the system, and $-\partial F/\partial H^\alpha = M^\alpha$ is the average magnetization, we obtain

$$M^\alpha = \langle \mathcal{M}^\alpha \rangle = \sum_f \mu_f \langle S_f^\alpha \rangle. \tag{8.27}$$

For a system consisting of N identical spins, we have

$$M^\alpha = M_0 \sigma_S^\alpha, \quad M_0 = N\mu S \quad (\mu_f = \mu), \tag{8.28}$$

$$\sigma_S^\alpha = S^{-1} \langle S_f^\alpha \rangle, \tag{8.29}$$

where M_0 is the total magnetic moment of the system, and σ_S is the relative magnetization per site.

In the special case when the Hamiltonian of the system is given by

$$\mathcal{H} = \mathcal{E}_0 + \sum_\nu E_\nu n_\nu \qquad (n_\nu = 0,\ 1,\ 2,\ \ldots), \tag{8.30}$$

where $\mathcal{E}_0$, E_ν are some functions of the external magnetic field H, the expression for the magnetization may be represented in yet a different form. For this purpose, we shall calculate the partition function for the Hamiltonian of Eq. (8.30):

$$Q = \sum_{\ldots, n_\nu, \ldots} e^{-\frac{1}{\vartheta}\left(\mathcal{E}_0 + \sum_\nu E_\nu n_\nu\right)} = e^{-\frac{\mathcal{E}_0}{\vartheta}} \prod_\nu \left(1 - e^{-\frac{E_\nu}{\vartheta}}\right)^{-1}. \tag{8.31}$$

According to Eq. (8.9), we obtain the following expression for the free energy:

$$F = -\vartheta \ln Q = \mathcal{E}_0 + \vartheta \sum_\nu \ln\left(1 - e^{-\frac{E_\nu}{\vartheta}}\right). \tag{8.32}$$

Differentiating Eq. (8.32) with respect to E_μ, we obtain the following expressions for the average values of the occupation numbers:

$$\langle n_\mu \rangle = \frac{\partial F}{\partial E_\mu} = \frac{\sum\limits_{\dots, n_\nu, \dots} n_\mu e^{-\frac{1}{\vartheta}\left(\mathscr{E}_0 + \sum\limits_\nu E_\nu n_\nu\right)}}{\sum\limits_{\dots, n_\nu, \dots} e^{-\frac{1}{\vartheta}\left(\mathscr{E}_0 + \sum\limits_\nu E_\nu n_\nu\right)}} \qquad (8.33)$$

or

$$\langle n_\mu \rangle = \left(e^{\frac{E_\mu}{\vartheta}} - 1\right)^{-1}. \qquad (8.34)$$

Finally, substituting Eq. (8.32) into Eq. (8.22), we obtain

$$M^\alpha = -\frac{\partial F}{\partial H^\alpha} = -\frac{\partial \mathscr{E}_0}{\partial H^\alpha} - \sum_\mu \langle n_\mu \rangle \frac{\partial E_\mu}{\partial H^\alpha} \qquad (\alpha = x,\ y,\ z). \qquad (8.35)$$

The formula (8.35) is convenient in actual calculations concerned with systems having a Hamiltonian of the type given by Eq. (8.30).

The specific heat at constant volume is given by

$$C_v = k\frac{\partial \langle \mathscr{H} \rangle}{\partial \vartheta} = -k\frac{\partial}{\partial \vartheta}\left(\vartheta^2 \frac{\partial}{\partial \vartheta}\right)\frac{F}{\vartheta} = k\frac{\partial}{\partial \vartheta}\left(\vartheta^2 \frac{\partial}{\partial \vartheta}\right)\ln Q, \qquad (8.36)$$

where k is Boltzmann's constant.

For systems with a Hamiltonian of the type given by Eq. (8.30) we obtain

$$C_v = \frac{k}{\vartheta^2}\sum_\nu E_\nu^2(\langle n_\nu^2 \rangle - \langle n_\nu \rangle^2) = \frac{k}{\vartheta^2}\sum_\nu E_\nu^2(\langle n_\nu \rangle + \langle n_\nu \rangle^2). \qquad (8.37)$$

Similarly, we can define the thermodynamic functions for the grand canonical Gibbs distribution.

Sec. 9. Density Matrices for Systems of Particles*

For simplicity, we shall consider a system of N identical particles. We shall use $\xi_1, \xi_2, \dots, \xi_N$ to denote the variables refer-

*See Bogolyubov's monograph (1949).

ring to the first, second, . . . , and N-th particles. Then the wave functions of the whole system will be functions of $\xi_1, \ldots, \xi_N$,

$$\psi_n(\xi, t) = \psi_n(\xi_1, \ldots, \xi_N, t), \tag{9.1}$$

and the operators acting on these functions will be matrices of the form

$$\mathcal{A} = \mathcal{A}(\xi_1, \ldots, \xi_N; \xi_1', \ldots, \xi_N'). \tag{9.2}$$

For example, in the coordinate representation, Schrödinger's equation for a system of particles subject to pair interaction,

$$\left\{\sum_i \left(-\frac{1}{2m}\Delta_{x_i}\right) + \right. \\ \left. + \frac{1}{2}\sum_{i \neq k} \Phi(x_i - x_k)\right\} \psi(x_1, \ldots, x_N) = \mathcal{E}\psi(x_1, \ldots, x_N),$$

where $\Phi(x_i - x_k)$ is the energy of interaction between the i-th and k-th particles, can be written in the form

$$\int \mathcal{H}(x_1, \ldots, x_N; x_1', \ldots, x_N') \psi(x_1', \ldots, x_N') dx_1' \ldots dx_N' = \\ = \mathcal{E}\psi(x_1', \ldots, x_N'),$$

where

$$\mathcal{H}(x_1, \ldots, x_N; x_1', \ldots, x_N') = \\ = \sum_i \delta(x_i - x_i')\left(-\frac{1}{2m}\Delta_{x_i'}\right)\prod_{l \neq i} \delta(x_l - x_l') + \\ + \sum_{\substack{i, k \\ (i \neq k)}} \delta(x_i - x_i')\delta(x_k - x_k')\Phi(x_i' - x_k')\prod_{l \neq i, k} \delta(x_l - x_l').$$

The equation for a particle in a central force field,

$$-\frac{1}{2m}\Delta_x\psi(x) + \Phi(x)\psi(x) = \mathcal{E}\psi(x)$$

in the momentum representation has the form

$$\frac{p^2}{2m}\psi(p) + \int \nu(p - p')\psi(p')\, dp' = \mathcal{E}\psi(p),$$

where $\nu(p)$ is the Fourier transform of the potential $\Phi(x)$. In the matrix form, the above equation is written as follows:

$$\int \mathcal{H}(p, p')\psi(p')\,dp' = \mathcal{E}\psi(p),$$

where

$$\mathcal{H}(p, p') = \delta(p - p')\frac{p'^2}{2m} + \nu(p - p').$$

We shall consider the density matrix

$$\rho(\xi_1, \ldots, \xi_N; \xi_1', \ldots, \xi_N'; t) = \\ = \sum_n W_n \psi_n(\xi_1, \ldots, \xi_N, t)\mathring{\psi}_n(\xi_1', \ldots, \xi_N', t). \qquad (9.3)$$

In the case of particles obeying the Bose–Einstein statistics, we have

$$P\psi_n(\xi_1, \ldots, \xi_N, t) = \psi_n(\xi_1, \ldots, \xi_N, t), \qquad (9.4)$$

where P is the permutation operator for the variables ξ_i ($i = 1, 2, \ldots, N$).

In the case of particles obeying the Fermi–Dirac statistics, we have

$$P\psi_n(\xi_1, \ldots, \xi_N, t) = (-1)^P \psi_n(\xi_1, \ldots, \xi_N, t); \qquad (9.5)$$

here, as usual, $(-1)^P = 1$ for the even permutation and $(-1)^P = -1$ for the odd permutation.

Therefore, in both cases,

$$P\rho = \rho P, \qquad (9.6)$$

or

$$P\rho P^{-1} = \rho. \qquad (9.7)$$

Since the ξ-representation of the operator $P\rho P^{-1}$ is obtained from the ξ-representation of the operator ρ by the same simultaneous permutation of the variables ξ and ξ', the density matrix is symmetrical with respect to the particle permutation.

In investigating actual dynamical systems, it is usually unnecessary to know the total density matrix. In fact, in calculating the average values of the dynamic variables, we deal with operators depending on the variables of one, two, . . . , s particles:

$$\mathcal{A}_1 = \sum_{1 \leq r \leq N} \mathcal{A}(r), \qquad \mathcal{A}_2 = \sum_{1 \leq r_1 < r_2 \leq N} \mathcal{A}(r_1, r_2), \ldots,$$
$$\mathcal{A}_s = \sum_{1 \leq r_1 < r_2 < \ldots r_s \leq N} \mathcal{A}(r_1, r_2, \ldots, r_s), \tag{9.8}$$

where the argument r_i denotes dependence on the coordinates of the i-th particle.

The average values of the dynamic variables are calculated by partial contraction of the matrix ρ.* In fact, using the condition of symmetry (9.7) for the matrix ρ with respect to the particle permutation, we obtain

$$\langle\mathcal{A}_1\rangle = N \operatorname*{Sp}_{(1)} \{\mathcal{A}(1)\rho_1(1)\},$$
$$\langle\mathcal{A}_2\rangle = \frac{N(N-1)}{2!} \operatorname*{Sp}_{(1,2)} \{\mathcal{A}(1, 2)\rho_2(1, 2)\},$$
$$\cdots\cdots\cdots\cdots\cdots\cdots \tag{9.9}$$
$$\langle\mathcal{A}_s\rangle = \frac{N(N-1)\ldots(N-s+1)}{s!} \times$$
$$\times \operatorname*{Sp}_{(1, 2, \ldots, s)} \{\mathcal{A}(1, 2, \ldots, s)\rho_s(1, 2, \ldots, s)\},$$

where the following notation is used:

$$\rho_1(1) = \operatorname*{Sp}_{(2, \ldots, N)} \rho(1, 2, \ldots, N),$$
$$\rho_2(1, 2) = \operatorname*{Sp}_{(3, \ldots, N)} \rho(1, 2, 3, \ldots, N), \tag{9.10}$$
$$\cdots\cdots\cdots\cdots\cdots\cdots$$
$$\rho_s(1, 2, \ldots, s) = \operatorname*{Sp}_{(s+1, \ldots, N)} \rho(1, 2, \ldots, s, s+1, \ldots, N);$$

for brevity we shall use the following notation:

$$\rho_2(1, 2) \equiv \rho_2(\xi_1, \xi_2; \xi_1', \xi_2'; t);$$
$$\operatorname*{Sp}_{(2)} \rho_2(1, 2) \equiv \operatorname*{Sp}_{(\xi_2)} \rho_2(\xi_1, \xi_2; \xi_1', \xi_2'; t) \text{ etc.}$$

The quantities ρ_s are called the density matrices of systems of s particles or s-particle density matrices.†

---

*That is, we take traces (spurs) of ρ for some of the variables.

† The density matrices for systems of particles were introduced into quantum statistical problems by Gurov (1946, 1947),

For the density matrices ρ_s, we have, on the basis of Eqs. (9.7) and (9.10), the following relationships:

$$P_s \rho_s P_s^{-1} = \rho_s \tag{9.11}$$

$$\rho_s(1, \ldots, s) = \underset{(s+1)}{\mathrm{Sp}}\ \rho_{s+1} \quad (1, \ldots, s, s+1), \tag{9.12}$$

where P_s is the permutation operator for particles of the system.

We shall replace the density matrices ρ_s with the operators F_s, using the formulas

$$F_s(1, \ldots, s) = N(N-1)\ldots(N-s+1)\rho_s(1, \ldots, s). \tag{9.13}$$

The latter operators will obey, according to Eq. (9.12), the relationships

$$F_s(1, \ldots, s) = \frac{1}{N-s} \underset{(s+1)}{\mathrm{Sp}}\, F_{s+1}(1, \ldots, s, s+1).$$

In the limit for fixed s and $N \to \infty$, we can neglect the quantity s compared with N in the factor on the right. Consequently, we obtain the following relationships for the operators F_s:

$$P_s F_s P_s^{-1} = F_s, \tag{9.14}$$

$$F_s(1, \ldots, s) = \frac{1}{N} \underset{(s+1)}{\mathrm{Sp}}\ F_{s+1} \quad (1, \ldots, s, s+1). \tag{9.15}$$

By analogy with classical statistics, the quantities F_s are called the distribution operators or – although not an exact name but a more usual one – the distribution functions.

Then, the expressions in Eq. (9.9) for the average values as-

---

Bogolyubov and Gurov (1947), and Born and Green (1947). In these and subsequent papers, a number of problems in statistical mechanics were considered, in particular, the problems arising in the theory of transport equations for quantum systems. Earlier suggestions to introduce density matrices for systems of particles were made by Husimi (1940).

sume the form

$$\langle \mathcal{A}_s \rangle = \frac{1}{s!} \mathop{\mathrm{Sp}}_{(1, \ldots, s)} \{\mathcal{A}(1, \ldots, s) F_s(1, \ldots, s)\}. \qquad (9.16)$$

We shall write down the operator $\mathcal{A}_s$ of Eq. (9.8) in the second quantization representation [cf., Eqs. (3.41) and (3.42)]:

$$\mathcal{A}_s = \sum_{1 \le r_1 < \ldots < r_s \le N} \mathcal{A}(r_1, \ldots, r_s) =$$

$$= \frac{1}{s!} \sum A(f_1, \ldots, f_s; f'_s, \ldots, f'_1)\, a^+_{f_1} \ldots a^+_{f_s} a_{f'_s} \ldots a_{f'_1}, \qquad (9.17)$$

where a^+_f, a_f are the particle creation and annihilation operators in the one-particle state f; $A_s(f_1, \ldots, f_s; f'_s, \ldots, f'_1)$ are the corresponding matrix elements of the operator $\mathcal{A}_s$. We shall take the average value of $\mathcal{A}_s$, given by Eq. (9.17), over an ensemble [cf., Eq. (8.4)]:

$$\langle \mathcal{A}_s \rangle = \frac{1}{s!} \sum A(f_1, \ldots, f_s; f'_s, \ldots, f'_1) \times$$

$$\times \left\langle a^+_{f_1} \ldots a^+_{f_s} a_{f'_s} \ldots a_{f'_1} \right\rangle. \qquad (9.18)$$

Comparing Eqs. (9.18) and (9.16), we obtain the following expression for the operator F_s in the second quantization representation:

$$F_s(1, \ldots, s) = F_s(f_1, \ldots, f_s; f'_s, \ldots, f'_1) =$$

$$= \left\langle a^+_{f_1} \ldots a^+_{f_s} a_{f'_s} \ldots a_{f'_1} \right\rangle. \qquad (9.19)$$

In principle, the problems of statistical mechanics are to a considerable extent solved if we know the operators ρ_s and F_s. The main difficulty is to calculate these quantities. One of the methods is to determine them from a chain (set) of coupled equations, each of which contains F_s and F_{s+1} [Gurov (1946, 1947), Bogolyubov and Gurov (1947), Bogolyubov (1949)]. We shall use a different method: we shall determine F_s from Green's functions [cf., Chaps. VII-VIII].

Sec. 10. Wick – Bloch – de Dominicis Statistical Theorem*

In calculating the averages of the products of the particle creation and annihilation operators, it is very useful to employ the statistical theorem of Wick, Bloch, and de Dominicis.† We shall consider separately the cases of fermion and boson systems.‡

Let us consider a system of N identical particles, and let the Hamiltonian of the system be

$$\mathscr{H} = \sum E_f n_f \qquad (n_f = a_f^+ a_f), \\ E_f = E_f' - \lambda, \tag{10.1}$$

where E'_f is the energy of a particle in the f-th state, and λ is the chemical potential, which is introduced in order to allow for the constancy of the number of particles and is given by the usual equation

$$\sum_f \langle n_f \rangle = N. \tag{10.2}$$

Here, the averaging is carried out over the density matrices

$$\rho = e^{\frac{\Omega - \mathscr{H}}{\theta}} = \prod_f \rho_1(f), \qquad \rho_1(f) = \frac{e^{-\frac{E_f}{\theta} n_f}}{\sum_{n_f} e^{-\frac{E_f}{\theta} n_f}}. \tag{10.3}$$

We shall call the average value of the two operators $\mathscr{A}_1$ and $\mathscr{A}_2$ the operator contraction:

$$\overline{\mathscr{A}_1 \mathscr{A}_2} = \langle \mathscr{A}_1 \mathscr{A}_2 \rangle, \tag{10.4}$$

where $\mathscr{A}$ represents either a creation or an annihilation operator.

*See Bloch and de Dominicis (1958).

† The proof given here was proposed by Luttinger (1961a).

‡ The generalization to the Pauli operators is given in a paper by Tyablikov and Moskalenko (1964).

For a product of several operators, we shall define a contraction system as a paired arrangement of operators, all in the form of contractions. A complete contraction system is a system of contractions in which none of the operators remains unpaired. Then, the expression obtained in the Fermi–Dirac case is given the sign $(-1)^P$, where P is the permutation transforming the original product of operators into a given product.

For example, let us consider the system of contractions

$$\mathcal{A}_1\mathcal{A}_2\mathcal{A}_3\mathcal{A}_4\mathcal{A}_5\mathcal{A}_6 =$$
$$= \overline{\mathcal{A}_1\mathcal{A}_3}\,\overline{\mathcal{A}_4\mathcal{A}_5}\,\overline{\mathcal{A}_2\mathcal{A}_6} \times \begin{cases} (-1)^P & \text{for Fermi–Dirac statistics;} \\ 1 & \text{for Bose–Einstein statisties;} \end{cases}$$

P is the permutation which changes the order of the indices from (123456) into(13 45 26).

Under the conditions given by Eqs. (10.1) and (10.3), the following theorem is satisfied:

The average value of a product of the creation and annihilation operators $\langle \mathcal{A}_1, \mathcal{A}_2 \ldots \mathcal{A}_{2p} \rangle$ for a system with a Hamiltonian of the type given by Eq. (10.1) is equal to the sum of all the possible complete systems of contractions of this product.

This is the statistical theorem of Wick, Bloch, and de Dominicis.

For example, in the Fermi–Dirac statistics case,

$$\langle \mathcal{A}_1\mathcal{A}_2\mathcal{A}_3\mathcal{A}_4 \rangle =$$
$$= \overline{\mathcal{A}_1\mathcal{A}_2}\,\overline{\mathcal{A}_3\mathcal{A}_4} + \mathcal{A}_1\mathcal{A}_2\mathcal{A}_3\mathcal{A}_4 + \overline{\mathcal{A}_1\overline{\mathcal{A}_2\mathcal{A}_3}\mathcal{A}_4} =$$
$$= \langle \mathcal{A}_1\mathcal{A}_2 \rangle \langle \mathcal{A}_3\mathcal{A}_4 \rangle - \langle \mathcal{A}_1\mathcal{A}_3 \rangle \langle \mathcal{A}_2\mathcal{A}_4 \rangle + \langle \mathcal{A}_1\mathcal{A}_4 \rangle \langle \mathcal{A}_2\mathcal{A}_3 \rangle;$$

and in the Bose–Einstein statistics case,

$$\langle \mathcal{A}_1\mathcal{A}_2\mathcal{A}_3\mathcal{A}_4 \rangle =$$
$$= \langle \mathcal{A}_1\mathcal{A}_2 \rangle \langle \mathcal{A}_3\mathcal{A}_4 \rangle + \langle \mathcal{A}_1\mathcal{A}_3 \rangle \langle \mathcal{A}_2\mathcal{A}_4 \rangle + \langle \mathcal{A}_1\mathcal{A}_4 \rangle \langle \mathcal{A}_2\mathcal{A}_3 \rangle.$$

We note first of all that if we have a product of two creation and annihilation operators, $\mathcal{A}_1$ and $\mathcal{A}_2$, then the only nonzero contractions will be $\overbrace{a_f^+ a_f}$ and $\overbrace{a_f a_f^+}$. In fact, any other combination of the operators a_f^+, a_f will not have diagonal elements, and its average value will be equal to zero, since the density matrix of Eq. (10.3) has only diagonal elements. Using Eqs. (10.4) and (10.3), we obtain:

for the Fermi–Dirac statistics

$$\begin{aligned} \overbrace{a_{f_1}^+ a_{f_2}} &= \Delta(f_1 - f_2)\, n_{f_1}^+ = \frac{\Delta(f_1 - f_2)}{1 + e^{\frac{E_f}{\vartheta}}}, \\ \overbrace{a_{f_2} a_{f_1}^+} &= \Delta(f_1 - f_2)\, n_{f_1}^- = \frac{\Delta(f_1 - f_2)}{1 + e^{-\frac{E_f}{\vartheta}}}, \\ \overbrace{a_{f_1} a_{f_2}} &= \overbrace{a_{f_1}^+ a_{f_2}^+} = 0 \quad (n_{f_1}^+ + n_{f_1}^- = 1) \end{aligned} \tag{10.5}$$

and for the Bose–Einstein statistics,

$$\begin{aligned} \overbrace{a_{f_1}^+ a_{f_2}} &= -\Delta(f_1 - f_2)\, n_{f_1}^+ = -\frac{\Delta(f_1 - f_2)}{1 - e^{\frac{E_f}{\vartheta}}}, \\ \overbrace{a_{f_2} a_{f_1}^+} &= -\Delta(f_1 - f_2)\, n_{f_1}^- = \frac{\Delta(f_1 - f_2)}{1 - e^{-\frac{E_f}{\vartheta}}}, \\ \overbrace{a_{f_1} a_{f_2}} &= \overbrace{a_{f_1}^+ a_{f_2}^+} = 0 \quad (n_{f_1}^+ + n_{f_1}^- = 1). \end{aligned} \tag{10.6}$$

Let us consider first the case of the Fermi–Dirac statistics.

Then, the formulas of Eq. (10.5) may be rewritten in the following form [$\mathcal{A}_f$ denotes either a particle creation (a_f^+) or a particle annihilation (a_f) operator]:

$$\langle \mathcal{A}_{f_1} \mathcal{A}_{f_2} \rangle = \frac{[\mathcal{A}_{f_1}, \mathcal{A}_{f_2}]}{1 + e^{\pm \frac{1}{\vartheta} E_{f_1}}} = n_{f_1}^{\pm} [\mathcal{A}_{f_1}, \mathcal{A}_{f_2}], \tag{10.7}$$

where $[\mathcal{A}_{f_1}, \mathcal{A}_{f_2}]$ is the anticommutator of the operators $\mathcal{A}_{f_1}$ and $\mathcal{A}_{f_2}$. We shall consider the average of the product

$$\langle \mathcal{A}_{f_1} \mathcal{A}_{f_2} \cdots \mathcal{A}_{f_{2p}} \rangle. \tag{10.8}$$

We shall write down the identity

$$\mathcal{A}_{f_1}\cdot\mathcal{A}_{f_2}\ldots\mathcal{A}_{f_{2p}}+\mathcal{A}_{f_2}\ldots\mathcal{A}_{f_{2p}}\cdot\mathcal{A}_{f_1}=$$
$$=[\mathcal{A}_{f_1},\ \mathcal{A}_{f_2}\mathcal{A}_{f_3}\ldots\mathcal{A}_{f_{2p}}]=$$
$$=[\mathcal{A}_{f_1},\mathcal{A}_{f_2}]\mathcal{A}_{f_3}\ldots\mathcal{A}_{f_{2p}}-\mathcal{A}_{f_2}[\mathcal{A}_{f_1},\mathcal{A}_{f_3}]\mathcal{A}_{f_4}\ldots\mathcal{A}_{f_{2p}}+\ldots \quad (10.9)$$

and calculate the average values of both sides. We note that

$$\langle\mathcal{A}_{f_2}\ldots\mathcal{A}_{f_{2p}}\mathcal{A}_{f_1}\rangle=\mathrm{Sp}\,(\mathcal{A}_{f_2}\ldots\mathcal{A}_{f_{2p}}\mathcal{A}_{f_1}\rho)=$$
$$=e^{\pm\frac{E_f}{\vartheta}}\,\mathrm{Sp}\,(\mathcal{A}_{f_2}\ldots\mathcal{A}_{f_{2p}}\rho\mathcal{A}_{f_1})=e^{\pm\frac{E_f}{\vartheta}}\langle\mathcal{A}_{f_1}\mathcal{A}_{f_2}\ldots\mathcal{A}_{f_{2p}}\rangle, \quad (10.10)$$

because

$$a_f^+\rho=\rho a_f^+e^{\frac{E_f}{\vartheta}},\quad a_f\rho=\rho a_f e^{-\frac{E_f}{\vartheta}}. \quad (10.11)$$

To prove the formulas of Eq. (10.11), we shall consider the operator

$$a_f(\alpha)=e^{\alpha n_f}a_f e^{-\alpha n_f},\quad a_f(0)\equiv a_f.$$

We shall differentiate both sides with respect to α:

$$\frac{da_f(\alpha)}{d\alpha}=e^{\alpha n_f}n_f a_f e^{-\alpha n_f}-e^{\alpha n_f}a_f n_f e^{-\alpha n_f}=$$
$$=e^{\alpha n_f}(n_f a_f-a_f n_f)\,e^{-\alpha n_f}=-e^{\alpha n_f}a_f e^{-\alpha n_f}.$$

Using the definition of the operator $a_f(\alpha)$, we obtain

$$\frac{da_f(\alpha)}{d\alpha}=-a_f(\alpha).$$

Let us integrate this equation. We can easily see that we obtain

$$a_f(\alpha)=a_f e^{-\alpha}$$

or, according to the definition of $a_f(\alpha)$,

$$e^{\alpha n_f}a_f e^{-\alpha n_f}=a_f e^{-\alpha}.$$

Multiplying both sides by $e^{-\alpha n_f}$, we obtain

$$a_f e^{-\alpha n_f}=e^{-\alpha n_f}a_f e^{-\alpha}$$

and then, since $$(\mathcal{A}\mathcal{B})^+ = \mathcal{B}^+\mathcal{A}^+,$$

$$a_f^+ e^{-\alpha n_f} = e^{-\alpha n_f} a_f^+ e^{\alpha}.$$

We shall assume that $\alpha = (1/\vartheta)E_f$ and that $\rho_1(f)$ differs from $\exp[-(1/\vartheta)E_f n_f]$ only by the normalization factor, and that $\rho_1(g)$ and a_f commute if $f \neq g$. Then the equations (10.11) follow from the last two formulas and from Eq. (10.3).

Consequently, we obtain

$$\begin{aligned}\left(1 + e^{\pm\frac{1}{\vartheta}E_{f_1}}\right)\langle \mathcal{A}_{f_1}\mathcal{A}_{f_2} \cdots \mathcal{A}_{f_{2p}}\rangle &= \\ = [\mathcal{A}_{f_1}, \mathcal{A}_{f_2}]\langle \mathcal{A}_{f_3}\mathcal{A}_{f_4} \cdots \mathcal{A}_{f_{2p}}\rangle &- \\ -[\mathcal{A}_{f_1}, \mathcal{A}_{f_3}]\langle \mathcal{A}_{f_2}\mathcal{A}_{f_4} \cdots \mathcal{A}_{f_{2p}}\rangle &+ \cdots, \end{aligned} \qquad (10.12)$$

where the plus sign in the exponential is used if $\mathcal{A}_{f_1} = a_{f_1}^+$, and the minus sign if $\mathcal{A}_{f_1} = a_{f_1}$. Dividing this expression by $\left(1 + e^{\pm\frac{1}{\vartheta}E_{f_1}}\right)$ and using Eq. (10.7), we find

$$\begin{aligned}\langle \mathcal{A}_{f_1}\mathcal{A}_{f_2} \cdots \mathcal{A}_{f_{2p}}\rangle &= \langle \mathcal{A}_{f_1}\mathcal{A}_{f_2}\rangle\langle \mathcal{A}_{f_3}\mathcal{A}_{f_4} \cdots \mathcal{A}_{f_{2p}}\rangle - \\ &- \langle \mathcal{A}_{f_1}\mathcal{A}_{f_3}\rangle\langle \mathcal{A}_{f_2}\mathcal{A}_{f_4} \cdots \mathcal{A}_{f_{2p}}\rangle + \cdots \end{aligned} \qquad (10.13)$$

Repeating the same process for the averages on the right-hand side of Eq. (10.13), we obtain a representation of the product given by Eq. (10.8) in terms of a sum of complete contraction systems with coefficients equal to ±1, depending on the parity or lack of parity of the permutation transforming the original arrangement of the operators into the new one.

We shall also prove that if two (or more) identical operators a_f (or a_f^+) occur next to one another, then the sum of all contractions is equal to zero. In fact, we have

$$\begin{aligned}&\langle \mathcal{A}_{f_1} \cdots \mathcal{A}_{f_l}\mathcal{A}_f\mathcal{A}_f\mathcal{A}_{f_{l+3}} \cdots \mathcal{A}_{f_{2p}}\rangle = \\ &= \cdots + (-1)^{l-1}\langle \mathcal{A}_{f_1}\mathcal{A}_{f_l}\rangle\langle \mathcal{A}_{f_2} \cdots \mathcal{A}_{f_{l-1}}\mathcal{A}_f\mathcal{A}_f\mathcal{A}_{f_{l+3}} \cdots \mathcal{A}_{f_{2p}}\rangle + \\ &\quad + (-1)^l\langle \mathcal{A}_{f_1}\mathcal{A}_f\rangle\langle \mathcal{A}_{f_2} \cdots \mathcal{A}_{f_l}\mathcal{A}_f\mathcal{A}_{f_{l+3}} \cdots \mathcal{A}_{f_{2p}}\rangle + \\ &\quad + (-1)^{l+1}\langle \mathcal{A}_{f_1}\mathcal{A}_f\rangle\langle \mathcal{A}_{f_2} \cdots \mathcal{A}_{f_l}\mathcal{A}_f\mathcal{A}_{f_{l+3}} \cdots \mathcal{A}_{f_{2p}}\rangle + \\ &+ (-1)^{l+2}\langle \mathcal{A}_{f_1}\mathcal{A}_{f_{l+3}}\rangle\langle \mathcal{A}_{f_2} \cdots \mathcal{A}_{f_l}\mathcal{A}_f\mathcal{A}_f\mathcal{A}_{f_{l+4}} \cdots \mathcal{A}_{f_{2p}}\rangle + \cdots\end{aligned}$$

Hence, we see that the terms on the right obtained by splitting the product $\mathcal{A}_f\mathcal{A}_f$ cancel mutually [they are the terms with $(-1)^i$ and $(-1)^{i+1}$]. All the remaining terms vanish because $\mathcal{A}_f\mathcal{A}_f=0$. Repeating the same treatment, we find that

$$\langle\mathcal{A}_{f_1}\ldots\mathcal{A}_{f_l}\mathcal{A}_f\mathcal{A}_f\mathcal{A}_{f_{l+3}}\ldots\mathcal{A}_{f_{2p}}\rangle=$$
$$=\ldots\langle\mathcal{A}_{f_l}\mathcal{A}_f\mathcal{A}_f\mathcal{A}_{f_{l+3}}\rangle\ldots=0. \qquad (10.14)$$

This concludes the proof for the Fermi–Dirac case.

We shall now consider the Bose–Einstein statistics.

We shall rewrite the formulas of Eq. (10.6) in the form

$$\langle\mathcal{A}_{f_1}\mathcal{A}_{f_2}\rangle=n^{\mp}_{f_1}[\mathcal{A}_{f_1},\ \mathcal{A}_{f_2}], \qquad (10.15)$$

where $[\mathcal{A}_{f_1},\ \mathcal{A}_{f_2}]$ is the commutator of the operators $\mathcal{A}_{f_1}$ and $\mathcal{A}_{f_2}$. To prove the representation of the average

$$\langle\mathcal{A}_{f_1}\ \ldots\ \mathcal{A}_{f_{2p}}\rangle$$

again in the form of a sum of the products of all the contractions, we shall use the identity for the product commutator:

$$\mathcal{A}_{f_1}\mathcal{A}_{f_2}\ldots\mathcal{A}_{f_{2p}}-\mathcal{A}_{f_2}\ldots\mathcal{A}_{f_{2p}}\mathcal{A}_{f_1}=[\mathcal{A}_{f_1},\mathcal{A}_{f_2}\ldots\mathcal{A}_{f_{2p}}]=$$
$$=[\mathcal{A}_{f_1},\ \mathcal{A}_{f_2}]\mathcal{A}_{f_3}\mathcal{A}_{f_4}\ldots\mathcal{A}_{f_{2p}}+\mathcal{A}_{f_2}[\mathcal{A}_{f_1},\mathcal{A}_{f_3}]\mathcal{A}_{f_4}\ldots\mathcal{A}_{f_{2p}} \qquad (10.16)$$

We shall also use the formulas

$$a_f^+\rho=\rho a_f^+e^{\frac{E_f}{\vartheta}},\quad a_f\rho=\rho a_fe^{-\frac{E_f}{\vartheta}}$$

(which are proved as in the preceding case).

Consequently,

$$\langle\mathcal{A}_{f_2}\ldots\mathcal{A}_{f_{2p}}\mathcal{A}_{f_1}\rangle=\mathrm{Sp}\,(\mathcal{A}_{f_2}\ldots\mathcal{A}_{f_{2p}}\mathcal{A}_{f_1}\rho)=$$
$$=\mathrm{Sp}\,(\mathcal{A}_{f_2}\ldots\mathcal{A}_{f_{2p}}\rho\mathcal{A}_{f_1})\,e^{\pm\frac{E_{f_1}}{\vartheta}}=$$
$$=e^{\pm\frac{E_{f_1}}{\vartheta}}\,\mathrm{Sp}\,(\mathcal{A}_{f_1}\mathcal{A}_{f_2}\ldots\mathcal{A}_{f_{2p}}\rho). \qquad (10.17)$$

As the result of this, Eq. (10.16) for the product commutator may be written in the form

$$\left(1 - e^{\pm \frac{1}{\theta} E_{f_1}}\right) \langle \mathcal{A}_{f_1} \mathcal{A}_{f_2} \ldots \mathcal{A}_{f_{2p}} \rangle = \\ = [\mathcal{A}_{f_1}, \mathcal{A}_{f_2}] \langle \mathcal{A}_{f_3} \mathcal{A}_{f_4} \ldots \mathcal{A}_{f_{2p}} \rangle + \\ + [\mathcal{A}_{f_1}, \mathcal{A}_{f_3}] \langle \mathcal{A}_{f_2} \mathcal{A}_{f_4} \ldots \mathcal{A}_{f_{2p}} \rangle + \ldots \qquad (10.18)$$

or, according to Eq. (10.15), in the form

$$\langle \mathcal{A}_{f_1} \mathcal{A}_{f_2} \mathcal{A}_{f_3} \ldots \mathcal{A}_{f_{2p}} \rangle = \langle \mathcal{A}_{f_1} \mathcal{A}_{f_2} \rangle \langle \mathcal{A}_{f_3} \mathcal{A}_{f_4} \ldots \mathcal{A}_{f_{2p}} \rangle + \\ + \langle \mathcal{A}_{f_1} \mathcal{A}_{f_3} \rangle \langle \mathcal{A}_{f_2} \mathcal{A}_{f_4} \ldots \mathcal{A}_{f_{2p}} \rangle + \ldots \qquad (10.19)$$

The right-hand side has averages of products containing smaller numbers (2p − 2) of operators and, therefore, repeating the same process, we obtain the representation of the average of the operator products in terms of a sum of the products of all the contractions. We thus prove the theorem for the Bose–Einstein case.

In conclusion, we must stress that the theorem in the form given here is valid only for systems which have a Hamiltonian of the Eq. (10.1) type.

If the Hamiltonian does not conserve the number of particles, for example, if it has the form

$$\mathcal{H} = \sum E_f a_f^+ a_f + \sum A_{f_1 f_2} a_{f_1}^+ a_{f_2}^+ + \sum \overset{*}{A}_{f_1 f_2} a_{f_1} a_{f_2}, \qquad (10.20)$$

then, in contrast to the case considered, the anomalous contractions or the anomalous averages will not be identically equal to zero:

$$\overbrace{a_{f_1} a_{f_2}} = \langle a_{f_1} a_{f_2} \rangle \neq 0. \qquad (10.21)$$

We can construct the corresponding analog of the theorem for this case as well; it will then include the anomalous contractions.

The anomalous averages were introduced first in the investigation of imperfect quantum gases [Bogolyubov (1947, 1958), Bogolyubov et al. (1958)]. In the theory of magnetism, they are met with in all those cases when the Hamiltonian does not conserve the number of particles (elementary excitations), for example, in the case of antiferromagnetism and ferromagnetism, in the presence of anisotropy, etc.

Sec. 11. Degeneracy of States and Quasi-Averages*

In statistical mechanics problems, degeneracy is always present, due to the additive integrals of motion. Therefore, in the investigation of various problems – and particularly in the applications of the perturbation theory – it is necessary to excercise some caution, since it is well known that the perturbation theory for a degenerate level differs from the theory for a nondegenerate level. It is found that in statistical mechanics problems involving degeneracy, we are not dealing with the usual averages, but with averages of a special type known as the quasi-averages. The quasi-averages and their role in problems involving degeneracy were investigated by Bogolyubov (1961). We shall quote here only those parts of his work which refer to the theory of magnetism.

In quantum mechanical problems involving degeneracy, several eigenfunctions of the total energy operator correspond to a single energy level. Therefore, the eigenfunctions are indeterminate to the extent of arbitrary constants.

In statistical mechanics, the usual average of any dynamic variable $\mathcal{A}$ is defined unambiguously:

$$\langle \mathcal{A} \rangle = \frac{\mathrm{Sp}\left(\mathcal{A} e^{-\frac{\mathcal{H}}{\vartheta}}\right)}{\mathrm{Sp}\left(e^{-\frac{\mathcal{H}}{\vartheta}}\right)}.$$

Therefore, the various quantities (including observable quantities) based on the usual averages should be determinate. Hence, one might assume that degeneracy plays no role in statistical mechanics and that it can be ignored, for example, in the applications of the perturbation theory.

In fact, the situation is much more complex. In order to deal with the situation existing in real matter, we shall consider the case of an isotropic ferromagnet.

*See Bogolyubov (1961).

We shall write the Hamiltonian of a system in the form

$$\tilde{\mathscr{H}} = -\frac{1}{2}\sum_{f_1 f_2} I(f_1 - f_2)(S_{f_1}, S_{f_2}), \tag{11.1}$$

where f is the number of a lattice site, I is the exchange integral, and S^{α}_{f} is the electron spin operator at the site f. To make the case definite, we shall assume that $I(f) \geq 0$ for all f.

We shall use S^{α} to denote the operator for the total spin of the system:

$$S^{\alpha} = \sum_{f} S^{\alpha}_{f} \qquad (\alpha = x, y, z) \tag{11.2}$$

and we note that each component of the total spin operator is an integral of motion, i.e.,

$$S^{\alpha}\tilde{\mathscr{H}} - \tilde{\mathscr{H}}S^{\alpha} = 0. \tag{11.3}$$

The following commutationrelationships apply to the operators S^{α}:

$$\begin{gathered} S^x S^y - S^y S^x = iS^z, \quad S^y S^z - S^z S^y = iS^x, \\ S^z S^x - S^x S^z = iS^y. \end{gathered} \tag{11.4}$$

We shall now calculate the average over an ensemble of any component of the total spin. For example,

$$\langle S^z \rangle = \frac{\mathrm{Sp}\left(S^z e^{-\frac{\tilde{\mathscr{H}}}{\vartheta}}\right)}{\mathrm{Sp}\left(e^{-\frac{\tilde{\mathscr{H}}}{\vartheta}}\right)}. \tag{11.5}$$

Using the commutative relationships of Eq. (11.4), we find

$$\mathrm{Sp}\left(S^z e^{-\frac{\tilde{\mathscr{H}}}{\vartheta}}\right) = -i\,\mathrm{Sp}\left([S^x S^y - S^y S^x]\, e^{-\frac{\tilde{\mathscr{H}}}{\vartheta}}\right).$$

Since S^x commutes with $\tilde{\mathscr{H}}$, we have

$$\mathrm{Sp}\left(S^y S^x e^{-\frac{\tilde{\mathscr{H}}}{\vartheta}}\right) = \mathrm{Sp}\left(S^y e^{-\frac{\tilde{\mathscr{H}}}{\vartheta}} S^x\right) = \mathrm{Sp}\left(S^x S^y e^{-\frac{\tilde{\mathscr{H}}}{\vartheta}}\right)$$

and, consequently,

$$\langle S^z \rangle = 0. \tag{11.6}$$

Similarly, we find that

$$\langle S^x \rangle = \langle S^y \rangle = 0.$$

We shall introduce an operator for the total magnetic moment:

$$\mathcal{M}^\alpha = \mu S^\alpha \qquad (\alpha = x, y, z), \tag{11.7}$$

where μ is the magnetic moment per atom.

It is evident that

$$\mathrm{Sp}\left(\mathcal{M}^\alpha e^{-\frac{\tilde{\mathcal{H}}}{\vartheta}}\right) = 0,$$

and, therefore, the average magnetic moment of the system is zero

$$M^\alpha = \langle \mathcal{M}^\alpha \rangle = 0 \qquad (\alpha = x, y, z). \tag{11.8}$$

Thus, we have found that the usual average for the magnetic moment of our system is zero. This is due to the isotropy of the Hamiltonian (10.1) with respect to the spin-rotation group.

The result of Eq. (11.8) is valid for all temperatures ϑ, including temperatures below the Curie point. However, it is known that at temperatures below the Curie point the magnetic moment of our system is not equal to zero, although its direction is not fixed. In this sense, the statistical equilibrium states considered here are degenerate.

On the application of an external field,

$$H^\alpha = h\gamma^\alpha, \quad h > 0, \quad \sum_\alpha (\gamma^\alpha)^2 = 1 \qquad (\alpha = x, y, z).$$

Then, the Hamiltonian (11.1) is replaced by the following expression:

$$\mathcal{H}_{h\gamma} = \tilde{\mathcal{H}} - h(\gamma, \mathcal{M}). \tag{11.9}$$

We shall assume tentatively that the field is directed along the z-axis. In this case, only the operator S^z or $\mathcal{M}^z$ commutes with the Hamiltonian (11.9). Repeating the treatment described above, we find that $\langle S^x \rangle = \langle S^y \rangle = 0$. In fact,

$$\mathrm{Sp}\left(S^x e^{-\frac{1}{\vartheta}\mathcal{H}_{h\gamma}}\right) = -i\,\mathrm{Sp}\left([S^y S^z - S^z S^y]\, e^{-\frac{1}{\vartheta}\mathcal{H}_{h\gamma}}\right).$$

The quantity S^z commutes with $\mathcal{H}_{h\gamma}$ and, therefore,

$$\mathrm{Sp}\left(S^z S^y e^{-\frac{1}{\vartheta}\mathcal{H}_{h\gamma}}\right) = \mathrm{Sp}\left(S^y e^{-\frac{1}{\vartheta}\mathcal{H}_{h\gamma}} S^z\right) = \mathrm{Sp}\left(S^y S^z e^{-\frac{1}{\vartheta}\mathcal{H}_{h\gamma}}\right).$$

Hence, we see that the average value $\langle S^x \rangle$ is equal to zero (and the same applies to $\langle S^y \rangle$). This does not apply to $\langle S^z \rangle$. Since, at temperatures below the Curie point ϑ_C, an isotropic ferromagnet has a spontaneous moment oriented along the external field, we may assume that $\langle \mathcal{M}^z \rangle = M_h$, where M_h is some number, such that

$$\lim_{h \to 0} M_h = M_0 \quad (M_0 \neq 0). \tag{11.10}$$

The quantity M_0 represents the magnetic moment for vanishingly weak fields at temperatures $\vartheta < \vartheta_C$.

For the Hamiltonian of Eq. (11.9), we now have

$$\langle \mathcal{M}^\alpha \rangle = \gamma^\alpha M_h \quad (\alpha = x,\ y,\ z); \tag{11.11}$$

in the limit when h tends to zero, we have

$$\langle \mathcal{M}^\alpha \rangle = \gamma^\alpha M_0 \quad (\alpha = x,\ y,\ z). \tag{11.12}$$

Thus, we observe here an instability of the usual averages under the action of an infinitely small perturbation,* which is $h(\gamma, \mathcal{M})$. When this perturbation is applied, the average value $\langle \mathcal{M}^\alpha \rangle$ increases by a finite amount, which is given by

$$\gamma^\alpha M_0 \quad \left(M_0 = \lim_{h \to 0} M_h\right). \tag{11.13}$$

*By going to the limit $h \to 0$, we understand the limit of statistical mechanics $N \to \infty$ (where N is the number of particles in the system), followed by $h \to 0$.

We shall now introduce the concept of the quasi-averages for systems described by the Hamiltonian of Eq. (11.1).

We shall consider some dynamic variable $\mathcal{A}$, which is a linear combination of the spin operator products

$$\mathcal{A} = S_{f_1}^{\alpha_1} \dots S_{f_p}^{\alpha_p}$$

and we shall define the quasi-averages $\{\mathcal{A}\}$ as follows:

$$\{\mathcal{A}\} = \lim_{h \to 0} \langle \mathcal{A} \rangle_{h\gamma}, \tag{11.14}$$

where $\langle \mathcal{A} \rangle_{h\gamma}$ is the normal average of $\mathcal{A}$ for the Hamiltonian $\mathcal{H}_{h\gamma}$ of Eq. (11.9).

Thus, in the presence of degeneracy, the quasi-averages depend only on an arbitrary unit vector γ.

We can easily see that the quasi-average $\{\mathcal{A}\}$ is related to the normal average $\langle \mathcal{A} \rangle$ by the expression

$$\langle \mathcal{A} \rangle = \int \{\mathcal{A}\} \, d\gamma. \tag{11.15}$$

The quasi-averages provide a better physical description of the statistical equilibrium states than do the normal averages. The normal averages represent the quasi-averages averaged out over all directions of the unit vector γ.

The normal averages

$$\left\langle S_{f_1}^{\alpha_1} \dots S_{f_p}^{\alpha_p} \right\rangle \tag{11.16}$$

should be invariant with respect to the spin-rotation transformation. The corresponding quasi-averages

$$\left\{ S_{f_1}^{\alpha_1} \dots S_{f_p}^{\alpha_p} \right\} \tag{11.17}$$

should be only covariant, i.e., the expression (11.17) should remain invariant when the spins are rotated only if the same rotation applies also to the vector γ.

The indeterminacy of the direction of the vector M or the unit vector γ represents the degeneracy of the statistical equilibrium states of the dynamic system we have considered. The degeneracy can be removed by fixing the direction of the vector γ – for example, by assuming that it lies along the z-axis. The quasi-averages then become definite numbers. This type of quasi-average is used in the theory of ferromagnetism.

Thus, to remove the degeneracy, it is necessary to include in the Hamiltonian a perturbation operator which is not invariant with respect to the group of transformations associated with the degeneracy. In the considered case of an isotropic ferromagnet, such a perturbation is the energy of interaction of the system with an external magnetic field.

We note that the anisotropy removes the degeneracy only partly, because the Hamiltonian remains invariant with respect to the transformations reversing all the spin directions.

Chapter IV

Method of Approximate Second Quantization

One of the methods of calculating the equilibrium properties of matter is to determine the thermodynamic functions in terms of the partition functions. To do this, it is necessary to know the energy spectrum of the system. The problem of the exact determination of the spectrum of a strongly magnetic substance is very complex and, therefore, we shall consider here a simple method for the approximate determination of the lower part of the energy spectrum or, in other words, a method for the determination of the weakly excited states of a system.

Sec. 12. Quasi-Classical Method and Approximate Second Quantization*

We shall now present a method of calculating the lower part of the energy spectrum which is known as the approximate second quantization method (ASQ method) or, when applied to strongly magnetic substances, as the method of spin waves. The basis of the method was given by Bloch (1930, 1932), who was able to represent the lower part of the spectrum of a spin system in the form of a set of single-particle excitations, which he called spin waves. His results are valid only at relatively low temperatures; the states close to the ground state are excited mainly at low temperatures. This justifies the approximation of replacing the exact spectrum with the spectrum of the weakly excited states.

*See Bogolyubov (1949), Bogolyubov and Tyablikov (1949c, 1957).

The principle of the method is very simple.* We shall assume that the operators S_f^α are replaced with classical vectors $\gamma_f^\alpha S_f$, where γ_f is the unit vector and S_f is the value of the spin at a site f. The determination of the ground state of a system then reduces to the determination of a minimum of the classical expression for $\mathscr{E}_0$, obtained from the Hamiltonian of the system, $\tilde{\mathscr{H}}$, by replacing the spin operators with the classical vectors. The condition for the first variation to vanish,

$$\delta\mathscr{E}_0 = 0$$

gives equations for the determination of the classical vectors γ_f. The sufficient conditions for a minimum are provided by the positive sign of the second variation:

$$\delta^2\mathscr{E}_0 > 0.$$

The expression for $\delta^2\mathscr{E}_0$ describes "small oscillations about the equilibrium position" and can be reduced to the diagonal form by a suitable canonical transformation. Therefore, the positive value of $\delta^2\mathscr{E}_0$ means that all the frequencies of these small oscillations are positive, and the spectrum of these frequencies represents the spectrum of the weakly excited states of the system.†

We have considered this situation in order to present more clearly the principle of the method. We shall now find the approximate eigenfunction which represents such a selection of the ground state and determine the conditions of applicability of the proposed method for determining the spectrum of the weakly excited states.

For simplicity, we shall not consider crystals with one-ion anisotropy.‡ We shall assume that the lattice of a crystal occupies a volume V and consists of N sites, at each of which the spin is S_f. Moreover, we shall assume that μ_f represents the mag-

*We shall consider here only the spin Hamiltonians; the more general case is considered in the work of Bogolyubov (1949) and Bogolyubov and Tyablikov (1949c, 1957).

† This fact was used for the first time in the theory of magnetism by Hulten (1936) in his investigation of isotropic antiferromagnets.

‡ For one-ion anisotropy see Sec. 18.

netic moment at the f-th site; $I(f_1 - f_2)$ is the exchange integral for the spins at the sites f_1 and f_2; H is an external magnetic field. Moreover, we shall consider only the simplest lattices; the generalization to more complex lattices does not present any great difficulty.

We shall arbitrarily write the Hamiltonian of our system in the form

$$\tilde{\mathcal{H}} = \tilde{\mathcal{H}}(S_f^\alpha). \qquad (12.1)$$

The ground state of the system will be found from Fock's minimum principle:

$$\mathcal{E}_0 = \min \frac{(C_0^*, \tilde{\mathcal{H}} C_0)}{(C_0^*, C_0)}, \qquad (12.2)$$

where C_0 is a trial wave function. We shall replace the spin variables S_f^α in the Hamiltonian (12.1) with the variables $S_f^\pm$, S_f^z, in accordance with the formulas (5.8)-(5.10). We shall select, as the function C_0, the vacuum wave function*:

$$C_0(\ldots, S_f^z, \ldots) = \prod_f \Delta(S_f^z - S_f). \qquad (12.3)$$

The variational method yields, in general, too high a value for the ground state energy $\mathcal{E}_0$ represented by Eq. (12.2). Therefore, we shall define the transformation coefficients of Eq. (5.8) to make the value of $\mathcal{E}_0$ as small as possible.

The Hamiltonian of Eq. (12.1) includes products of the operators S_f^α for various sites. Therefore, the expression (12.2) contains only the average values of the operators (5.8) applied to the function $C_0(S_f^z) = \Delta(S_f^z - S_f)$. Obviously,

$$(C_0^*(S_f^z), S_f^\alpha C_0(S_f^z)) - \gamma_f^\alpha S_f. \qquad (12.4)$$

Consequently, the expression for $\mathcal{E}_0$, given by Eq. (12.2), is obtained from $\tilde{\mathcal{H}}$ of Eq. (12.1) by replacing the operators S_f^α with

*The vacuum state of a system is that state in which all the spins are oriented along the same direction and the z component of each of the spins has its maximum value. Other states are regarded as unstable.

with the classical vectors $\gamma_f^\alpha S_f$, and by finding the minimum with respect to γ_f:

$$\mathcal{E}_0 = \min \mathcal{E}_0(\gamma_f S_f), \quad \sum_\alpha (\gamma_f^\alpha)^2 = 1. \tag{12.5}$$

The parameters which need to be determined are the components of the unit vectors γ_f. From the condition for the minimum of the form given by Eq. (12.5), we obtain the following expressions for these factors:

$$\frac{\partial}{\partial \gamma_f^\alpha}\left\{\mathcal{E}_0(\gamma_f S_f) - \sum_f S_f \lambda_f (\gamma_f^2 - 1)\right\} = \frac{\partial \mathcal{E}_0(\gamma_f S_f)}{\partial \gamma_f^\alpha} - 2S_f \lambda_f \gamma_f^\alpha = 0,$$
$$\sum_\alpha (\gamma_f^\alpha)^2 = 1 \quad (\alpha = x, y, z), \tag{12.6}$$

where the quantities λ_f are Lagrange's indeterminate multipliers.

Thus, the determination of the ground-state energy (in the zeroth approximation) using the classical form given by Eq. (12.5) represents the so-called quasi-classical method. In this method, the electron spin operators are replaced with classical vectors and the energy of the interaction between the operators is taken to be equal to the exchange energy (cf., Sec. 1).

We have coupled here the quasi-classical method for determining the ground-state energy with the variational method due to Fock (1930). We have established that a trial wave function has zero values of the spin deviations: $S_f - S_f^z = 0$. Next, we must solve two problems: first, we have to estimate the degree of accuracy of our calculation of the ground state and, secondly, we have to find the energy levels of at least the weakly excited states and establish the criteria of the applicability of the method.

We know that if the quantities γ_f are found from the condition for a minimum of the ground-state energy in the quasi-classical approximation, the terms in $\tilde{\mathcal{H}}$ which contain the lowest powers of the operators $S_f^\pm$, S_f^z will be linear in the spin deviations $S_f - S_f^z$ and quadratic in $S_f^\pm$. In fact, it follows that

$$\mathcal{E}_0(\gamma_f S_f) = [\tilde{\mathcal{H}}]_{S_f^\alpha \to \gamma_f^\alpha S_f}. \tag{12.7}$$

We shall write the transformation of Eq. (5.8) in the form

$$S_f^\alpha = \gamma_f^\alpha S_f + \delta S_f^\alpha, \tag{12.8}$$

where

$$\delta S_f^\alpha = -\gamma_f^\alpha (S_f - S_f^z) + A_f^\alpha S_f^+ + A_f^{*\alpha} S_f^- \tag{12.9}$$

is considered to be a "small" operator correction to the numbers $\gamma_f^\alpha S_f$. Substituting Eq. (12.8) into $\tilde{\mathscr{H}}$, we obtain

$$\begin{aligned}\tilde{\mathscr{H}} = \tilde{\mathscr{H}}(S_f^\alpha) = \mathscr{H}(\gamma_f^\alpha S_f + \delta S_f^\alpha) = \\ = \mathscr{E}_0(\gamma_f S_f) + \sum_\alpha \frac{\partial \mathscr{E}_0(\gamma_f S_f)}{\partial \gamma_f^\alpha} S_f^{-1} \delta S_f^\alpha + \\ + \frac{1}{2} \sum_{\alpha_1, \alpha_2} \frac{\partial^2 \mathscr{E}_0(\gamma_f S_f)}{\partial \gamma_{f_1}^{\alpha_1} \partial \gamma_{f_2}^{\alpha_2}} S_{f_1}^{-1} S_{f_2}^{-1} \delta S_{f_1}^{\alpha_1} \delta S_{f_2}^{\alpha_2} + \dots\end{aligned}$$

We shall take the scalar product of Eq. (12.6) with $S_f^{-1} \delta S_f^\alpha$, sum over α, and apply the conditions of Eq. (5.9) to the vectors γ_f and A_f. This gives the following equation:

$$\sum_\alpha \frac{\partial \mathscr{E}_0(\gamma_f S_f)}{\partial \gamma_f^\alpha} S_f^{-1} \delta S_f^\alpha = 2\lambda_f (\gamma_f, \delta S_f) = -2\lambda_f (S_f - S_f^z). \tag{12.10}$$

Similarly,

$$\sum_\alpha \frac{\partial \mathscr{E}_0(\gamma_f S_f)}{\partial \gamma_f^\alpha} S_f^{-1} \gamma_f^\alpha = 2\lambda_f. \tag{12.11}$$

Consequently, $\tilde{\mathscr{H}}$ transforms into

$$\begin{aligned}\tilde{\mathscr{H}} = \mathscr{E}_0(\gamma_f S_f) - \sum_f 2\lambda_f (S_f - S_f^z) + \\ + \frac{1}{2} \sum_{f_1, f_2, \alpha_1, \alpha_2} \frac{\partial^2 \mathscr{E}_0(\gamma_f S_f)}{\partial \gamma_{f_1}^{\alpha_1} \partial \gamma_{f_2}^{\alpha_2}} S_{f_1}^{-1} S_{f_2}^{-1} \delta S_{f_1}^{\alpha_1} \delta S_{f_2}^{\alpha_2} + \dots\end{aligned} \tag{12.12}$$

According to Eq. (5.7), we may assume that $S_f - S_f^z$ is of the order of $S_f^- S_f^+$.* Consequently, if we expand $\tilde{\mathscr{H}}$ in powers of the

*In the quasi-classical approximation, a similar result follows from the normalization condition (5.2).

spin operators $S_f - S_f^z$, $S_f^{\pm}$, we then obtain

$$\tilde{\mathcal{H}} = \mathcal{E}_0 + \mathcal{H}_2 + \mathcal{H}_3 + \ldots = \\ = \mathcal{E}_0(\gamma_f S_f) - \sum_f 2\lambda_f (S_f - S_f^z) + \\ + \frac{1}{2} \sum_{f_1, f_2, \alpha_1, \alpha_2} \frac{\partial^2 \mathcal{E}_0 (\gamma_f S_f)}{\partial \gamma_{f_1}^{\alpha_1} \partial \gamma_{f_2}^{\alpha_2}} S_{f_1}^{-1} S_{f_2}^{-1} \tau_{f_1}^{\alpha_1} \tau_{f_2}^{\alpha_2} + \mathcal{H}_3 + \ldots, \tag{12.13}$$

where

$$\tau_f^{\alpha} = A_f^{\alpha} S_f^{+} + A_f^{*\alpha} S_f^{-}, \tag{12.14}$$

$\mathcal{H}_3$, ... are the third, etc., forms containing powers of the operators; $\mathcal{E}_0$ is found in accordance with Eq. (12.7); γ_f, λ_f are found by solving the system of equations given in Eq. (12.6); A_f are the transformation coefficients of Eq. (5.8).

The spectrum of the quadratic form $\mathcal{H}_2$ approximates the lower part of the spectrum of the total Hamiltonian $\tilde{\mathcal{H}}$. Therefore, the ground state $\mathcal{E}_0$ is stable if all the eigenvalues of $\mathcal{H}_2$ are positive.

To determine the ground state, we have used the variational method. In this method, the wave functions of the ground state were selected so that the numerical values of the spin deviations were zero:

$$S_f - S_f^z = 0. \tag{12.15}$$

We shall now assume that the condition of Eq. (12.15) is satisfied only approximately in the sense that the probability of finding the values $S_f - S_f^z \neq 0$ is small. Consequently, we shall consider the average values of the products of the spin deviations as quantities of increasing order of smallness:

$$1 \gg \{\langle S_f - S_f^z \rangle\} \gg \{\langle (S_{f_1} - S_{f_1}^z)(S_{f_2} - S_{f_2}^z) \rangle_{f_1 \neq f_2}\} \gg \ldots \tag{12.16}$$

When this condition is satisfied, the main term in the Hamiltonian will be the form $\mathcal{H}_2$ and the spectrum can then be described by the expression

$$\tilde{\mathcal{H}} \approx \mathcal{E}_0 + \mathcal{H}_2, \tag{12.17}$$

where $\mathcal{E}_0$ is the ground-state energy of Eq. (12.7) and $\mathcal{H}_2$ is the form given by Eq. (12.13).

In accordance with the assumption that the probability of nonzero spin deviations is small, we shall neglect the difference between the right-hand parts of the commutation relationships of Eq. (5.5) and their values for the ground state. In other words, we shall assume approximately that the Bose-type commutation relationships apply to the operators $S_f - S_f^z$, $S_f^{\pm}$:

$$\begin{aligned} S_f^+ S_g^- - S_g^- S_f^+ &\approx 2S_f \Delta(f-g), \\ S_f^{\pm}(S_g - S_g^z) - (S_g - S_g^z) S_f^{\pm} &\approx \pm S_f^{\pm} \Delta(f-g). \end{aligned} \tag{12.18}$$

Applying the usual normalization to the operators, we can write

$$S_f^+ = \sqrt{2S_f} b_f, \quad S_f^- = \sqrt{2S_f} b_f^+, \quad S_f - S_f^z = n_f = b_f^+ b_f, \tag{12.19}$$

where b_f, b_f^+ are operators which satisfy the pure Bose commutation relationships of the type given by Eq. (3.27). Consequently, the problem of the approximate calculation of the lower part of the spectrum of the spin system reduces to an investigation of a form which is quadratic in terms of the Bose operators and is given by Eqs. (12.13), (12.14), and (12.19).

In the important special case of a magnetically isotropic crystal, the spin system is described by the Hamiltonian (6.14). By repeating our previous calculations, we can easily obtain the following expression for the Hamiltonian:

$$\tilde{\mathscr{H}} = \mathscr{E}_0 + \mathscr{H}_2 + \mathscr{H}_3 + \mathscr{H}_4, \tag{12.20}$$

where

$$\mathscr{E}_0 = -\sum_f \mu_f S_f (\gamma_f, H) - \frac{1}{2} \sum_{f_1, f_2} S_{f_1} S_{f_2} I(f_1 - f_2)(\gamma_{f_1}, \gamma_{f_2}), \tag{12.21}$$

$$\mathscr{H}_2 = -\sum_f 2\lambda_f (S_f - S_f^z) - \frac{1}{2} \sum_{f_1, f_2} I(f_1 - f_2)(\tau_{f_1}, \tau_{f_2}), \tag{12.22}$$

$$\mathscr{H}_3 = -\sum_{f_1, f_2} I(f_1 - f_2)(\gamma_{f_1}, \tau_{f_2})(S_{f_1} - S_{f_1}^z), \tag{12.23}$$

$$\mathscr{H}_4 = -\frac{1}{2} \sum_{f_1, f_2} I(f_1 - f_2)(\gamma_{f_1}, \gamma_{f_2})(S_{f_1} - S_{f_1}^z)(S_{f_2} - S_{f_2}^z); \tag{12.24}$$

τ_f are the operators given by Eq. (12.14); γ_f, λ_f are found from the condition for a minimum of the ground-state energy given in Eq. (12.6):

$$-\sum_{f'} I(f-f')S_{f'}\gamma_{f'}^{\alpha} - \mu_f H^{\alpha} = 2\lambda_f \gamma_f^{\alpha},$$
$$\sum_{\alpha} (\gamma_f^{\alpha})^2 = 1. \tag{12.25}$$

In the calculation of λ_f, it is convenient to use the notation of Eq. (12.11):

$$2\lambda_f = -\mu_f(\gamma_f, H) - \sum_{f'} I(f-f')S_{f'}(\gamma_f, \gamma_{f'}). \tag{12.26}$$

According to Eq. (12.17), the ground-state energy and the approximate form of the lower part of the spectrum are given, respectively, by the classical form $\mathcal{E}_0$ of Eq. (12.21) and the operator form $\mathcal{H}_2$ of Eq. (12.22). The spin operators are given by the approximate formulas of Eq. (12.19). Consequently, for a magnetically isotropic crystal, the problem reduces to an investigation of the quadratic form:

$$\begin{aligned}\mathcal{H} \approx \mathcal{E}_0 + \mathcal{H}_2 = \mathcal{E}_0 - \sum_f 2\lambda_f n_f - \\ - \sum_{f_1, f_2} \sqrt{S_{f_1}S_{f_2}} I(f_1 - f_2)\{2(A^*_{f_1}, A_{f_2}) b^+_{f_1} b_{f_2} + \\ + (A_{f_1}, A_{f_2}) b_{f_1} b_{f_2} + (A^*_{f_1}, A^*_{f_2}) b^+_{f_1} b^+_{f_2}\}.\end{aligned} \tag{12.27}$$

where b_f, b_f^+ are the Bose operators, $n_f = b_f^+ b_f$, and the quantities γ_f, λ_f are found from Eq. (12.25).

In the approximate second quantization method, the ground state is defined by the condition that all the spin deviations for this state are zero. In other words, for the wave function of the ground state, we have

$$(S_f - S_f^z) C_0 = 0. \tag{12.28}$$

We shall now consider the state C_{f_0}, in which the spin deviation at a site f_0 is equal to unity:

$$[S_f - \Delta(f - f_0) - S_f^z] C_{f_0} = 0. \tag{12.29}$$

The function C_{f_0} can be regarded as the result of the application of the operator $S^-_{f_0}$ to the function C_0. Due to the translational degeneracy, such a deviation is equally likely at any of the equivalent lattice sites. Therefore, the wave function of the state with a single "excited" spin will have the form

$$C_\nu = \frac{1}{\sqrt{N}} \sum_f e^{i(f,\nu)} C_f = \frac{1}{\sqrt{N}} \sum_f e^{i(f,\nu)} S^-_f C_0. \tag{12.30}$$

The function C_ν represents a state in which only one spin wave with a wave vector ν is excited, and is the eigenfunction of Eq. (12.27) when $S_f = S$. For this reason, the approximate method of calculating the weakly excited states of a spin system, described above, is also known as the method of spin waves, and the transition from the total Hamiltonian $\mathscr{H}$ to the Hamiltonian $\mathscr{H}_2$, followed by the replacement of the spin operators with the Bose operators, is known as the transition to the spin-wave approximation.

Basically, the approximate second quantization method rests on the possibility of considering the quadratic form in the Hamiltonian as its main part and on the approximate replacement of the spin operators with the Bose operators.

Sec. 13. Diagonalization of Quadratic Forms*

In the second quantization method, the approximate Hamiltonians are frequently represented as forms quadratic in the Bose operators. We shall now consider the problem of the diagonalization of such forms.†

Let us consider a quadratic form of the type

$$\mathscr{H} = \frac{1}{2} \sum_{\alpha,\beta} R_{\alpha\beta} x^+_\alpha x^+_\beta + \sum_{\alpha,\beta} S_{\alpha\beta} x^+_\alpha x_\beta + \frac{1}{2} \sum_{\alpha,\beta} R^*_{\alpha\beta} x_\alpha x_\beta, \tag{13.1}$$

*See Tyablikov (1947), Bogolyubov (1949), and Bogolyubov and Tyablikov (1949c).

† The extension of the results to the diagonalization of forms quadratic in the Fermi operators does not present any difficulties.

whose coefficients satisfy the conditions

$$S_{\alpha\beta} = S^*_{\beta\alpha}, \quad R_{\alpha\beta} = R_{\beta\alpha}, \tag{13.2}$$

and in which the variables x_α may be either the Bose operators or ordinary numbers (not operators).

From the conditions of Eq. (13.2), it follows that $\mathscr{H}$ is a Hermitian form.

We shall use E_ν and $u_{\alpha\nu}$, $v_{\alpha\nu}$ to denote the eigenvalues and eigenfunctions of the system of equations:

$$\begin{aligned} E_\nu u_{\alpha\nu} &= \sum_\beta S_{\alpha\beta} u_{\beta\nu} + \sum_\beta R_{\alpha\beta} v_{\beta\nu}, \\ -E_\nu v_{\alpha\nu} &= \sum_\beta S^*_{\alpha\beta} v_{\beta\nu} + \sum_\beta R^*_{\alpha\beta} u_{\beta\nu}. \end{aligned} \tag{13.3}$$

For the functions $u_{\alpha\nu}$, $v_{\alpha\nu}$ we have the following orthogonality and normalization conditions:

$$\begin{aligned} &\sum_\alpha (u_{\alpha\nu} u^*_{\alpha\mu} - v_{\alpha\nu} v^*_{\alpha\mu}) = \Delta(\nu - \mu), \\ &\sum_\alpha (u_{\alpha\nu} v_{\alpha\mu} - u_{\alpha\mu} v_{\alpha\nu}) = 0 \end{aligned} \tag{13.4}$$

and, correspondingly,

$$\begin{aligned} &\sum_\nu (u_{\alpha\nu} u^*_{\beta\nu} - v_{\beta\nu} v^*_{\alpha\nu}) = \Delta(\alpha - \beta), \\ &\sum_\nu (u_{\beta\nu} v^*_{\alpha\nu} - u_{\alpha\nu} v^*_{\beta\nu}) = 0. \end{aligned} \tag{13.5}$$

If x_α, x^+_α represent the Bose operators, the canonical transformation

$$x_\alpha = \sum_\nu (u_{\alpha\nu} \xi_\nu + v^*_{\alpha\nu} \xi^+_\nu), \tag{13.6}$$

where ξ_ν, ξ^+_ν are new Bose operators, alters the quadratic form $\mathscr{H}$ of Eq. (13.1) to the diagonal type

$$\mathscr{H} = \Delta\mathscr{E}_0 + \sum_\nu E_\nu \xi^+_\nu \xi_\nu, \tag{13.7}$$

$$\Delta\mathscr{E}_0 = -\sum_{\alpha,\nu} E_\nu |v_{\alpha\nu}|^2. \tag{13.8}$$

If x_α, x_α^+ are ordinary numbers, a canonical transformation of the type of Eq. (13.6), where ξ_ν, ξ_ν^+ are also ordinary numbers, reduces the quadratic form of Eq. (13.1) to

$$\mathcal{H} = \sum_\nu E_\nu \xi_\nu^+ \xi_\nu. \tag{13.9}$$

The inverse transformation has the form

$$\xi_\nu = \sum_\alpha \left(u_{\alpha\nu}^* x_\alpha - v_{\alpha\nu}^* x_\alpha^+\right). \tag{13.10}$$

We shall now explain the results given by Eqs. (13.3)-(13.8). The operator form of Eq. (13.1) is considered to be a Hamiltonian $\mathcal{H}$ of some system represented by the Bose operators. The equations of motion for the operators x_α, x_α^+ then have the form

$$i\frac{dx_\alpha}{dt} = [x_\alpha, \mathcal{H}], \quad i\frac{dx_\alpha^+}{dt} = [x_\alpha^+, \mathcal{H}]$$

or, in more detail,

$$\begin{aligned} i\frac{dx_\alpha}{dt} &= \sum_\beta S_{\alpha\beta} x_\beta + \sum_\beta R_{\alpha\beta} x_\beta^+, \\ -i\frac{dx_\alpha^+}{dt} &= \sum_\beta S_{\alpha\beta}^* x_\beta^+ + \sum_\beta R_{\alpha\beta}^* x_\beta. \end{aligned} \tag{13.11}$$

We shall apply to Eq. (13.11) the transformation of variables given by (13.6), where ξ_ν, ξ_ν^+ are the new Bose operators, and $u_{\alpha\nu}$, $v_{\alpha\nu}$ are some functions which have to be determined.

We shall now require the new Bose operators ξ_ν, ξ_ν^+ to satisfy the following equations of motion:

$$i\frac{d\xi_\nu}{dt} = E_\nu \xi_\nu, \quad i\frac{d\xi_\nu^+}{dt} = -E_\nu \xi_\nu^+. \tag{13.12}$$

Substituting Eq. (13.6) into Eq. (13.11), and using Eq. (13.12), we obtain

$$\sum_\nu \left(E_\nu u_{\alpha\nu} \xi_\nu - E_\nu v_{\alpha\nu}^* \xi_\nu^+\right) = \sum_{\beta,\nu} \left(S_{\alpha\beta} u_{\beta\nu} \xi_\nu + S_{\alpha\beta} v_{\beta\nu}^* \xi_\nu^+ + R_{\alpha\beta} u_{\beta\nu}^* \xi_\nu^+ + R_{\alpha\beta} v_{\beta\nu} \xi_\nu\right).$$

Since the operators ξ_ν, $\overset{+}{\xi}_\nu$ are linearly independent, their coefficients should vanish independently of one another. This gives us Eqs. (13.3) from which to determine the functions $u_{\alpha\nu}$, $v_{\alpha\nu}$.

We shall now require the transformation (13.6) to be canonical, i.e., we shall require the operators x_α, $\overset{+}{x}_\alpha$, defined by this formula, to satisfy the Bose commutation relationships if ξ_ν, $\overset{+}{\xi}_\nu$ are the Bose operators.

For this purpose, we shall use the commutators

$$x_\alpha x_\beta^+ - x_\beta^+ x_\alpha, \quad x_\alpha x_\beta - x_\beta x_\alpha$$

and substitute in them Eq. (13.6). As a result of some simple calculations, we then obtain

$$x_\alpha x_\beta^+ - x_\beta^+ x_\alpha = \sum_\nu (u_{\alpha\nu} u_{\beta\nu}^* - v_{\beta\nu} v_{\alpha\nu}^*),$$

$$x_\alpha x_\beta - x_\beta x_\alpha = \sum_\nu (v_{\alpha\nu}^* u_{\beta\nu} - v_{\beta\nu}^* u_{\alpha\nu}).$$

Equating the left-hand parts of these expressions to $\Delta(\alpha - \beta)$ and 0, respectively, we obtain the orthonormalization conditions of Eq. (13.5) for the functions $u_{\alpha\nu}$, $v_{\alpha\nu}$.

We shall now derive formulas (13.4)-(13.5) and (13.7)-(13.9) using Eq. (13.3) and the conditions (13.2), which apply to their coefficients.

We note, first of all, that the eigenvalues E_ν of the system of equations (13.3) are real. In fact, it follows from Eq. (13.3) that

$$E_\nu \sum_\alpha (u_{\alpha\nu} u_{\alpha\nu}^* - v_{\alpha\nu} v_{\alpha\nu}^*) =$$
$$= \sum_{\alpha,\beta} (S_{\alpha\beta} u_{\beta\nu} u_{\alpha\nu}^* + S_{\alpha\beta}^* v_{\beta\nu} v_{\alpha\nu}^* + R_{\alpha\beta} v_{\beta\nu} u_{\alpha\nu}^* + R_{\alpha\beta}^* u_{\beta\nu} v_{\alpha\nu}^*).$$

We can easily see that the sum on the left-hand side of the above equation is real by definition and the double sum of the right-hand part is real by virtue of Eq. (13.2). Consequently,

$$E_\nu^* = E_\nu. \tag{13.13}$$

We shall now establish the validity of the orthogonality and normalization conditions of Eq. (13.4). Using Eq. (13.3), we obtain

$$E_\nu \sum_\alpha (u_{\alpha\nu} u^*_{\alpha\mu} - v_{\alpha\nu} v^*_{\alpha\mu}) =$$
$$= \sum_{\alpha,\beta} (S_{\alpha\beta} u_{\beta\nu} u^*_{\alpha\mu} + R_{\alpha\beta} v_{\beta\nu} u^*_{\alpha\mu} + S^*_{\alpha\beta} v_{\beta\nu} v^*_{\alpha\mu} + R^*_{\alpha\beta} u_{\beta\nu} v^*_{\alpha\mu}).$$

We shall interchange the indices ν and μ and use the conjugate expression, so that we finally have

$$E_\mu \sum_\alpha (u^*_{\alpha\mu} u_{\alpha\nu} - v^*_{\alpha\mu} v_{\alpha\nu}) =$$
$$= \sum_{\alpha,\beta} (S^*_{\alpha\beta} u^*_{\beta\mu} u_{\alpha\nu} + R^*_{\alpha\beta} v^*_{\beta\mu} u_{\alpha\nu} + S_{\alpha\beta} v^*_{\beta\mu} v_{\alpha\nu} + R_{\alpha\beta} u^*_{\beta\mu} u_{\alpha\nu}).$$

Calculating the difference of these two expressions, we obtain

$$(E_\nu - E_\mu) \sum_\alpha (u_{\alpha\nu} u^*_{\alpha\mu} - v_{\alpha\nu} v^*_{\alpha\mu}) = 0.$$

Hence, in the usual way, we establish the validity of the first of the relationships in Eq. (13.4).

Next, from Eq. (13.3), we obtain the equation

$$E_\nu \sum_\alpha (u_{\alpha\nu} v_{\alpha\mu} - v_{\alpha\nu} u_{\alpha\mu}) =$$
$$= \sum_{\alpha,\beta} (S_{\alpha\beta} u_{\beta\nu} v_{\alpha\mu} + R_{\alpha\beta} v_{\beta\nu} v_{\alpha\mu} + S^*_{\alpha\beta} v_{\beta\nu} u_{\alpha\mu} + R^*_{\alpha\beta} u_{\beta\nu} u_{\alpha\mu})$$

and, by the interchange $\mu \rightleftharpoons \nu$, we obtain the analogous equation

$$E_\mu \sum_\alpha (u_{\alpha\mu} v_{\alpha\nu} - v_{\alpha\mu} u_{\alpha\nu}) =$$
$$= \sum_{\alpha,\beta} (S_{\alpha\beta} u_{\beta\mu} v_{\alpha\nu} + R_{\alpha\beta} v_{\beta\mu} v_{\alpha\nu} + S^*_{\alpha\beta} v_{\beta\mu} u_{\alpha\nu} + R^*_{\alpha\beta} u_{\beta\mu} u_{\alpha\nu}).$$

Their difference gives the following expression:

$$(E_\nu + E_\mu) \sum_\alpha (u_{\alpha\nu} v_{\alpha\mu} - v_{\alpha\nu} u_{\alpha\mu}) = 0.$$

Hence, we find the second of the relationships in Eq. (13.4).

We shall now show that by virtue of Eq. (13.4) the transformation which is inverse to Eq. (13.6) has the form of Eq. (13.10).

We shall take the arbitrary system of quantities $\{X_\alpha\}$ and represent it in the form of an expansion:

$$X_\alpha = \sum_\nu (u_{\alpha\nu}\Xi_\nu + v^*_{\alpha\nu}\Xi^*_\nu); \tag{13.14}$$

and, similarly,

$$X^*_\alpha = \sum_\nu (u^*_{\alpha\nu}\Xi^*_\nu + v_{\alpha\nu}\Xi_\nu). \tag{13.15}$$

Multiplying Eq. (13.14) by $u^*_{\alpha\mu}$ and Eq. (13.15) by $v^*_{\alpha\mu}$, we sum with respect to α and subtract one expression from the other:

$$\sum_\alpha (u^*_{\alpha\mu}X_\alpha - v^*_{\alpha\mu}X^*_\alpha) =$$
$$= \sum_\nu \Xi_\nu \sum_\alpha (u_{\alpha\nu}u^*_{\alpha\mu} - v_{\alpha\nu}v^*_{\alpha\mu}) + \sum_\nu \Xi^*_\nu \sum_\alpha (v^*_{\alpha\nu}u^*_{\alpha\mu} - u^*_{\alpha\nu}v^*_{\alpha\mu}).$$

Hence, using the relationships of Eq. (13.4), we obtain

$$\Xi_\nu = \sum_\alpha (X_\alpha u^*_{\alpha\nu} - X^*_\alpha v^*_{\alpha\nu}). \tag{13.16}$$

The transformation (13.16) is the inverse of the transformation given by Eq. (13.14). If the quantities X_α, Ξ_ν are replaced with the operators x_α, ξ_ν, we obtain the inverse transformations (13.6) and (13.10).

Substituting the values of the "coefficients" Ξ_ν of Eq. (13.16) into the expansion given by Eq. (13.14), we obtain the equation

$$X_\alpha = \sum_{\beta,\nu} \{(X_\beta u^*_{\beta\nu} - X^*_\beta v^*_{\beta\nu}) u_{\alpha\nu} + (X^*_\beta u_{\beta\nu} - X_\beta v_{\beta\nu}) v^*_{\alpha\nu}\},$$

which should be valid for any complex values of the quantities X_α. Setting the coefficients of X_β and $X_\beta{}^*$ equal to zero, we obtain the relationships of Eq. (13.5).

We shall show that the transformation of Eq. (13.6) is canonical. In other words, if the operators ξ^+_ν, ξ_ν satisfy the commutation relationships of the Bose–Einstein statistics, the operators x^+_α, x_α also satisfy the commutation relationships of this statistics.

To prove this, we shall use commutators. Employing the orthonormalization conditions of Eq. (13.5), we obtain

$$x_\alpha x_\beta^+ - x_\beta^+ x_\alpha =$$
$$= \sum_{\nu,\mu} \{ (u_{\alpha\nu}\xi_\nu + v_{\alpha\nu}^*\xi_\nu^+)(u_{\beta\mu}^*\xi_\mu^+ + v_{\beta\mu}\xi_\mu) -$$
$$- (u_{\beta\mu}^*\xi_\mu^+ + v_{\beta\mu}\xi_\mu)(u_{\alpha\nu}\xi_\nu + v_{\alpha\nu}^*\xi_\nu^+) \} =$$
$$= \sum_{\nu,\mu} (u_{\alpha\nu}u_{\beta\mu}^* - v_{\beta\mu}v_{\alpha\nu}^*)\,\Delta(\nu-\mu) = \Delta(\alpha-\beta).$$
$$x_\alpha x_\beta - x_\beta x_\alpha = \sum_\nu (v_{\beta\nu}^* u_{\alpha\nu} - v_{\alpha\nu}^* u_{\beta\nu}) = 0.$$

Hence, we see that if ξ are the Bose operators, then x are also the Bose operators. Similarly, the converse of this is proved.

To diagonalize the quadratic form of Eq. (13.1), we shall subject it to the canonical transformation of Eq. (13.6). After making the obvious transformations, we obtain

$$\begin{aligned} \mathscr{H} = &\sum_{\nu,\mu} \xi_\nu\xi_\mu \sum_{\alpha,\beta} \left(\frac{1}{2} R_{\alpha\beta}v_{\alpha\nu}v_{\beta\mu} + S_{\alpha\beta}v_{\alpha\nu}u_{\beta\mu} + \frac{1}{2} R_{\alpha\beta}^* u_{\alpha\nu}u_{\beta\mu}\right) + \\ &+ \sum_{\nu,\mu} \xi_\nu^+\xi_\mu^+ \sum_{\alpha,\beta} \left(\frac{1}{2} R_{\alpha\beta}^* v_{\alpha\nu}^* v_{\beta\mu}^* + S_{\alpha\beta}u_{\alpha\nu}^* v_{\beta\mu}^* + \frac{1}{2} R_{\alpha\beta}u_{\alpha\nu}^* u_{\beta\mu}^*\right) + \\ &+ \sum_{\nu,\mu} \xi_\nu^+\xi_\mu \sum_{\alpha,\beta} \left(\frac{1}{2} R_{\alpha\beta}u_{\alpha\nu}^* v_{\beta\mu} + S_{\alpha\beta}u_{\alpha\nu}^* u_{\beta\mu} + \frac{1}{2} R_{\alpha\beta}^* v_{\alpha\nu}^* u_{\beta\mu}\right) + \\ &+ \sum_{\nu,\mu} \xi_\nu\xi_\mu^+ \sum_{\alpha,\beta} \left(\frac{1}{2} R_{\alpha\beta}v_{\alpha\nu}u_{\beta\mu}^* + S_{\alpha\beta}v_{\alpha\nu}v_{\beta\mu}^* + \frac{1}{2} R_{\alpha\beta}^* u_{\alpha\nu}v_{\beta\mu}^*\right). \end{aligned} \quad (13.17)$$

We shall transform the sums involving α and β, using Eq. (13.3). For example, for the first sum we obtain

$$\sum_{\alpha,\beta} \left(\frac{1}{2} R_{\alpha\beta}v_{\alpha\nu}v_{\beta\mu} + S_{\alpha\beta}v_{\alpha\nu}u_{\beta\mu} + \frac{1}{2} R_{\gamma\beta}^* u_{\alpha\nu}u_{\beta\mu}\right) =$$
$$= \frac{1}{2}\sum_\alpha v_{\alpha\nu} \sum_\beta (R_{\alpha\beta}v_{\beta\mu} + S_{\alpha\beta}u_{\beta\mu}) +$$
$$+ \frac{1}{2}\sum_\beta u_{\beta\mu} \sum_\alpha (S_{\alpha\beta}v_{\alpha\nu} + R_{\alpha\beta}^* u_{\alpha\nu}) =$$
$$= \frac{1}{2}(E_\mu - E_\nu)\sum_\alpha u_{\alpha\mu}v_{\alpha\nu} = Q_{\mu\nu}.$$

On the basis of the second condition of Eq. (13.4), the quantity $Q_{\mu\nu}$ is antisymmetric with respect to the interchange of indices,

$$Q_{\mu\nu} + Q_{\nu\mu} = 0$$

and, therefore,

$$\sum_{\nu,\mu} \xi_\nu\xi_\mu Q_{\mu\nu} = 0.$$

In exactly the same way, we can show that the second term in Eq. (13.17) is equal to zero.

The sum with respect to α, β in the third term of Eq. (13.17) transforms into

$$\sum_{\alpha,\beta}\left(\frac{1}{2}R_{\alpha\beta}u^{*}_{\alpha\nu}v_{\beta\mu}+S_{\alpha\beta}u^{*}_{\alpha\nu}u_{\beta\mu}+\frac{1}{2}R^{*}_{\alpha\beta}v^{*}_{\alpha\nu}u_{\beta\mu}\right)=$$
$$=\frac{1}{2}(E_{\mu}+E_{\nu})\sum_{\alpha}u^{*}_{\alpha\nu}u_{\alpha\mu}$$

and the last term becomes

$$-\frac{1}{2}(E_{\mu}+E_{\nu})\sum_{\alpha}v_{\alpha\nu}v^{*}_{\alpha\mu}.$$

Consequently, the expression (13.17) becomes

$$\mathscr{H}=\frac{1}{2}\sum_{\nu,\mu}\Big\{\xi^{+}_{\nu}\xi_{\mu}(E_{\nu}+E_{\mu})\sum_{\alpha}u^{*}_{\alpha\nu}u_{\alpha\mu}-$$
$$-\xi_{\nu}\xi^{+}_{\mu}(E_{\nu}+E_{\mu})\sum_{\alpha}v_{\alpha\nu}v^{*}_{\alpha\mu}\Big\}. \qquad (13.18)$$

In the above equation, we shall interchange the operators ξ_{ν} and ξ^{+}_{μ} and the indices ν and μ in the second term of the expression in braces. Then $\mathscr{H}$ becomes

$$\mathscr{H}=-\sum_{\alpha,\nu}E_{\nu}|v_{\alpha\nu}|^{2}+\frac{1}{2}\sum_{\nu,\mu}\xi^{+}_{\nu}\xi_{\mu}(E_{\nu}+E_{\mu})\sum_{\alpha}\left(u^{*}_{\alpha\nu}u_{\alpha\mu}-v^{*}_{\alpha\nu}v_{\alpha\mu}\right)=$$
$$=-\sum_{\alpha,\nu}E_{\nu}|v_{\alpha\nu}|^{2}+\sum_{\nu}E_{\nu}\xi^{+}_{\nu}\xi_{\nu}, \qquad (13.19)$$

which is identical with Eqs. (13.7)-(13.8).

If the quantities x (and, correspondingly, ξ) are ordinary numbers, the quadratic form in Eq. (13.1) is again reduced to the diagonal form by the canonical transformation of Eq. (13.6). The difference between this case and the case when these quantities are operators lies in the fact that, here, the quantities ξ, ξ^{+} commute with one another and, therefore, the form of Eq. (13.18) reduces to an expression of the type given by Eq. (13.9):

$$\mathscr{H}=\frac{1}{2}\sum_{\nu,\mu}\xi^{+}_{\nu}\xi_{\mu}(E_{\nu}+E_{\mu})\sum_{\alpha}\left(u^{*}_{\alpha\nu}u_{\alpha\mu}-v^{*}_{\alpha\nu}v_{\alpha\mu}\right)=\sum_{\nu}E_{\nu}\xi^{+}_{\nu}\xi_{\nu}. \qquad (13.20)$$

which does not contain a term similar to $\Delta\mathscr{E}_0$ of Eq. (13.8).

We note that the number of eigenfunctions of the linear system of equations given by Eq. (13.3) is even, because the functions u, v occur in pairs. If E_ν is an eigenvalue, E_μ is also an eigenvalue. As assumed, the quadratic form of Eq. (13.1) describes the states of a dynamic system close to the ground state of the system. A sufficient condition for this is the positive sign of the form in Eq. (13.1). Therefore, in our calculations, we shall use only the solutions with $E_\nu > 0$. The solutions for which $E_\nu < 0$ should be rejected.

Let us assume that there is a solution $(u_{\alpha\nu}, v_{\alpha\nu})$ of the system of Eq. (13.3), which corresponds to an eigenvalue $E_\nu > 0$, and which satisfies the normalization conditions of Eq. (13.4). We shall carry out the substitution

$$(u,\ v)\rightarrow(v^{*\prime},\ u^{*\prime});\quad E\rightarrow -E'. \tag{13.21}$$

We shall then obtain

$$E'_\nu u'_{\alpha\nu}=\sum_\beta S_{\alpha\beta}u'_{\beta\nu}+\sum_\beta R_{\alpha\beta}v'_{\beta\nu},$$

$$-E'_\nu v'_{\alpha\nu}=\sum_\beta S^*_{\alpha\beta}v'_{\beta\nu}+\sum_\beta R^*_{\alpha\beta}u'_{\beta\nu}. \tag{13.22}$$

The above system of equations is of the same type as the original system of Eq. (13.3), but it has an eigenvalue $E'_\nu = -E_\nu$. The normalization conditions for the new functions can be obtained as before. They will be similar to the conditions of Eq. (13.4). Applying the transformation of Eq. (13.21), we obtain

$$\sum_\alpha\left(u'_{\alpha\nu}u^{*\prime}_{\alpha\mu}-v'_{\alpha\nu}v^{*\prime}_{\alpha\mu}\right)=\sum_\alpha\left(v^*_{\alpha\nu}v_{\alpha\mu}-u^*_{\alpha\nu}u_{\alpha\mu}\right)=-\Delta(\nu-\mu). \tag{13.23}$$

Hence, we see that the normalization conditions of Eq. (13.23) for the new functions differ by their sign from the normalization conditions given by Eq. (13.4) for the original functions. Thus, the selection of the solutions corresponding to $E_\nu > 0$ and the normalization conditions of Eq. (13.4) excludes the solutions with $E_\nu < 0$.

Sec. 14. Conditions for the Application of the ASQ Method

We shall consider in detail the conditions for the application of the method of approximate second quantization, which was presented in Sec. 12.

The first condition is the condition of the smallness of the forms which include operators in powers higher than two.

The Hamiltonian $\mathscr{H}_2$ of Eq. (12.13) is quadratic in the "Bose operators" b, b^+, as given by Eq. (13.1). The canonical transformation of Eq. (13.6) transforms Eq. (13.1) into the diagonal form of Eqs. (13.7)-(13.8):

$$\mathscr{H}_2 = \Delta\mathscr{E}_0 + \sum_\nu E_\nu n_\nu \qquad (n_\nu = \xi_\nu^+ \xi_\nu).$$

In this form, the Hamiltonian is similar to the Hamiltonian of a system of noninteracting bosons. The quantity ν is now the number of a single-particle state, and E_ν denotes the energy of a particle in a state ν ($n_\nu = 0, 1, 2, \ldots$). The total number of particles is not conserved. Consequently, each eigenvalue of $\mathscr{H}_2$ is represented in the form of a sum of the energies of individual "elementary excitations."* The operators ξ_ν^+, ξ_ν are called the creation or annihilation operators for elementary excitations. The state in which there is one elementary excitation of number ν (one spin wave with a wave vector ν) is equivalent to the one-particle states considered earlier in Eq. (12.30).

In principle, the problem of determining the eigenvalues of the "zeroth-approximation" Hamiltonian $\mathscr{H}_2$ may be regarded as solved. The conditions of smallness of the cubic, quartic,... forms, compared with $\mathscr{H}_2$ in the total Hamiltonian of the system, or the conditions for the application of the ASQ method given by Eq. (12.16), are satisfied if the average values of the occupation numbers (or spin deviations $S_f - S_f^z$) are small:

$$\langle n_f \rangle = \langle b_f^+ b_f \rangle \ll 1. \tag{14.1}$$

*They are also known as "quasi-particles" or, in the case of spin systems, "spin waves."

Using Eq. (13.6), we shall rewrite these conditions in the form

$$\langle n_f \rangle = \sum_\nu \{|u_{f\nu}|^2 + |v_{f\nu}|^2\} \overline{N}_\nu + \sum_\nu |v_{f\nu}|^2 \ll 1, \qquad (14.2)$$

where $\overline{N}_\nu$ represents the average values of the occupation numbers of quasi-particles or spin waves:

$$\overline{N}_\nu = \langle \xi_\nu^+ \xi_\nu \rangle = \left\{ \exp \frac{E_\nu}{\vartheta} - 1 \right\}^{-1}. \qquad (14.3)$$

The conditions of Eq. (14.2) are observed when the following inequalities are obeyed:

$$D_f = \sum_\nu |v_{f\nu}|^2 \ll 1, \quad \overline{N}_\nu \ll 1. \qquad (14.4)$$

Since $E_\nu \geq 0$, the second of the inequalities in Eq. (14.4) is always obeyed at sufficiently low temperatures. The first of the inequalities of Eq. (14.4) is more rigorous and it gives the basic criterion for the applicability of the ASQ method. If the coefficients R in the form $\mathscr{H}_2$ vanish, then $v_{f\nu} \equiv 0$ and the method is always applicable; if $R \not\equiv 0$, an additional investigation of the conditions of Eq. (14.4) is necessary.

In some cases, the accuracy of this approximation is insufficient and one needs to find corrections to this approximation. These corrections will be of two types: first, the corrections due to forms of higher order in powers of the operators in the total Hamiltonian, which describe the interaction of quasi-particles with one another; they may be calculated using the standard perturbation theory. The allowance for these forms in higher approximations is spoken of as "the allowance for the dynamic interaction (of spins or spin waves)." Secondly, there will be corrections due to the non-Bose properties of the operators which occur in the Hamiltonian. The origin of these properties is associated with the features of the commutation relationships applying to the spin operators, i.e., due to the kinematic properties of these operators. In this case, one speaks of the problem of "allowing for the kinematic interaction." The special difficulties of the quantum theory of magnetism are associated with this type of correction.*

*A detailed analysis of the difficulties associated with the allow-

We shall consider first the more usual case of the spin $S = 1/2$. Here, the spin operators can be conveniently written in terms of the Pauli operators. According to Eq. (5.15), we have

$$\begin{gathered} S_f^+ = b_f, \quad S_f^- = b_f^+, \quad S_f - S_f^z = n_f, \\ n_f = b_f^+ b_f, \quad b_f^2 = b_f^{+2} = 0 \quad (n_f^0 = 0,\ 1). \end{gathered} \tag{14.5}$$

The Pauli operators are defined in the subspace of functions of the occupation numbers $n_f^0 = 0, 1$ (in other words, in the subspace of functions for which the spin at a site assumes the values $S_f^z = \pm 1/2$). At any one site, these operators obey the Fermi commutation relationships of Eq. (5.13). Therefore, when the operators b_f^+, b_f act on a wave function $C(n_f^0)$ of the occupation numbers, we have

$$\begin{gathered} b_f C(n_f^0) = \Delta(n_f^0 - 1) C(n_f^0 - 1), \\ b_f^+ C(n_f^0) = \Delta(n_f^0) C(n_f^0 + 1). \end{gathered} \tag{14.6}$$

Thus, we see that the subspace of the functions $\{C(0), C(1)\}$ transforms into itself when the operators b_f, b_f^+ are applied. Similarly, the application of the operators b_f, b_f^+ leaves invariant the subspace of the functions $\{C(2), C(3), \ldots\}$. The Hamiltonian of the system has the same properties because it consists of products of the Pauli operators. Following Dyson (1956a, b), the former subspace will be called physical, and the latter nonphysical (see also Sec. 5).

The problem is to determine the eigenfunctions of $\mathscr{H}$ in the physical space. However, on going over to the spin-wave approximation, the nonphysical states are automatically included (due to the existence of spin-wave occupation numbers which can be as large as we please). The inclusion of the nonphysical space leads to unpredictable errors, due to the fact that at a site the projection of the spin onto the quantization axis may assume the values $|S_f^z| > 1/2$, thereby contradicting premises of the problem. Consequently, we are faced with the very difficult problem of eliminat-

ance for the kinematic interaction was given by Dyson (1956a, b) in connection with an investigation of low-temperature expansions for an isotropic ferromagnet.

ing the contributions of the nonphysical states. This can be regarded also as the problem of allowing for the conditions which are imposed upon the occupation numbers when they are transformed to the Bose operators (or allowing for the conditions imposed on the occupation numbers of spin waves). The conditions imposed on the occupation numbers are equivalent to some interaction (for details see Sec. 37).

For the spins $S \geq 1$, the greatest difficulties are met with, as before, in allowing for the kinematic effects. However, in this case, the physical space is more extensive than that for the spin $S = 1/2$, and we would expect the restrictions on the occupation numbers to be less rigorous because $n_f^0 = 0, 1, \ldots, 2S$. At low temperatures ($\vartheta \to 0$), the contribution of the nonphysical states will decrease asymptotically for $S \gg 1$, because the nonphysical states will be those states with $n_f^0 \geq 2S + 1$, the probability of whose excitation is low at low temperatures [for details, see Izyumov (1959), Oguchi (1960), and Sec. 5].

The methods of allowing for the dynamic and kinematic interactions will now be considered using an isotropic ferromagnet as the example.

The Hamiltonian of the spin system will be taken in the form given by Eqs. (12.20)-(12.24). We shall assume that the lattice consists of N equivalent sites and that all these sites are occupied by atoms of the same type with the spin S, and that $I(f) \geq 0$ for all $|f|$. The magnetic field will be assumed to be directed along the z-axis.

We can easily see that the form given by Eq. (12.21) has a minimum when all the classical vectors γ_f are oriented parallel to the external field:

$$\gamma_f \| H, \quad H^\alpha = H\delta_{\alpha z}. \tag{14.7}$$

However, then γ_f is independent of f and it follows from Eq. (5.9) that

$$(A^*_{f_1}, A_{f_2}) = \frac{1}{2}, \quad (A_{f_1}, A_{f_2}) = (A_{f_1}, \gamma_{f_2}) = 0. \tag{14.8}$$

Consequently, the Hamiltonian of the system transforms into

$$\mathcal{H}=\mathcal{E}_0+\mathcal{H}_2+\mathcal{H}_4, \tag{14.9}$$

where

$$\mathcal{E}_0=-N\mu HS-\frac{1}{2}NS^2J(0), \tag{14.10}$$

$$\mathcal{H}_2=(\mu H+SJ(0))\sum_f(S-S_f^z)-\frac{1}{2}\sum_{f_1,f_2}I(f_1-f_2)S_{f_1}^-S_{f_2}^+, \tag{14.11}$$

$$\mathcal{H}_4=-\frac{1}{2}\sum_{f_1,f_2}I(f_1-f_2)(S-S_{f_1}^z)(S-S_{f_2}^z) \tag{14.12}$$

$$\left(J(\nu)=\sum_f I(f)\,e^{i(f,\nu)}\right).$$

In the calculations, it is convenient to represent the spin operators as operators with very simple commutative properties. We shall consider only those representations of the spin operators which have been discussed in Sec. 5.

A Hamiltonian of the type employed in the ASQ method (cf., Sec. 12) is used as the zeroth approximation. For this purpose, we take from the total Hamiltonian of Eq. (14.9) the quadratic form of (14.11) and we replace the spin operators in it with the Bose operators, using Eq. (12.9). Consequently, we obtain

$$\tilde{\mathcal{H}}_0=\mathcal{E}_0+\mathcal{H}_2'=$$
$$=\mathcal{E}_0+(\mu H+SJ(0))\sum_f n_f-\sum_{f_1,f_2}SI(f_1-f_2)\,b_{f_1}^+b_{f_2}. \tag{14.13}$$

The allowance for the dynamic and kinematic interaction corrections depends on the type of representation used for the spin operators. We shall briefly consider each of these representations.

<u>Pauli Operators of Eq. (14.5) for Spin S = $^1/_2$.</u> After substitution, the formulas (14.10)-(14.12) become

$$\mathcal{E}_0=-\frac{1}{2}N\mu H-\frac{1}{8}NJ(0), \tag{14.14}$$

$$\mathcal{H}_2=\left(\mu H+\frac{1}{2}J(0)\right)\sum_f n_f-\frac{1}{2}\sum_{f_1,f_2}I(f_1-f_2)\,b_{f_1}^+b_{f_2}, \tag{14.15}$$

$$\mathcal{H}_4 = -\frac{1}{2} \sum_{f_1, f_2} I(f_1 - f_2) n_{f_1} n_{f_2}. \tag{14.16}$$

Using the Hamiltonian of Eqs. (14.9), (14.15)-(14.16) at low temperatures, we can approximately assume that the operators b_f, b_f^+ are of the Bose type, and that $\mathcal{H}_2$ of Eq. (14.15) is the zeroth-approximation Hamiltonian. The dynamic interaction is described by the operator $\mathcal{H}_4$ of Eq. (14.16) and the kinematic interaction enters through the commutation relationships (5.13) for the Pauli operators in the form of the conditions imposed on the occupation numbers. The assumption that the kinematic and dynamic interactions can be considered to be small pertubations may be checked by using the equations of motion for the operators b_f, b_f^+ (see Sec. 37).

<u>Holstein - Primakoff - Izyumov Representation of Eqs. (5.16)-(5.17) for Spins $S \geq 1/2$.</u> The formulas (14.11)-(14.12) become

$$\mathcal{H}_2 = (\mu H + SJ(0)) \sum_f n_f - \sum_{f_1, f_2} SI(f_1 - f_2)\, a_{f_1}^+ \varphi(n_{f_1}) \varphi(n_{f_2})\, a_{f_2}, \tag{14.17}$$

$$\mathcal{H}_4 = -\frac{1}{2} \sum_{f_1, f_2} I(f_1 - f_2) n_{f_1} n_{f_2}, \tag{14.18}$$

where $n_f = a_f^+ a_f$, $\varphi(n_f) = [1 - (n_f/2S)]^{1/2}$, and the ground-state energy $\mathcal{E}_0$ is defined in accordance with Eq. (14.10). At low temperatures, the contribution of the nonphysical states decreases asymptotically for $S \gg 1$. Therefore, we can allow for the kinematic effects by expanding the operators $\varphi(n_f)$ into a series,

$$\varphi(n_f) = 1 - \frac{n_f}{4S} - \frac{n_f^2}{32S^2} - \ldots \tag{14.19}$$

and neglecting the non-Bose nature of the operators a_f, a_f^+ of Eq. (5.18). Substituting Eq. (14.19) into the Hamiltonian of Eqs. (14.17) and (14.18), we obtain

$$\mathcal{H} = \mathcal{E}_0 + \mathcal{H}_2 + \mathcal{H}' + \mathcal{H}'', \tag{14.20}$$

where

$$\mathscr{H}_2 = (\mu H + SJ(0)) \sum_f n_f - \sum_{f_1, f_2} SI(f_1 - f_2)\, a^+_{f_1} a_{f_2}, \tag{14.21}$$

$$\mathscr{H}' = -\frac{1}{2} \sum_{f_1, f_2} I(f_1 - f_2)\, n_{f_1} n_{f_2}, \tag{14.22}$$

$$\begin{aligned}\mathscr{H}'' = {} & \frac{1}{4} \sum_{f_1, f_2} I(f_1 - f_2)\left(1 + \frac{1}{8S}\right)(a^+_{f_1} a^+_{f_1} a_{f_1} a_{f_2} + a^+_{f_1} a^+_{f_2} a_{f_2} a_{f_2}) + \\ & + \sum_{f_1, f_2} I(f_1 - f_2)\frac{1}{32S}(a^+_{f_1} a^+_{f_1} a^+_{f_1} a_{f_1} a_{f_1} a_{f_2} + a^+_{f_1} a^+_{f_2} a^+_{f_2} a_{f_2} a_{f_2} a_{f_2}) - \\ & - \sum_{f_1, f_2} I(f_1 - f_2)\frac{1}{8S}\, a^+_{f_1} a^+_{f_1} a^+_{f_2} a_{f_2} a_{f_2} a_{f_1} + O(S^{-2}).\end{aligned} \tag{14.23}$$

The operator $\mathscr{H}_2$ of Eq. (14.21) describes the noninteracting spin waves, $\mathscr{H}'$ of Eq. (14.22) represents the ordinary dynamic interaction of spin waves, and $\mathscr{H}''$ of Eq. (14.23) gives the kinematic interaction. At low temperatures, $\mathscr{H}'$ and $\mathscr{H}''$ can be regarded as small perturbations compared with $\mathscr{H}_0$. The higher the value of S, the more exactly the Hamiltonian of Eqs. (14.20)-(14.23) describes the behavior of the system.

Dyson's Representation of Eq. (5.23). In this case, the Hamiltonian of Eqs. (14.9), (14.11)-(14.12) transforms into

$$\tilde{\mathscr{H}} = \mathscr{E}_0 + \mathscr{H}_2 + \mathscr{H}' + \mathscr{H}'', \tag{14.24}$$

where

$$\mathscr{H}_2 = (\mu H + SJ(0)) \sum_f n_f - \sum_{f_1, f_2} SI(f_1 - f_2)\, a^+_{f_1} a_{f_2}, \tag{14.25}$$

$$\mathscr{H}' = -\frac{1}{2} \sum_{f_1, f_2} I(f_1 - f_2)\, n_{f_1} n_{f_2}, \tag{14.26}$$

$$\mathscr{H}'' = \frac{1}{2} \sum_{f_1, f_2} I(f_1 - f_2)\, a^+_{f_1} n_{f_2} a_{f_2}, \tag{14.27}$$

and $\mathscr{E}_0$ is defined by Eq. (14.10).

The operator $\mathscr{H}_2$ of Eq. (14.25) describes, as before, the noninteracting spin waves, $\mathscr{H}'$ of Eq. (14.26) gives the dynamic interaction of the spin waves, and $\mathscr{H}''$ of Eq. (14.27) gives the kinematic interaction of the spin waves. It should be mentioned that, in this case, as indicated by Eq. (14.27), the Hamiltonian is not Hermitian in the new variables.

At low temperatures, the operators $\mathscr{H}'$, $\mathscr{H}''$ can be regarded as small perturbations and we can use the perturbation theory methods. From the equivalence of the Dyson and Holstein–Primakoff–Izyumov representations, it follows that at low temperatures the contribution of the nonphysical states decreases asymptotically when $S \gg 1$.

Chapter V

Applications of the Method of Approximate Second Quantization

In the present chapter, we consider the examples of an isotropic ferromagnet and an isotropic antiferromagnet; we shall calculate their spectra of elementary excitations and their magnetizations as a function of the temperature and the external magnetic field. In the concluding part, some further applications of the method will be indicated.

Sec. 15. Isotropic Ferromagnets

Ferromagnet with a Simple Lattice. We shall consider, as a straightforward example, the problem of the approximate determination of the lower part of the energy spectrum of an isotropic ferromagnet and the temperature dependence of its magnetization. We shall assume that the lattice consists of N equivalent sites occupied by atoms of the same kind and, therefore, $S_f = S$, $\mu_f = \mu$. The volume of the lattice will be denoted by V. As far as the exchange integrals are concerned, we shall assume that

$$I(f_1 - f_2) \geq 0 \quad \text{for all} \quad |f_1 - f_2|.$$

To make the case definite, we shall take the external magnetic field to be directed along the z-axis.

Under these assumptions, the spin system is described approximately by the Hamiltonian (14.13):

$$\mathcal{H} = \mathcal{E}_0 + \mathcal{H}_2. \tag{15.1}$$

where

$$\mathcal{E}_0 = -N\mu SH - \frac{1}{2} NS^2 J(0),$$

$$\mathcal{H}_2 = (\mu H + SJ(0)) \sum_f n_f - \sum_{f_1, f_2} SI(f_1 - f_2)\, b^+_{f_1} b_{f_2}; \qquad (15.3)$$

b^+_f, b_f are the Bose operators.

The quadratic form of Eq. (15.3) is diagonalized by the canonical transformation*

$$b_f = \frac{1}{\sqrt{N}} \sum_\nu e^{i(f,\nu)} \xi_\nu, \qquad (15.4)$$

where ξ^+_ν, ξ_ν are new Bose operators. Substituting Eq. (15.4) into Eq. (15.3), we obtain the required formula for the energy of weakly excited states,

$$\mathcal{H}_2 = \sum_\nu E_S(\nu) N_\nu \quad (N_\nu = \xi^+_\nu \xi_\nu), \qquad (15.5)$$

where $E_S(\nu)$ is the energy of an elementary excitation or a spin wave with a wave vector ν:

$$E_S(\nu) = \mu H + SJ(0)(1 - \gamma_\nu) = \mu H + SJ(0)\,\mathcal{E}_\nu, \qquad (15.6)$$

where

$$\gamma_\nu = J(\nu)/J(0), \quad J(\nu) = \sum_f I(f)\, e^{i(f,\nu)}. \qquad (15.7)$$

The operators ξ_ν, ξ^+_ν are also called the operators for the annihilation and creation of spin waves. In this approximation, the free energy of the system is equal to

$$F = \mathcal{E}_0 + \vartheta \sum_\nu \ln\left(1 - e^{-\frac{1}{\vartheta} E_S(\nu)}\right). \qquad (15.8)$$

Hence, by differentiation with respect to H, we obtain an expression for the magnetization (8.35):

---

*The transformation of Eq. (15.4) is a special case of the general transformation of Eq. (13.6), when $u_{f\nu} = N^{-1/2} \exp i(f, \nu)$; $v_{f\nu} = 0$.

$$M = -\frac{\partial F}{\partial H} = N\mu S - \sum_{\nu} \frac{\partial E_S(\nu)}{\partial H} \bar{N}_{\nu}^{(S)} = M_0\left(1 - \frac{1}{S} P_S\right), \quad (15.9)$$

where $M_0 = N\mu S$ is the saturation magnetization of a sample, $\bar{N}_{\nu}^{(S)}$ are the average occupation numbers for spin waves, P_S is the sum of the occupation numbers:

$$P_S = \frac{1}{N} \sum_{\nu} \bar{N}_{\nu}^{(S)}, \quad \bar{N}_{\nu}^{(S)} = \langle N_{\nu} \rangle = \left(e^{\frac{1}{\vartheta} E_S(\nu)} - 1\right)^{-1}. \quad (15.10)$$

In accordance with Eq. (A2.14), which is given in Appendix 2, we shall go over in Eq. (15.10) from a sum to an integral, assuming that

$$\frac{1}{N} \sum_{\nu} \bar{N}_{\nu}^{(S)} \rightarrow \frac{v}{(2\pi)^3} \int \bar{N}_{\nu}^{(S)} d\nu, \quad (15.11)$$

where $v = V/N$ is the volume per site. Integration with respect to ν is carried out in the first reduced Brillouin zone for the wave vectors.

Since $E_S(\nu) \geq 0$, then at low temperatures $\exp[-(1/\vartheta)E_S(\nu)]$ can be considered to be a small quantity and the expression for P_S may be rewritten in the form

$$\begin{aligned} P_S &= \sum_{n=1}^{\infty} \frac{v}{(2\pi)^3} \int e^{-\frac{n}{\vartheta} E_S(\nu)} d\nu = \\ &= \sum_{n=1}^{\infty} e^{-\frac{n}{\vartheta}\mu H} \frac{v}{(2\pi)^3} \int e^{-\frac{n}{\vartheta} SJ(0)\mathcal{E}_{\nu}} d\nu. \end{aligned} \quad (15.12)$$

As $\vartheta \rightarrow 0$, the main contribution to the magnetization is given by the spin waves having small wave vectors, since mainly low-energy spin waves are excited. Therefore, we can expand $E_S(\nu)$ of Eq. (15.6) as a series in powers of ν and we need take terms only up to the order of ν^2, inclusive. We note that $E_S(\nu)$ is an even function of ν, and that the expansions begin with the terms ν^2.

In fact,

$$J(0) - J(\nu) = \sum I(f)(1 - e^{i(f,\nu)}) \approx$$
$$\approx \frac{1}{2}\sum (f, \nu)^2 I(f) = \frac{\nu^2}{6}\sum f^2 I(f).$$

Substituting this expression into Eq. (15.6), we obtain an approximate expression for the spin-wave energy:

$$E_S(\nu) \approx \mu H + \frac{1}{6} SJ(0)\,\overline{\delta^2}\cdot\nu^2,$$
$$\overline{\delta^2} = \frac{\sum f^2 I(f)}{\sum I(f)} \quad \left[J(0) = \sum I(f)\right]. \tag{15.13}$$

We shall substitute Eq. (15.13) into the integral (15.12) and extend the integration to all values of ν. As a result of this, we can easily calculate an approximate expression for the quantity P_S:

$$P_S \approx \sum_{n=1}^{\infty} e^{-\frac{n}{\vartheta}\mu H} \frac{v}{(2\pi)^3}\int e^{-\frac{n}{6\vartheta} SJ(0)\overline{\delta^2}\cdot\nu^2} d\nu =$$
$$= \frac{v}{(\overline{\delta^2})^{3/2}}\left(\frac{\vartheta}{\frac{2\pi}{3} SJ(0)}\right)^{3/2} \sum_{n=1}^{\infty}\frac{e^{-\frac{n}{\vartheta}\mu H}}{n^{3/2}}. \tag{15.14}$$

Using Eq. (15.14), we can transform Eq. (15.9) into the following expression:

$$\sigma_S = 1 - S^{-1}P_S \approx 1 - \frac{S^{-1}v}{(\overline{\delta^2})^{3/2}}\left(\frac{\vartheta}{\frac{2\pi}{3} SJ(0)}\right)^{3/2} Z_{3/2}\left(\frac{\mu H}{\vartheta}\right), \tag{15.15}$$

where $\sigma_S = M/M_0$ is the magnetization per site of Eq. (8.29) and, for H = 0,

$$\sigma_S = 1 - \frac{S^{-1}v}{(\overline{\delta^2})^{3/2}}\left(\frac{\vartheta}{\frac{2\pi}{3} SJ(0)}\right)^{3/2} \zeta\left(\frac{3}{2}\right), \tag{15.16}$$

where

$$Z_p(x) = \sum_{n=1}^{\infty} n^{-p} e^{-nx}, \quad \zeta(p) = \sum_{n=1}^{\infty} n^{-p} \tag{15.17}$$

[ζ (p) is Riemann's zeta-function].

Formulas of the (15.15) and (15.16) type, which give the dependence of the magnetization of an isotropic ferromagnet on the temperature and field, or only on the temperature, were first obtained by Bloch (1930, 1932). It is evident from Eq. (15.16) that for H = 0 at low temperatures, the magnetization of an isotropic ferromagnet varies with temperature as $\vartheta^{3/2}$, which is "Bloch's three-halves power law." It can be easily seen that the law is a consequence of the quadratic dependence of $E_S(\nu)$ on ν for small values of ν.

We shall now allow for the contribution to the magnetization of higher terms in the expansion of the spin-wave energy in powers of the wave vector.

We shall consider only the nearest-neighbor approximation. Then,

$$I(f)=\begin{cases} I, & f=\delta, \\ 0, & f\neq\delta, \end{cases}\quad J(\nu)=Iz\gamma_\nu,\quad \gamma_\nu=\frac{1}{z}\sum_\delta e^{i(\nu,\delta)}, \qquad (15.18)$$

where δ is a vector joining the nearest sites; summation over δ represents summation over all the nearest neighbors; z is the number of nearest neighbors.

For cubic lattices (including simple cubic, bcc, and fcc), the expansion of the magnetization in powers of temperature has the form [Dyson (1956b)]

$$\sigma_S=1-S^{-1}P_S=1-S^{-1}\Big\{\rho Z_{3/2}\left(\frac{h}{\tau}\right)\tau^{3/2}+ \\ +\frac{3\pi}{4}\rho Z_{5/2}\left(\frac{h}{\tau}\right)\tau^{5/2}+\pi^2\omega\rho Z_{7/2}\left(\frac{h}{\tau}\right)\tau^{7/2}+\dots\Big\}, \qquad (15.19)$$

where the following dimensionless quantities are used:

$$\tau=\frac{\vartheta}{\frac{2\pi}{3}SIz},\qquad h=\frac{\mu H}{\frac{2\pi}{3}SIz},\qquad \rho=\frac{v}{\delta^3}, \qquad (15.20)$$

and ω is a number which depends on the lattice geometry:

$$\left.\begin{aligned} &\text{simple cubic lattice:}\quad z=6,\ \omega=\frac{33}{32};\\ &\text{bcc lattice:}\quad z=8,\ \omega=\frac{281}{288};\\ &\text{fcc lattice:}\quad z=12,\ \omega=\frac{15}{16}. \end{aligned}\right\} \qquad (15.21)$$

In the same approximation, the specific heat of a spin system is given, according to Eqs. (8.36) and (15.8), by the following expression:

$$C = k\left\{\frac{15}{4}\zeta\left(\frac{5}{2}\right)\tau^{3/2} + \frac{105}{16}\pi\rho\zeta\left(\frac{7}{2}\right)\tau^{5/2} + \right.$$
$$\left. + \frac{63}{4}\pi^2\omega\zeta\left(\frac{9}{2}\right)\tau^{7/2} + \ldots\right\}. \quad (15.22)$$

The expansions (15.19) and (15.22) were obtained in the spin-wave approximation. Therefore, we shall call them the Bloch expansions for an isotropic ferromagnet. We must remember that the corrections for the dynamic and kinematic interactions of spin waves are not included in the expansions (15.19) and (15.22). Dyson's investigations (1956a, b) have shown that the corrections for the dynamic and kinematic interactions balance each other out to a considerable degree. Thus, for example, the formula for the magnetization (15.19) is valid to terms of the order of $\vartheta^{7/2}$, inclusive.

The criterion for the application of the approximate second quantization method, given by Eq. (14.4), is satisfied identically by an isotropic ferromagnet. In fact, the canonical transformation of Eq. (15.4) may be regarded as the special case of the transformation (13.6) for $v_{f\nu} = 0$. Consequently, $D_f = 0$. This is because the wave function of the state with the parallel orientation of all spins in an isotropic ferromagnet,

$$C_0 = \prod_f \Delta(S - S_f^z)$$

is the exact eigenfunction of the ground state of the Hamiltonian:

$$\mathcal{H}C_0 = \mathcal{E}_0 C_0,$$

where $\mathcal{E}_0$, $\mathcal{H}$ are given by the expressions (14.9)-(14.12).

The spin system of real ferromagnets can relatively rarely be modeled by the isotropic exchange Hamiltonian. Among the factors which distort the model to a greater or lesser extent are the magnetic anisotropy, striction, influence of the conduction electrons in conducting ferromagnets, etc. Therefore, the model given here is mainly of heuristic importance.

The requirements of the model are relatively well satisfied by the compound $CrBr_3$. Gossard et al (1961) measured its nuclear magnetic resonance and found that, in the range 1-4°K ($T_C \approx 38$°K), the temperature dependence of the magnetization of $CrBr_3$ was described well by the three-term formula (15.19) (accurate to values of the order of $\vartheta^{5/2}$, inclusive).

Another substance which satisfies the requirements of the model is EuS. The direct exchange between the 4f-electrons is hardly possible and obviously the indirect exchange applies. The compound EuS has the NaCl-type ionic lattice. The ground state of the Eu^{2+} ion is ^8S; the value of the europium-ion spin in the lattice is close to $S = 7/2$; the Debye temperature is $T_D \approx 200$°K. According to McCollum and Callaway (1962), the experimental temperature dependence of the specific heat of EuS in the temperature range 1-4°K is quite accurately described by the first two terms of a formula of the type given by Eq. (15.22); the Curie temperature of EuS is $T_C = 17 \pm 1$°K. According to Charap and Boyd (1964), a formula of the type of Eq. (15.16), obtained allowing for the first and second nearest neighbors and for the spin–spin (dipole) interaction, describes well the experimental temperature dependence of the magnetization; they found that T_C of EuS was 16.4 ± 1°K.

Spin Waves in Complex Lattices.* The spin waves in a Heisenberg ferromagnet and its lattice vibrations are very similar, because both are discrete normal-mode vibrations. Therefore, some of the data on the nature of the vibration spectrum of a lattice may be applied to the spin-wave spectrum. In particular, in complex magnetic lattices, the spin-wave spectrum may have acoustic-mode and optical-mode vibration branches. Since, in some effects, the nature of the spectrum is of great importance, we shall consider this problem in more detail.

Let a lattice consist of N unit cells, each of which contains n sites occupied by magnetically active atoms. We shall denote the number of a unit cell and its radius vector by the index f, and the

*See Saénz (1962) and Wallace (1962).

number of an atom in a cell and its radius vector by the index j. The coordinate of each atom (or spin) in the lattice will be written as $r_{fj} = f + j$. The spin at a site r_{fj} will be denoted by S_{fj}, and we shall assume that the value of the spin depends only on the index j; the values of the spin can be arbitrary: $S_j \geq 1/2$.

We shall assume uniaxial anisotropy, which may be represented by some internal field H_a, directed along the z-axis; an external magnetic field H will also be assumed to be directed along the z axis. We shall consider a configuration in which the spins are collinear with this axis and assume, therefore, that the configuration is stable. We shall introduce a matrix $\varepsilon = (\varepsilon_j \delta_{jj'})$, where $\varepsilon_j = +1$ (−1) if S_{fj} is directed parallel (antiparallel) to the z-axis.

We shall write the Hamiltonian of the spin system in the form

$$\mathcal{H} = -\sum_{f,j} A_j S^z_{fj} - \frac{1}{2} \sum_{f,f',j,j'} I(f, j; f', j')(S_{fj}, S_{f'j'}), \quad (15.23)$$

where $I(f, j; f', j')$ are exchange integrals depending only on the difference between the coordinates of the sites r_{fj} and $r_{f'j'}$; A_j represents linear and homogeneous functions of the external and anisotropy fields ($A_j = H + \varepsilon_j H_{aj}$). Since we have confined ourselves to the spin-wave approximation, we can represent the operators S_{fj} approximately in terms of the Bose operators (12.19):

$$S^x_{fj} = \sqrt{S_j}\,\frac{a^+_{fj} + a_{fj}}{\sqrt{2}}, \quad S^y_{fj} = i\varepsilon_j \sqrt{S_j}\,\frac{a^+_{fj} - a_{fj}}{\sqrt{2}}, \quad S^z_{fj} = \varepsilon_j (S_j - n_{fj}). \quad (15.24)$$

We shall substitute Eq. (15.24) into Eq. (15.23) and carry out a canonical transformation of the variables:

$$a_{fj} = \frac{1}{\sqrt{N}} \sum_{\nu} a_{\nu j} e^{i(f,\nu)}. \quad (15.25)$$

Consequently, the expression for $\mathcal{H}$ becomes

$$\mathcal{H} = \mathcal{E}_0 + \sum_{\nu, j} \left[\varepsilon_j A_j + \sum_{j'} \varepsilon_j \varepsilon_{j'} S_{j'} I_{jj'}(0) \right] a^+_{\nu j} a_{\nu j} -$$

$$-\frac{1}{2}\sum_{\nu,\,j,\,j'}\varepsilon_j\varepsilon_{j'}\sqrt{S_jS_{j'}}\,I_{jj'}(\nu)\Big(\frac{\varepsilon_j\varepsilon_{j'}-1}{2}a^+_{\nu j}a^+_{-\nu j'}+ \tag{15.26}$$

$$+2\frac{\varepsilon_j\varepsilon_{j'}+1}{2}a^+_{\nu j}a_{\nu j'}+\frac{\varepsilon_j\varepsilon_{j'}-1}{2}a_{-\nu j'}a_{\nu j}\Big),$$

where

$$\mathscr{E}_0=-\sum_{f,\,j}\Big[\varepsilon_jA_jS_j+\frac{1}{2}\sum_{j'}\varepsilon_j\varepsilon_{j'}S_jS_{j'}I_{jj'}(0)\Big], \tag{15.27}$$

$$I_{jj'}(\nu)=\sum_f I(f,\,j;\,f',\,j')\,e^{-i(\nu,\,f-f')}. \tag{15.28}$$

Since the exchange integral is real and depends only on the difference between the coordinates of the sites,

$$I^*_{jj'}(\nu)=I_{jj'}(\nu)=I_{jj'}(-\nu). \tag{15.29}$$

Consequently, the quadratic form (15.26) is Hermitian and its eigenvalues are real.

The states of a system are stable if the energy of the excited states is positive. Therefore, the form (15.26) should not be negative. We shall assume that this condition is satisfied. *

We shall write the transformed Hamiltonian $\mathscr{H}$ in the standard form:

$$\mathscr{H}=\mathscr{E}_0+\frac{1}{2}\sum\{R_{jj'}(\nu)\,a^+_{\nu j}a^+_{-\nu j'}+$$

$$+2S_{jj'}(\nu)\,a^+_{\nu j}a_{\nu j'}+R^*_{jj'}(\nu)\,a_{-\nu j'}a_{\nu j}\}, \tag{15.30}$$

where

$$S_{jj'}(\nu)=\Big[\varepsilon_jA_j+\sum_{j''}\varepsilon_j\varepsilon_{j''}S_{j''}I_{jj''}(0)\Big]\Delta(j-j')-$$

$$-\varepsilon_j\varepsilon_{j'}\sqrt{S_jS_{j'}}\,I_{jj'}(\nu)\frac{\varepsilon_j\varepsilon_{j'}+1}{2},$$

$$R_{jj'}(\nu)=-\varepsilon_j\varepsilon_{j'}\sqrt{S_jS_{j'}}\,I_{jj'}(\nu)\frac{\varepsilon_j\varepsilon_{j'}-1}{2}. \tag{15.31}$$

* For details of this condition of stability in the theory of magnetism, see Sec. 12, and the papers by Van Kranendonk and Van Vleck (1958) and by Kaplan (1960).

According to Eqs. (13.1)-(13.8), the form (15.30) is diagonalized,

$$\mathcal{H} = \mathcal{E}_0 + \Delta\mathcal{E}_0 + \sum_{\nu,\,p} E^{(p)}(\nu)\,\xi^{+}_{\nu p}\xi_{\nu p},$$
$$\Delta\mathcal{E}_0 = -\sum_{\nu,\,p,\,j} E^{(p)}(\nu)\,|\,v_{jp}(\nu)\,|^2, \tag{15.32}$$

by the canonical transformation

$$a_{\nu j} = \sum_p \{u_{jp}(\nu)\,\xi_{\nu p} + v^{*}_{jp}(\nu)\,\xi^{+}_{\nu p}\} \quad (p = 1,\,2,\,\ldots,\,n), \tag{15.33}$$

where $\xi^{+}_{\nu p}$, $\xi_{\nu p}$ are the new Bose operators, (u, v) are the eigenvectors, and $E^{(p)}$ are the eigenvalues of the system of equations

$$E^{(p)}u_{jp}(\nu) = \sum_{j'} S_{jj'}(\nu)\,u_{j'p}(\nu) + \sum_{j'} R_{jj'}(\nu)\,v_{j'p}(\nu),$$
$$-E^{(p)}v_{jp}(\nu) = \sum_{j'} S^{*}_{jj'}(\nu)\,v_{j'p}(\nu) + \sum_{j'} R^{*}_{jj'}(\nu)\,u_{j'p}(\nu) \tag{15.34}$$
$$(j,\,j' = 1,\,2,\,\ldots,\,n);$$

and the vectors (u, v) obey the following orthogonality and normalization conditions:

$$\sum_j \{u_{jp}(\nu)\,u^{*}_{jp'}(\nu) - v_{jp}(\nu)\,v^{*}_{jp'}(\nu)\} = \Delta(p - p'),$$
$$\sum_j \{u_{jp}(\nu)\,v_{jp'}(\nu) - u_{jp'}(\nu)\,v_{jp}(\nu)\} = 0. \tag{15.35}$$

In the system of 2n linear equations of (15.24), the quantity ν is a parameter. For a fixed value of ν, the system has 2n linearly independent solutions, corresponding to 2n eigenvalues: $\pm E^{(p)}$, $p = 1, 2, \ldots, n$. The quasi-continuous index ν assumes N values corresponding to the points in the first Brillouin zone in the reciprocal lattice space of the vectors f, and each of the eigenvalues $E^{(p)}$ assumes N values $E^{(p)}(\nu)$. Since the form (15.26) is positive, we have n eigenvalues which are positive at all values of ν. We shall assume that $E^{(p)}(\nu) \geq 0$, $p = 1, 2, \ldots, n$. We shall call the quantities $E^{(p)}(\nu)$ the branches of the spectrum of elementary excitations or spin waves, and p will be the number of a given branch. The number of different branches may, in general, be less than n, since some of them may be equal (degeneracy in the quantum-mechanical sense). The other n branches of the spectrum $[-E^{(p)}(\nu) \leq 0]$ are rejected in the final result (cf., Sec. 13).

We shall show that when $A_j = 0$ [cf., Eq. (15.23)] we have at least one acoustical branch, i.e., we have $E^{(p)}(0) = 0$.

We shall write the system of equations (15.34) in the matrix form:

$$\begin{pmatrix} S(\nu) & R(\nu) \\ R^*(\nu) & S^*(\nu) \end{pmatrix} \begin{pmatrix} u \\ v \end{pmatrix} = \begin{pmatrix} E(\nu) & O \\ O & -E(\nu) \end{pmatrix} \begin{pmatrix} u \\ v \end{pmatrix}, \tag{15.36}$$

where $S(\nu)$, $R(\nu)$ are square matrices of the n-th order with the elements given by Eq. (15.31); O is a square matrix of the n-th order in which all the elements are equal to zero, and $E(\nu)$ is a diagonal matrix of the n-th order.

We shall use primes to denote the values in the case $A_j = 0$ [for example, $S'(\nu) = S(\nu)|_{A_j = 0}$, etc.]. We shall assume that $\nu = 0$. Since $E'^{(p)}$ are the eigenvalues of the system of equations (15.36), then

$$(-1)^n \prod_p \{E'^{(p)}(0)\}^2 = \text{Det}\begin{vmatrix} S'(0) & R'(0) \\ S^{*\prime}(0) & R^{*\prime}(0) \end{vmatrix}. \tag{15.37}$$

We shall multiply each of the columns in the determinant (15.37) by $\sqrt{S_j}$, and from the sum of the first n columns we shall subtract the sum of the last n columns. Consequently, for any j, we have the identity

$$\sum_{j'} S'_{jj'}(0)\sqrt{S_{j'}} - \sum_{j'} R'_{jj'}(0)\sqrt{S_{j'}} = 0 \tag{15.38}$$

[this becomes obvious on substituting Eq. (15.31) into Eq. (15.38)]. Hence, we see that the determinant in Eq. (15.37) vanishes and, consequently, we have at least one acoustical branch.

Sec. 16. Ground State of an Antiferromagnet*

The Simplest Case. The ferromagnetic ordering of spins in a magnetically isotropic substance is a simple consequence of the positive sign of the exchange integral for all pairs of sites. We shall now consider the case when the exchange integral is positive for some pairs of sites and negative for other

*See Hulten (1936) and Tyablikov (1956a).

pairs, and we shall show that, under certain conditions, this gives rise to antiferromagnetic ordering.

Let us consider an isotropic crystal. For simplicity, we shall assume that there are only two equivalent sublattices, one inside the other. The sites in the two sublattices will be denoted by the letters f and g. We shall also assume that the exchange integrals for the spins in the sublattices f or g are positive, but that they are negative for the interaction between the two sublattices:

$$I(f_1-f_2)\geqslant 0,\quad I(g_1-g_2)\geqslant 0,\quad I(f-g)\leqslant 0.$$

The number of sites in each sublattice will be denoted by N:

$$\sum_f 1=\sum_g 1=N.$$

We shall consider only the case when the spin of each atom is $S=1/2$ and the magnetic moment is μ. In the spin-wave approximation, the Hamiltonian of our model has the form of Eq. (12.27), if we assume that the summation indices apply to both sublattices:

$$\tilde{\mathcal{H}}=\mathcal{E}_0+\mathcal{H}_2, \tag{16.1}$$

$$\mathcal{E}_0=-\frac{1}{2}\sum(\gamma_h,H)-\frac{1}{8}\sum I(h_1-h_2)(\gamma_{h_1},\gamma_{h_2}). \tag{16.2}$$

$$\begin{aligned}\mathcal{H}_2=-\sum 2\lambda_h n_h+\frac{1}{2}\sum R^*(h_1,h_2)\,b_{h_1}b_{h_2}+\\+\sum S(h_1,h_2)\,b^+_{h_1}b_{h_2}+\frac{1}{2}\sum R(h_1,h_2)\,b^+_{h_1}b^+_{h_2},\end{aligned} \tag{16.3}$$

where the following notation is used:

$$\begin{aligned}R(h_1,h_2)&=-I(h_1-h_2)\left(A^*_{h_1},A^*_{h_2}\right),\\S(h_1,h_2)&=-I(h_1-h_2)\left(A^*_{h_1},A_{h_2}\right)\quad(h=f,g).\end{aligned} \tag{16.4}$$

The classical vectors γ_h and the Lagrange multipliers λ_h are found from the condition (12.25):

$$\begin{aligned}-\frac{1}{2}\sum_{h'} I(h-h')\,\gamma^\alpha_{h'}-\mu H^\alpha=2\lambda_h\gamma^\alpha_h,\\\sum_\alpha(\gamma^\alpha_h)^2=1\quad(\alpha=x,y,z).\end{aligned} \tag{16.5}$$

The classical vectors A_h are defined in terms of the vectors γ_h using Eq. (5.10).

We shall now consider the ground state. We shall assume that in the zeroth approximation each of the magnetic sublattices f and g is magnetized to saturation. Consequently, the vectors S_f (and S_g) have a common quantization axis. Therefore

$$\gamma_f^\alpha = \gamma_1^\alpha, \quad \gamma_g^\alpha = \gamma_2^\alpha \quad (\alpha = x, y, z), \quad \lambda_f = \lambda_1, \quad \lambda_g = \lambda_2. \tag{16.6}$$

Since the spin Hamiltonian of our system is isotropic, we can assume, without affecting the generality of our considerations, that the magnetic field is directed along the z-axis. Consequently, the expressions in (16.5) assume the form

$$\left.\begin{aligned} -\left(\frac{1}{2} J_{11}(0) + 2\lambda_1\right)\gamma_1^\alpha - \frac{1}{2} J_{12}(0)\,\gamma_2^\alpha = \mu H \delta_{\alpha, z}, \\ -\frac{1}{2} J_{21}(0)\,\gamma_1^\alpha - \left(\frac{1}{2} J_{22}(0) + 2\lambda_2\right)\gamma_2^\alpha = \mu H \delta_{\alpha, z}, \end{aligned}\right\} \tag{16.7}$$

$$\sum_\alpha (\gamma_1^\alpha)^2 = 1, \quad \sum_\alpha (\gamma_2^\alpha)^2 = 1, \tag{16.8}$$

where the following notation is used:

$$\begin{gathered} J_{11}(0) = \sum_{f'} I(f - f'), \quad J_{22}(0) = \sum_{g'} I(g - g'), \\ J_{12}(0) = J_{21}(0) = \sum_g I(f - g). \end{gathered} \tag{16.9}$$

Equations (16.7)-(16.8) have two types of solution. One of these represents the case of the zero determinant Δ of the system (16.7), and the other the case when the determinant Δ is not equal to zero. Let $\Delta = 0$. From the condition for solubility of the equations, we find the Lagrange multipliers: $2\lambda_1 = -\frac{1}{2}J_{11}(0) + \frac{1}{2}J_{21}(0)$, $2\lambda_2 = -\frac{1}{2}J_{22}(0) + \frac{1}{2}J_{12}(0)$. Consequently, from Eqs. (16.7) we obtain

$$\gamma_1^\alpha + \gamma_2^\alpha = 0 \quad (\alpha = x, y), \quad \gamma_1^z = \gamma_2^z = \frac{\mu H}{|J_{12}(0)|}.$$

Let $\Delta \neq 0$. Then it follows from Eq. (16.7) that

$$\gamma_1^\alpha = \gamma_2^\alpha = 0 \quad (\alpha = x, y), \quad \gamma_1^z, \ \gamma_2^z = \pm 1.$$

From the minimum condition of the form of $\mathcal{E}_0$ given by Eq. (16.2), we have, depending on the value of the external field, one of the following solutions.

1. Weak fields: $H \leq H_c = (1/\mu)|J_{12}(0)|$,

$$\begin{gathered}
\gamma_1^z = \gamma_2^z = \frac{H}{H_c}, \quad \gamma_1^\alpha + \gamma_2^\alpha = 0 \quad (\alpha = x,\ y), \\
\mathcal{E}_0' = -\frac{N}{8}\left[J_{11}(0) + J_{22}(0) - 2J_{12}(0)\right] - \frac{N}{2}|J_{12}(0)|\left(\frac{H}{H_c}\right)^2, \\
2\lambda_i = -\frac{1}{2}J_{ii}(0) + \frac{1}{2}J_{12}(0) \quad (i = 1,\ 2).
\end{gathered} \tag{16.10}$$

In vanishingly weak fields, this solution corresponds to the antiparallel distribution of the sublattice spins, which are at right angles to the direction of the vector **H**.

2. Strong fields: $H > H_c = (1/\mu)|J_{12}(0)|$,

$$\begin{gathered}
\gamma_1^z = \gamma_2^z = 1, \quad \gamma_1^\alpha = \gamma_2^\alpha = 0 \quad (\alpha = x,\ y), \\
\mathcal{E}_0'' = -\frac{N}{8}\left[J_{11}(0) + J_{22}(0) - 2J_{12}(0)\right] + \\
+ \frac{N}{2}|J_{12}(0)|\left(1 - 2\frac{H}{H_c}\right), \\
2\lambda_i = -\frac{1}{2}J_{ii}(0) - \frac{1}{2}J_{12}(0) - |J_{12}(0)|\frac{H}{H_c}.
\end{gathered} \tag{16.11}$$

The spins of the two sublattices are oriented along the external field and the spin system possesses ferromagnetic ordering. We can, therefore, expect the system to behave like a ferromagnet.

The value of the external field at the transition from one type of solution to the other will be called the critical field:

$$H_c = \frac{1}{\mu}|J_{12}(0)|. \tag{16.12}$$

The order of magnitude of the field is $|J_{12}(0)| \approx \vartheta_N$, where $\vartheta_N = kT_N$ (T_N is the Néel temperature). Since, for typical antiferromagnets, $T_N \approx$ 10–100°K, $\mu \approx 10^{-20}$, the critical field is $H_c \approx 10^5$–10^6 Oe. Thus, the alignment of the magnetic moments of the sublattices can be observed only in sufficiently strong fields.

The quantity $\mathcal{E}_0$ represents the energy of the ground state in the zeroth approximation. In this approximation, the magnetization and susceptibility of an antiferromagnet are given by

$$M_0 = -\frac{\partial \mathscr{E}_0}{\partial H} = \begin{cases} N\mu \frac{H}{H_c}, & H \leq H_c, \\ N\mu, & H > H_c, \end{cases} \tag{16.13}$$

$$\chi_0 = \frac{\partial M_0}{\partial H} = \begin{cases} N\frac{\mu}{H_c}, & H \leq H_c, \\ 0, & H > H_c. \end{cases} \tag{16.14}$$

When $H = H_c$, the magnetization M_0 changes suddenly and the susceptibility χ_0 exhibits a discontinuity.

The well-known conclusion of Hulten (1936), that in an isotropic antiferromagnet the spins are aligned at right angles to the field, follows from Eq. (16.10) for the case of vanishingly weak fields. When the field intensity is increased ($H \leq H_c$), the moments of both sublattices rotate in the direction of the field and, finally, become parallel to the field when the field reaches the critical value ($H = H_c$). In strong fields ($H > H_c$), the moments of the sublattices are oriented along the field and the antiferromagnet behaves like a ferromagnet.

The directions of the projections of the sublattice magnetization vectors in a plane (x, y) perpendicular to the direction of the field H remain indeterminate and all we can say is that the component of the magnetization in the (x, y) plane is proportional to $[1-(H/H_c)_2]$. This is due to the degeneracy of the directions under simultaneous rotation of both sublattices about the directions of the vector H (cf., Sec. 11). The degeneracy is partly lifted by allowing for the anisotropy – for example, by introducing into the Hamiltonian terms of the type.

$$-\sum I_a(f_1 - f_2)\, S^x_{f_1} S^x_{f_2} - \sum I_a(g_1 - g_2)\, S^x_{g_1} S^x_{g_2},$$

where the I_a are positive quanities.

Our results, in fact, apply to uniaxially anisotropic antiferromagnets in an external field applied at right angles to the anisotropy axis.

Some Configurations in H = 0. The earliest representations of antiferromagnets as systems of antiparallel spins are due to Néel (1932, 1936, 1948), who proposed a checkerboard antiparallel distribution, and to Landau (1934), who proposed a layered distribution.

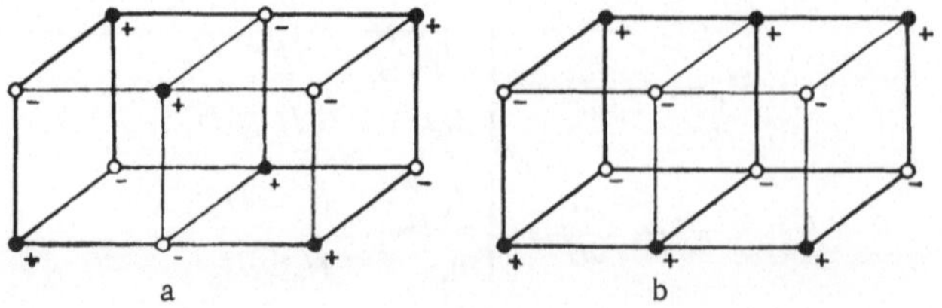

Fig. 7. Simple cubic lattice: a) $r < 1/4$; b) $r > 1/4$.

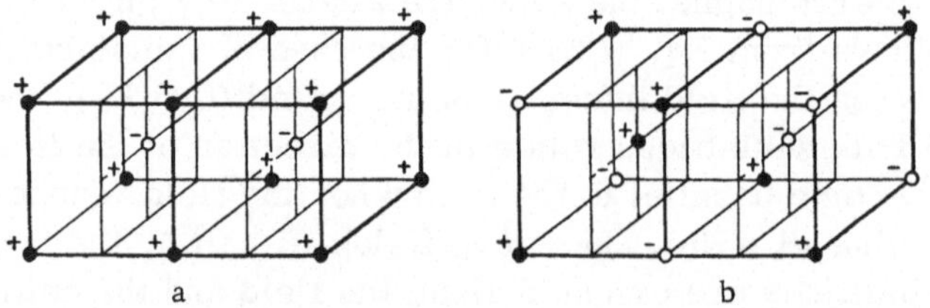

Fig. 8. Bcc lattice: a) $r < 2/3$; b) $r > 2/3$.

In general, spin configurations are governed by the signs and values of the exchange integrals in various environments and geometries of the crystal lattice. Adopting this point of view, several authors have considered the problem of the possible spin configurations in simple cubic lattices [Van Vleck (1941, 1951), Anderson (1950b), Luttinger (1951), Smart (1952)] and in hcp lattices [Li (1950) and Smart (1953)].

We shall quote here the results of Luttinger (1951) for cubic lattices. We shall describe the spin system by means of the Hamiltonian (Ising model)

$$\mathcal{H} = -\frac{1}{2} I_1 \sum_{f,\,\delta_1} S_f^z S_{f+\delta_1}^z - \frac{1}{2} I_2 \sum_{f,\,\delta_2} S_f^z S_{f+\delta_2}^z; \qquad (16.15)$$

in the first term the summation is only over the nearest neighbors while in the second term it is only over the second-nearest neighbors; I_1 is the value of the exchange integral for the nearest neighbors and I_2 is the same integral for the second-nearest neighbors; it is assumed that I_1, $I_2 < 0$; S^z_f are operators which assume the values ± 1.

Depending on the type of lattice and the value of $r = I_2/I_1$, the energy minimum of the form (16.15) corresponds to one of the configurations shown in Figs. 7, 8, 9. The simple cubic lattice can have two configurations, shown in Fig. 7. The bcc lattice also has two configurations (Fig. 8). The fcc lattice can be represented as an assembly of four simple cubic lattices, inserted into each other. The possible fcc configurations [see also Van Vleck (1951)] are shown in Fig. 9.

These considerations are based on a very much simplified Heisenberg model. The problem of the magnetic structures of antiferromagnets is presented more fully in "Antiferromagnetism" (1956), and in the reviews by Nagamiya et al. (1955) and Borovik-Romanov (1962).

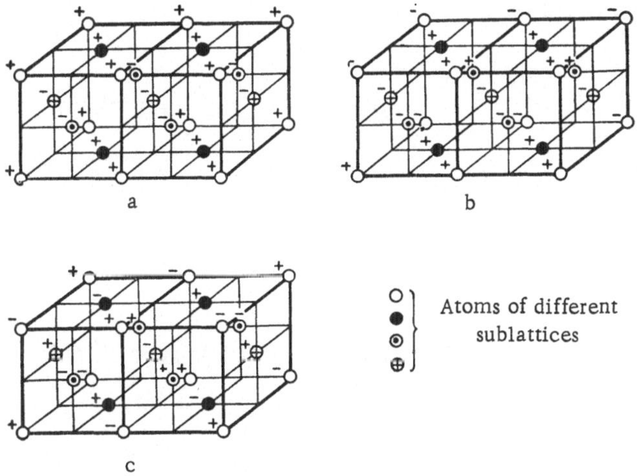

Fig. 9. Fcc lattice; a) $r < 1/3$; b) $r < 1/2$; c) $r > 1/2$.

Sec. 17. Elementary Excitations in Antiferromagnets*

We shall now consider the spectrum of elementary excitations (spin waves) in our model.

The quadratic form (16.3) is reduced to the diagonal form (13.7)-(13.8) by the transformation (13.6), where ξ^+, ξ are the new Bose operators; u, v are the eigenfunctions of the system of equations (13.3). In this case, the quantities R, S are defined by the formulas (16.4), and λ is found from the solution of the system (16.7)-(16.8).

We shall consider separately the cases of weak and strong fields. For convenience, we shall measure the field in units of H_c:

$$h = \frac{H}{H_c}, \quad H_c = \mu^{-1} |J_{12}(0)|. \tag{17.1}$$

1. Weak Fields: $h \leq 1$. To determine the spectrum of elementary excitations, we shall calculate the functions R, S from the formulas (16.4), (5.10). Using the expressions (16.10) for γ and λ, we obtain

$$\begin{aligned} A_f^x &= -\frac{h}{2}, \quad A_f^y = \frac{i}{2}, \quad A_f^z = \frac{1}{2}\sqrt{1-h^2}, \\ A_g^x &= \frac{h}{2}, \quad A_g^y = -\frac{i}{2}, \quad A_g^z = \frac{1}{2}\sqrt{1-h^2}. \end{aligned} \tag{17.2}$$

Consequently,

$$\begin{gathered} R(f_1, f_2) = R(g_1, g_2) = 0, \quad R(f, g) = -\frac{1}{2}(1-h^2) I(f-g), \\ S(f_1, f_2) = -\frac{1}{2} I(f_1 - f_2),\ S(g_1, g_2) = -\frac{1}{2} I(g_1 - g_2),\ (17.3) \\ S(f, g) = \frac{1}{2} h^2 I(f-g). \end{gathered} \tag{17.3}$$

Thus, Eqs. (13.3) assume the form

*See Bogolyubov and Tyablikov (1949c) and Tyablikov (1956a).

$$(E_\nu + 2\lambda_1) u_{f\nu} + \sum_{f'} \frac{1}{2} I(f - f') u_{f'\nu} - h^2 \sum_{g} \frac{1}{2} I(f - g) u_{g\nu} + \\ + (1 - h^2) \sum_{g} \frac{1}{2} I(f - g) v_{g\nu} = 0,$$
$$(-E_\nu + 2\lambda_1) v_{f\nu} + \sum_{f'} \frac{1}{2} I(f - f') v_{f'\nu} - h^2 \sum_{g} \frac{1}{2} I(f - g) v_{g\nu} + \\ + (1 - h^2) \sum_{g} \frac{1}{2} I(f - g) u_{g\nu} = 0; \tag{17.4}$$

there are two more similar equations, which are obtained by the interchange $f \rightleftharpoons g$, $\lambda_1 \rightleftharpoons \lambda_2$.

The solution of the system of equations (17.4) will be sought in the form

$$u_{f\nu} = \frac{u_{1\nu}}{\sqrt{N}} e^{i(f,\nu)}, \quad u_{g\nu} = \frac{u_{2\nu}}{\sqrt{N}} e^{i(g,\nu)},$$
$$v_{f\nu} = \frac{v_{1\nu}}{\sqrt{N}} e^{i(f,\nu)}, \quad v_{g\nu} = \frac{v_{2\nu}}{\sqrt{N}} e^{i(g,\nu)}. \tag{17.5}$$

To determine the values of $u_{1\nu}$, $u_{2\nu}$, $v_{1\nu}$, $v_{2\nu}$, we shall use a system of four algebraic equations:

$$(2E_\nu - A_1) u_{1\nu} - h^2 J_{12}(\nu) u_{2\nu} + (1 - h^2) J_{12}(\nu) v_{2\nu} = 0,$$
$$-h^2 J_{21}(\nu) u_{1\nu} + (2E_\nu - A_2) u_{2\nu} + (1 - h^2) J_{21}(\nu) v_{1\nu} = 0,$$
$$(1 - h^2) J_{12}(\nu) u_{2\nu} - (2E_\nu + A_1) v_{1\nu} - h^2 J_{12}(\nu) v_{2\nu} = 0,$$
$$(1 - h^2) J_{21}(\nu) u_{1\nu} - h^2 J_{21}(\nu) v_{1\nu} - (2E_\nu + A_2) v_{2\nu} = 0, \tag{17.6}$$

where the following notation is used:

$$A_i = -4\lambda_i - J_{ii}(\nu) = J_{ii}(0) - J_{12}(0) - J_{ii}(\nu) \quad (i = 1, 2),$$
$$J_{11}(\nu) = \sum_{f'} I(f - f') e^{i(f - f', \nu)}, \ \ldots \tag{17.7}$$

From the condition of solubility of this system of equations, we determine the spectrum of the eigenvalues E_ν:

$$E_\nu^{(1,2)} = \frac{1}{4} \Big\{ 2A_1^2 + 2A_2^2 - 4(1 - 2h^2) J_{12}^2(\nu) \pm \\ \pm 2\sqrt{(A_1 - A_2)^2 [(A_1 + A_2)^2 - 4(1 - 2h^2) J_{12}^2(\nu)] + 16h^4 A_1 A_2 J_{12}^2(\nu)} \Big\}^{1/2}. \tag{17.8}$$

In this case, we have two positive roots – two modes in the elementary excitation spectrum. The other two roots, which have the minus sign in front of the whole expression, are rejected. *

The operator form of Eq. (16.3) becomes

$$\mathscr{H}_2 = \Delta\mathscr{E}_0 + \sum_\nu E_\nu^{(1)} N_\nu^{(1)} + \sum_\nu E_\nu^{(2)} N_\nu^{(2)}, \tag{17.9}$$

where $N_\nu^{(1)}$, $N_\nu^{(2)}$ are the occupation number operators of spin waves (or elementary excitations) for the first and second modes in the energy spectrum.

Assuming that h = 0, we obtain the expression [Bogolyubov and Tyablikov (1949c)]

$$E_\nu^{(1,2)} = \frac{1}{4}\left\{\pm (A_1 - A_2) + \sqrt{(A_1 + A_2)^2 - 4J_{12}^2(\nu)}\right\}, \tag{17.10}$$

where A_1, A_2 are defined by Eq. (17.7).

We shall consider long-wave excitations ($\nu \to 0$). An approximate expression is obtained for E_ν by expanding the quantities in Eq. (17.10) as a series in powers of ν. We shall introduce the notation

$$\begin{aligned} J_{11}(0) - J_{11}(\nu) &= \alpha_{11}\nu^2 + \beta_{11}\nu^4 + \ldots, \\ -J_{12}(0) + J_{12}(\nu) &= \alpha_{12}\nu^2 + \beta_{12}\nu^4 + \ldots, \end{aligned} \tag{17.11}$$

where

$$\alpha_{11} = \frac{1}{6}\sum_{f'} |f - f'|^2 I(f - f'), \quad \alpha_{12} = -\frac{1}{6}\sum_{g} |f - g| I(f - g). \tag{17.12}$$

Substituting Eq. (17.11) into Eq. (17.10), we obtain the following approximate expression for the spectrum of long-wave elementary excitations:

$$E_\nu^{(1,2)} = w\nu + O(\nu^2), \tag{17.13}$$

$$w = \frac{1}{2}\sqrt{|J_{12}(0)|(\alpha_{11} + \alpha_{22} + \alpha_{12})}. \tag{17.14}$$

We thus find that, if h = 0, the energy of long-wave elementary excitations in an isotropic antiferromagnet depends linearly on

*See the remark at the end of Sec. 13 about the rejection of solutions with $E_\nu < 0$.

the wave vector, while in an isotropic ferromagnet the corresponding dependence is quadratic. The spectrum is then doubly degenerate.

Moreover, if h = 0, the system of equations (17.6) splits into two independent subsystems for the functions (u_1, v_2) and (u_2, v_1), respectively. Then, using the normalization conditions (13.4),

$$u_{1\nu}^2 - v_{2\nu}^2 = 1, \quad u_{2\nu}^2 - v_{1\nu}^2 = 1,$$

it is easy to find the explicit form of the functions (u, v). We shall consider the solution $u_{1\nu} \neq 0$, $v_{2\nu} \neq 0$, $u_{2\nu} = v_{1\nu} = 0$. Simple calculations give

$$\begin{aligned} u_{1\nu} &= \frac{2E_\nu^{(1)} + A_2}{\left[(2E_\nu^{(1)} + A_2)^2 - J_{12}^2(\nu)\right]^{1/2}}, \\ v_{2\nu} &= \frac{J_{12}(\nu)}{\left[(2E_\nu^{(1)} + A_2)^2 - J_{12}^2(\nu)\right]^{1/2}}, \end{aligned} \tag{17.15}$$

where $E_\nu^{(1)}$ is a mode of the spectrum (17.10) with two plus signs on the right-hand side. The second solution $u_{2\nu} \neq 0$, $v_{1\nu} \neq 0$, $u_{1\nu} = v_{2\nu} = 0$ is obtained from the first by the index interchange $1 \rightleftharpoons 2$ in Eq. (17.15) [obviously, $E_\nu^{(1)}$ then becomes $E_\nu^{(2)}$].

Using the expressions in Eq. (17.15) for $v_{1\nu}$, $v_{2\nu}$, we shall find how well the condition for the application of the approximate second quantization method, Eq. (14.4), is satisfied by antiferromagnets. Using Eqs. (17.5) and (17.15), we shall write the condition (14.4) in the form

$$D_1 = \frac{1}{N} \sum_\nu \frac{J_{12}^2(\nu)}{(2E_\nu^{(1)} + A_2)^2 - J_{12}^2(\nu)} \ll 1 \tag{17.16}$$

(the other condition is obtained by the index interchange $1 \rightleftharpoons 2$). We note, first of all, that if we neglect the interaction between the sublattices, and assume that $J_{12}(\nu) = 0$, Eq. (17.16) is satisfied identically: $D_1 = 0$. This result is self-evident, since we obtain a set of two independent ferromagnetic lattices. Moreover, since D_1 is a continuous function of J_{12}, $D_1 \ll 1$ when $|J_{12}| \ll |J_{11}|$. The situation is somewhat more complicated when $|J_{12}| \gtrsim |J_{11}|$.

We shall consider the case when the interaction is only between the sublattices. Then,

$$J_{11}=J_{22}=0,\quad A_1=A_2=|J_{12}(0)|,\quad E_\nu^{(1,2)}=\frac{1}{2}\sqrt{J_{12}^2(0)-J_{12}^2(\nu)}$$

and the expression for $D_{1,2}$ can be written in the form

$$D_i=\frac{1}{N}\sum_\nu\frac{\gamma_\nu^2}{(1+\sqrt{1-\gamma_\nu^2})^2-\gamma_\nu^2}=\frac{1}{2N}\sum_\nu\frac{\gamma_\nu^2}{\sqrt{1-\gamma_\nu^2}(1+\sqrt{1-\gamma_\nu^2})},$$

$$\gamma_\nu=\frac{J_{12}(\nu)}{J_{12}(0)}\qquad (i=1,\ 2). \tag{17.17}$$

Replacing $\sqrt{1-\gamma_\nu^2}$ in the denominator of Eq. (17.17) with zero or unity ($0 \leq \sqrt{1-\gamma_\nu^2} \leq 1$), we obtain the upper and lower limits for D_i:

$$\frac{1}{4}\varkappa<D_i<\frac{1}{2}\varkappa\quad (i=1,\ 2), \tag{17.18}$$

where $\varkappa$ is a number which depends on the lattice geometry:

$$\varkappa=\frac{1}{N}\sum_\nu\frac{\gamma_\nu^2}{\sqrt{1-\gamma_\nu^2}}. \tag{17.19}$$

In the nearest-neighbor approximation, the order of magnitude of the above quantity is $\varkappa \sim 1/z$, which, in general, gives a favorable estimate of the possibility of applying the second quantization method. * In this case,

$$\frac{1}{4z}<D_i<\frac{1}{2z}\quad (i=1,\ 2) \tag{17.20}$$

(z is the number of nearest neighbors).

* For the bcc lattice, Davis (1962) calculated the integrals

$$B_1=\frac{1}{(2\pi)^3}\int\int_{-\pi}^{\pi}\int d\nu\left\{\frac{1}{\sqrt{1-\gamma_\nu^2}}-1\right\},$$

$$B_2=\frac{1}{(2\pi)^3}\int\int_{-\pi}^{\pi}\int d\nu\sqrt{1-\gamma_\nu^2},$$

$$B_1=0.118\ldots,\qquad B_2=0.926\ldots$$

Hence, we find that $\varkappa=1+B_1 \approx 0.19$, and the estimate $\varkappa\sim 1/z$ is quite close to this value.

Allowance for the interaction within each of the sublattices leads to an additional ordering in the spin system. Therefore, when J_{11}, $J_{22} \neq 0$, the criterion for the application of the method should be satisfied even better.

We shall now consider the case of equivalent sublattices. For these sublattices we have, by definition,

$$\begin{gathered} J_{11}(\nu) = J_{22}(\nu), \quad A_1 = A_2, \\ A_1 = J_{11}(0) - J_{12}(0) - J_{11}(\nu). \end{gathered} \tag{17.21}$$

Then, the expression (17.8) for the energy of elementary excitations simplifies to

$$E_\nu^{(1,2)} = \frac{1}{2}\sqrt{A_1^2 - J_{12}^2(\nu) + 2h^2 J_{12}(\nu)\left[J_{12}(\nu) \pm A_1\right]}. \tag{17.22}$$

When h = 0, the two modes merge and the spectrum becomes doubly degenerate. If we consider only the case of small wave vectors, we then obtain for both modes an expression of the type given by Eq. (17.13) with the unimportant difference that $\alpha_{11} = \alpha_{22}$.

So far, our treatment has not been restricted to any particular type of lattice. To make the case definite, we shall now consider an antiferromagnet in which the magnetically active atoms are distributed at sites of a simple cubic lattice. Since, in this case, I_{11}, $I_{22} > 0$, $I_{12} < 0$, the spin configuration in H = 0 has the form shown in Fig. 7a. The magnetic lattice constant is equal to double the crystal lattice constant. Let us use b^r (r = 1, 2, 3) to denote the reciprocal magnetic lattice vector. Since the distance between the nearest sites in the spin sublattices is equal to half the lattice constant, it follows that

$$J_{12}(2\pi b - \nu) = -J_{12}(\nu), \quad J_{11}(2\pi b - \nu) = J_{11}(\nu).$$

Consequently, the roots of Eq. (17.22) differ by $2\pi b$. Thus, we need consider only one of them, for example, $E_\nu^{(1)}$.

In the case $\nu \ll 2\pi b$, we obtain the following expression by substituting the expansion (17.11) into Eq. (17.22):

$$E_\nu^{(1)} \approx \nu w \sqrt{1 - h^2}, \tag{17.23}$$

where w is given by the expression (17.14).

In the case $2\pi b - \nu \ll 2\pi b$, we similarly obtain, using $E^{(1)}_{2\pi b-\nu} = E^{(2)}_{\nu}$,

$$E^{(2)}_{2\pi b-\nu} \approx \sqrt{h^2 J^2_{12}(0) + w^2\left(1 + h^2 \frac{\alpha_{11} - 3\alpha_{12}}{\alpha_{11} + \alpha_{12}}\right)\nu^2}. \tag{17.24}$$

When $h \to 0$, Eqs. (17.23) and (17.24) reduce to Eq. (17.13).

We shall now calculate the magnetization. Using Eq. (8.35), we can write the expression for the magnetization in the form

$$M = M_0 + M_{\vartheta}, \tag{17.25}$$

where M_0 is the temperature-independent part of the magnetization, which is given by Eq. (16.13), and M_{ϑ} is the temperature-dependent part:

$$M_{\vartheta} = -\sum_{\nu} \frac{\partial E^{(1)}_{\nu}}{\partial H} \bar{N}^{(1)}_{\nu} - \sum_{\nu} \frac{\partial E^{(2)}_{\nu}}{\partial H} \bar{N}^{(2)}_{\nu}. \tag{17.26}$$

We shall now transform sums into integrals [cf., (A2.14)] and take into account the fact that our treatment is restricted to low temperatures. Then, the integral with respect to ν for the first mode of the spectrum [$E^{(1)}_{\nu}$] can be replaced by two integrals near the points $\nu = 0$ and $\nu = 2\pi b$, using the approximate expressions for $E^{(1)}_{\nu}$ given by Eqs. (17.23) and (17.24), respectively. Since the multiplier $\bar{N}^{(1)}_{\nu}$ is effectively different from zero only at sufficiently low values of ν (or $2\pi b - \nu$), the range of integration can be extended to all values of ν. The second mode of the spectrum, $E^{(2)}_{\nu}$, differs from the first only by the shift $2\pi b$ and, therefore, it can be allowed for in calculations by doubling the result for $E^{(1)}_{\nu}$. Simple calculations give

$$M_{\vartheta} = N \frac{v\mu}{\pi^2} \left(\frac{J_{12}(0)}{w}\right)^3 h\varphi(h, \tau), \tag{17.27}$$

where the following notation is used:

$$\left.\begin{aligned} \varphi(h, \tau) &= \int\limits_0^{\infty} \frac{1}{r_1} \frac{x^4\, dx}{e^{\frac{r_1}{\tau}} - 1} - \int\limits_0^{\infty} \frac{1 + \gamma x^2}{r_2} \frac{x^2\, dx}{e^{\frac{r_2}{\tau}} - 1}, \\ r_1 &= x\sqrt{1 - h^2}, \quad r_2 = \sqrt{h^2 + (1 + \gamma h^2) x^2}, \quad \gamma = \frac{\alpha_{11} - 3\alpha_{12}}{\alpha_{11} + \alpha_{12}}, \end{aligned}\right\} \tag{17.28}$$

$$\tau = \frac{\vartheta}{|J_{12}(0)|}, \quad h = \frac{H}{H_c}, \quad V = Nv, \quad x = \frac{w\nu}{|J_{12}(0)|}. \qquad (17.29)$$

When h=0, we have

$$\varphi(0, \tau) = -\int_0^\infty \frac{x\,dx}{e^{\frac{x}{\tau}} - 1} + (1-\gamma)\int_0^\infty \frac{x^3\,dx}{e^{\frac{x}{\tau}} - 1} \approx -\frac{\tau^2\pi^2}{6} + O(\tau^4). \qquad (17.30)$$

Using Eqs. (16.13), (17.27), and (17.30), we obtain an expression for the transverse magnetization in the limiting case of vanishingly weak fields:

$$M = M_0\left\{1 - \frac{v}{6}\left(\frac{|J_{12}(0)|}{w}\right)^3 \tau^2\right\} \quad (M_0 = N\mu h), \qquad (17.31)$$

where h is the external field in units of H_c, given by Eq. (17.1); τ is the temperature in dimensionless units; v is the volume per site; w is the quantity which is given by Eq. (17.14) and represents the velocity of propagation of long-wave vibtations.

Similarly, we obtain an expression for the transverse susceptibility in vanishingly weak fields:

$$\chi = \frac{\partial M}{\partial H} = \frac{N\mu^2}{|J_{12}(0)|}\left\{1 - \frac{v}{6}\left(\frac{|J_{12}(0)|}{w}\right)^3 \tau^2\right\}. \qquad (17.32)$$

If we allow for the interaction of only the nearest neighbors, then

$$v = a^3, \quad w = \frac{a}{\sqrt{12}}|J_{12}(0)|, \quad \frac{v}{6w^3}|J_{12}(0)|^3 = \sqrt{48}$$

and we have the following formulas for $E_\nu^{(1)}$, M, χ:

$$E_\nu^{(1)} = w\sqrt{1-h^2}\,\nu,$$

$$M = N\mu h\left(1 - \sqrt{48}\,\tau^2\right), \quad \chi = \frac{N\mu^2}{|J_{12}(0)|}\left(1 - \sqrt{48}\,\tau^2\right). \qquad (17.33)$$

Allowing for the nearest neighbors only, Hulten (1936) was the first to obtain the main term in the energy of elementary excitations, $E_\nu^{(1)}$, and the formulas for the magnetization and susceptibility given by Eq. (17.33). Hulten used the quasi-classical approximation to find the equilibrium configuration of spins and considered and quantized small oscillations of the spin vectors about the equilibrium configuration. We can easily see that this procedure represents the approximate second quantization method.

2. Strong Fields: $h > 1$. Substituting the quantity γ of Eq. (16.11) into Eq. (5.10), we find the vectors A and then, using Eq. (16.4), the coefficients of the quadratic form (16.3):

$$\begin{gathered} S(f_1, f_2) = -\frac{1}{2} I(f_1 - f_2), \quad S(g_1, g_2) = -\frac{1}{2} I(g_1 - g_2), \\ S(f, g) = \frac{1}{2} I(f - g), \quad R(f_1, f_2) = R(g_1, g_2) = R(f, g) = 0. \end{gathered} \tag{17.34}$$

Consequently, we obtain the following system of equations for the determination of the functions u, v:

$$\begin{gathered} (2E_\nu + A_1) u_{1,\nu} - J_{12}(\nu) u_{2,\nu} = 0, \\ -J_{21}(\nu) u_{1,\nu} + (2E_\nu + A_2) u_{2,\nu} = 0, \\ v_{l,\nu} = 0 \quad (l = 1, 2), \end{gathered} \tag{17.35}$$

where

$$A_l = 4\lambda_l + J_{ll}(\nu) = -J_{ll}(0) + J_{ll}(\nu) - J_{12}(0) - 2\mu H. \tag{17.36}$$

A similar system of equations for the functions v differs from Eq. (17.35) only in the sign of E_ν. We have two solutions: $u \neq 0$, $v = 0$, and $u = 0$, $v \neq 0$. The second solution corresponds to negative values of E_ν and can therefore be rejected.

The spectrum of elementary excitations is found, as usual, from the condition of the solubility of the system of equations (17.35):

$$E_\nu^{(1,2)} = -\frac{1}{4}(A_1 + A_2) \pm \frac{1}{4}\sqrt{(A_1 - A_2)^2 + 4J_{12}^2(\nu)}. \tag{17.37}$$

For the equivalent sublattices, $A_1 = A_2$, $J_{11}(\nu) = J_{22}(\nu)$. Using Eqs. (17.36) and (17.1), we obtain

$$\begin{aligned} E_\nu^{(1,2)} = \frac{1}{2}\{J_{11}(0) - J_{11}(\nu) + 2(h-1)|J_{12}(0)| + \\ + |J_{12}(0)| \pm J_{12}(\nu)\}. \end{aligned} \tag{17.38}$$

We note that, as in the preceding case of weak fields, $E_\nu^{(2)} = E^{(1)}_{2\pi b - \nu}$. Therefore, we shall consider only the first mode of the spectrum of elementary excitations. Using the expansions (17.11), we find that

$$E_{\nu}^{(1)} \approx \begin{cases} (h-1)|J_{12}(0)| + \frac{1}{2}(\alpha_{11}+\alpha_{12})\nu^2, & \nu \ll 2\pi b, \\ h|J_{12}(0)| + \frac{1}{2}(\alpha_{11}-\alpha_{12})(2\pi b-\nu)^2, & |2\pi b-\nu| \ll 2\pi b. \end{cases} \tag{17.39}$$

The temperature-independent part of the magnetization is given by Eq. (16.13) and the temperature-dependent part by an expression of the (17.26) type, where $E_{\nu}^{(1,2)}$ is given by Eq. (17.38). Since $E_{\nu}^{(2)} = E_{2\pi b-\nu}^{(1)}$, the contributions of the two modes of the spectrum to the magnetization are equal. We shall now go over, in accordance with Eq. (A2.14), from summation over ν to integration, and we shall use the approximate expressions for $E_{\nu}^{(1)}$, which are valid at $\nu \approx 0$ and $\nu \approx 2\pi b$. Using dimensionless variables, we obtain the following expression for the temperature-dependent part of the magnetization:

$$M_{\vartheta} = Nv\frac{\mu}{\pi^2}\left(\frac{|J_{12}(0)|}{w}\right)^3 \psi(h, \tau), \tag{17.40}$$

where

$$\psi(h, \tau) = -\int\limits_0^{\infty} \frac{x^2\,dx}{e^{\frac{h-1+x^2}{\tau}}-1} - \int\limits_0^{\infty} \frac{x^2\,dx}{e^{\frac{h+\alpha x^2}{\tau}}-1},$$

$$\alpha = \frac{\alpha_{11}-\alpha_{12}}{\alpha_{11}+\alpha_{12}}, \tag{17.41}$$

and h, τ, w are given by Eqs. (17.29) and (17.14); v is the volume per site.

Since, in the present case, $h \geq 1$, the second term in ψ can be dropped:

$$\psi(h, \tau) \approx -\int\limits_0^{\infty} \frac{x^2\,dx}{e^{\frac{h-1+x^2}{\tau}}-1} = -\tau^{3/2}\frac{\sqrt{\pi}}{4} Z_{3/2}\left(\frac{h-1}{\tau}\right). \tag{17.42}$$

Using Eqs. (17.25) and (16.13), (17.40), and (17.42), we obtain the final formulas for the magnetization and susceptibility:

$$M = N\mu\left\{1 - \frac{v}{4\pi^{3/2}}\left(\frac{|J_{12}(0)|}{w}\right)^3 \tau^{3/2} Z_{3/2}\left(\frac{h-1}{\tau}\right)\right\}, \tag{17.43}$$

$$\chi = N \frac{\mu^2 v}{4\pi^{3/2} |J_{12}(0)|} \left(\frac{|J_{12}(0)|}{w} \right)^3 \tau^{1/2} Z_{1/2} \left(\frac{h-1}{\tau} \right),$$

where Z_p is defined in accordance with Eq. (15.17).

We note that, in the region $h \approx 1$, the method of spin waves, in the form used here, gives incorrect results for the temperature-dependent contributions to the magnetization and susceptibility. This is because the determination of the critical field from the condition for the minimum of the ground-state energy is insufficiently accurate. The correct results may be obtained for strong fields by determining the critical field from the condition for the minimum of the free energy. This problem was discussed in detail by Turov and Irkhin (1958).

Sec. 18. Spin Waves in Helical Structures*

We shall now consider the elementary excitations in substances with helical magnetic structures. We shall deal only with the simple and ferromagnetic helices (cf., Figs. 6c, d); to simplify the calculations, the external field will be assumed to be zero.

It is usual to assume that in such substances the magnetocrystalline anisotropy is partly of the one-ion type. Therefore, we shall first indicate the changes which have to be made in the standard approximate second quantization method, presented in Sec. 12. Let the Hamiltonian of the system be

$$\mathcal{H} = \mathcal{H}_U + \mathcal{H}_A, \tag{18.1}$$

where $\mathcal{H}_U$ is the Hamiltonian which depends on the product of the spin operators for various sites, and $\mathcal{H}_A$ is the Hamiltonian of the one-ion anisotropy:

$$\mathcal{H}_A = -\sum \frac{I_{n\alpha}(f)}{n!} (S_f^\alpha)^n. \tag{18.2}$$

We shall find in $\mathcal{H}$ the terms which describe approximately the weakly excited states of the system. According to Eq. (12.13),

*See Kaplan (1961), Cooper et al. (1962), and Bar'yakhtar et al. (1964a).

$$\mathcal{H} \approx \mathcal{E}_0 - \sum 2\lambda_f (S_f - S_f^z) + \frac{1}{2} \sum \frac{\partial^2 \mathcal{E}_0}{\partial \gamma_{f_1}^{\alpha_1} \partial \gamma_{f_2}^{\alpha_2}} S_{f_1}^{-1} S_{f_2}^{-1} \tau_{f_1}^{\alpha_1} \tau_{f_2}^{\alpha_2}, \qquad (18.3)$$

where $\mathcal{E}_0$ is the classical form obtained from the operator form of $\mathcal{H}$ by replacing the operators S_f^{α} with the classical vectors $\gamma_f^{\alpha} S_f$ (γ_f^{α} are the components of the direction vector of the quantization axis of the f-th spin) on going over to the characteristic representation of the spin operators given by Eq. (5.8). The quantities λ_f, γ_f^{α} are found from Eqs. (12.6), and τ_f^{α} are given by the formulas (12.14). We shall write the form $\mathcal{E}_0$ as the sum of two terms, one of which corresponds to the operator form $\mathcal{H}_U$ of Eq. (18.1), and the other to $\mathcal{H}_A$:

$$\mathcal{E}_0 = \mathcal{E}_U + \mathcal{E}_A = \mathcal{E}_U - \sum \frac{I_{n\alpha}(f)}{n!} (S_f \gamma_f^{\alpha})^n. \qquad (18.4)$$

Substituting Eq. (18.4) into Eq. (18.3), we obtain

$$\mathcal{H} = \mathcal{E}_0 - \sum 2\lambda_f (S_f - S_f^z) + \frac{1}{2} \sum \frac{\partial^2 \mathcal{E}_U}{\partial \gamma_{f_1}^{\alpha_1} \partial \gamma_{f_2}^{\alpha_2}} S_{f_1}^{-1} S_{f_2}^{-1} \tau_{f_1}^{\alpha_1} \tau_{f_2}^{\alpha_2} - \\ - \sum \frac{I_{n\alpha}(f)}{2(n-2)!} (S_f \gamma_f^{\alpha})^{n-2} (\tau_f^{\alpha})^2. \qquad (18.5)$$

In the third term of this expression, the sites f_1 and f_2 are not identical, and the term can be written in the usual form:

$$\frac{1}{2} \sum \frac{\partial^2 \mathcal{E}_U}{\partial \gamma_{f_1}^{\alpha_1} \partial \gamma_{f_2}^{\alpha_2}} S_{f_1}^{-1} S_{f_2}^{-1} \{ A_{f_1}^{\alpha_1} A_{f_2}^{\alpha_2} S_{f_1}^{+} S_{f_2}^{+} + 2 A_{f_1}^{*\alpha_1} A_{f_2}^{\alpha_2} S_{f_1}^{-} S_{f_2}^{+} + \\ + A_{f_1}^{*\alpha_1} A_{f_2}^{*\alpha_2} S_{f_1}^{-} S_{f_2}^{-} \}. \qquad (18.6)$$

Using the commutation relationship for the spin operators, the fourth term in Eq. (18.5) is written as follows:

$$-\sum \frac{I_{n\alpha}(f)}{2(n-2)!} (S_f \gamma_f^{\alpha})^{n-2} \{ 2S_f |A_f^{\alpha}|^2 + \\ + 2(S_f + S_f^z)(S_f - S_f^z) |A_f^{\alpha}|^2 + (A_f^{\alpha})^2 (S_f^{+})^2 + (A_f^{*\alpha})^2 (S_f^{-})^2 \}.$$

Noting that in this approximation

$$(S_f + S_f^z)(S_f - S_f^z) \approx 2S_f (S_f - S_f^z)$$

and that, according to Eq. (5.9),

$$|A_f^{\alpha}|^2 = \frac{1}{4} [1 - (\gamma_f^{\alpha})^2].$$

we then obtain

$$-\sum \frac{S_f I_{n\alpha}(f)}{4(n-2)!}(S_f \gamma_f^{\alpha})^{n-2}[1-(\gamma_f^{\alpha})^2] - \\ -\sum \frac{S_f I_{n\alpha}(f)}{2(n-2)!}(S_f \gamma_f^{\alpha})^{n-2}[1-(\gamma_f^{\alpha})^2](S_f - S_f^z) - \\ -\sum \frac{I_{n\alpha}(f)}{2(n-2)!}(S_f \gamma_f^{\alpha})^{n-2}\{(A_f^{\alpha})^2(S_f^{+})^2 + (A_f^{*\alpha})^2(S_f^{-})^2\}. \tag{18.7}$$

Substituting Eqs. (18.6) and (18.7) into Eq. (18.5), and grouping the terms having the same powers of operators, we obtain

$$\mathcal{H} = \mathcal{E}_0 + \delta\mathcal{E}_0 + \mathcal{H}_2, \tag{18.8}$$

where

$$\delta\mathcal{E}_0 = -\sum \frac{S_f I_{n\alpha}(f)}{4(n-2)!}(S_f \gamma_f^{\alpha})^{n-2}[1-(\gamma_f^{\alpha})^2], \tag{18.9}$$

$$\mathcal{H}_2 = -\sum\left\{2\lambda_f + \frac{S_f I_{n\alpha}(f)}{2(n-2)!}(S_f \gamma_f^{\alpha})^{n-2}[1-(\gamma_f^{\alpha})^2]\right\}(S_f - S_f^z) + \\ +\frac{1}{2}\sum \frac{\partial^2 \mathcal{E}_U}{\partial \gamma_{f_1}^{\alpha_1}\partial \gamma_{f_2}^{\alpha_2}} S_{f_1}^{-1} S_{f_2}^{-1}\{A_{f_1}^{\alpha_1} A_{f_2}^{\alpha_2} S_{f_1}^{+} S_{f_2}^{+} + \\ + 2A_{f_1}^{*\alpha_1} A_{f_2}^{\alpha_2} S_{f_1}^{-} S_{f_2}^{+} + A_{f_1}^{*\alpha_1} A_{f_2}^{*\alpha_2} S_{f_1}^{-} S_{f_2}^{-}\} - \\ -\sum \frac{I_{n\alpha}(f)}{2(n-2)!}(S_f \gamma_f^{\alpha})^{n-2}\{(A_f^{\alpha})^2(S_f^{+})^2 + (A_f^{*\alpha})^2(S_f^{-})^2\} \tag{18.10}$$

and where λ_f, γ_f^{α} are found from the system of equations

$$\frac{\partial \mathcal{E}_0}{\partial \gamma_f^{\alpha}} - 2S_f \lambda_f \gamma_f^{\alpha} = 0, \quad \sum_{\alpha}(\gamma_f^{\alpha})^2 = 1. \tag{18.11}$$

In general, the quantities λ_f and γ_f^{α} should be determined from the condition for the minimum of $\mathcal{E}_0 + \delta\mathcal{E}_0$. We shall assume that $S^{-1} \ll 1$ (S is the spin). The order of magnitude is then given by $\delta\mathcal{E}_0/\mathcal{E}_0 \sim S^{-1}$. Considering $\delta\mathcal{E}_0$ now as a small quantity of the order of S^{-1}, we obtain an expression for the ground-state energy accurate to values of the order of S^{-1}, inclusive, using the conditions for the minimum of only $\mathcal{E}_0$. The first corrections to the coefficients of the quadratic form $\mathcal{H}_2$ are of the order of S^{-1}. We shall ignore these corrections, since we shall, in general, leave out corrections of the order of S^{-1} and higher orders, which appear when, in accordance with the ASQ method, the spin operators are replaced with the Bose operators [cf., Eqs. (12.19) and (5.22)].

In terms of the Bose operators of Eq. (12.19), the quadratic form (18.10) becomes

$$\begin{aligned}\mathcal{H}_2 = &-\sum\left\{2\lambda_f + \frac{S_f I_{n\alpha}(f)}{2(n-2)!}\left(S_f\gamma_f^\alpha\right)^{n-2}\left[1-(\gamma_f^\alpha)^2\right]\right\}n_f + \\ &+\sum\frac{\partial^2\mathcal{E}_U}{\partial\gamma_{f_1}^{\alpha_1}\partial\gamma_{f_2}^{\alpha_2}}\left(S_{f_1}S_{f_2}\right)^{-1/2}\left\{A_{f_1}^{\alpha_1}A_{f_2}^{\alpha_2}b_{f_1}b_{f_2} + \right. \\ &\left. + 2A_{f_1}^{*\alpha_1}A_{f_2}^{\alpha_2}b_{f_1}^{+}b_{f_2} + A_{f_1}^{*\alpha_1}A_{f_2}^{*\alpha_2}b_{f_1}^{+}b_{f_2}^{+}\right\} - \\ &-\sum\frac{S_f I_{n\alpha}(f)}{(n-2)!}\left(S_f\gamma_f^\alpha\right)^{n-2}\left\{(A_f^\alpha)^2 b_f^2 + (A_f^{*\alpha})^2 b_f^{+2}\right\}.\end{aligned} \quad (18.12)$$

We shall now consider the two simplest examples of helical structures.

<u>Simple Helix.</u> We shall consider a spin system which is described by a Hamiltonian of the type [Cooper and Elliott (1963)]

$$\mathcal{H} = -\frac{1}{2}\sum I(f_1 - f_2)(S_{f_1}, S_{f_2}) + \frac{K_2}{2}\sum (S_f^z)^2. \quad (18.13)$$

Obviously, the z-axis is a symmetry axis of the magnetic lattice. We shall compare the operator form of Eq. (18.13) with the classical form

$$\mathcal{E}_0 = -\frac{1}{2}\sum S^2 I(f_1 - f_2)(\gamma_{f_1}, \gamma_{f_2}) + \frac{1}{2}S^2 K_2 \sum (\gamma_f^z)^2. \quad (18.14)$$

We shall regard all sites in the magnetic lattice as equivalent. Then the conditions of Eq. (18.11) for the minimum of the form (18.14) are written as follows:

$$\begin{aligned}-\sum SI(f-f')\gamma_{f'}^z + SK_2\gamma_f^z - 2\lambda\gamma_f^z &= 0, \\ -\sum SI(f-f')\gamma_{f'}^{\pm} - 2\lambda\gamma_f^{\pm} &= 0,\end{aligned} \quad (18.15)$$

where the following notation is used:

$$\gamma_f^{\pm} = \gamma_f^x \pm i\gamma_f^y, \qquad S = S_f. \quad (18.16)$$

We shall seek a solution such that the spins in each basal plane are oriented parallel to one another (ferromagnetic ordering) and lie in that plane, but their directions change from one basal plane to another. Accordingly, we shall assume that

$$\gamma_f^z = \gamma_z = \text{const}, \quad \gamma_f^{\pm} = \rho e^{\pm i(q, f)} \left[\rho = \sqrt{1-(\gamma_z)^2}\right]. \tag{18.17}$$

where q is a vector parallel to the z-axis. Then, Eq. (18.15) becomes

$$(-2\lambda - SJ(0) + SK_2)\gamma_z = 0,$$
$$(-2\lambda - SJ(q))\rho = 0.$$

Of the three possible solutions,

$$\begin{aligned} &\text{I.}\ \ \rho = 1, && \gamma_z = 0, && -2\lambda = SJ(q), \\ &\text{II.}\ \ \rho = 0, && \gamma_z = 1, && -2\lambda = SJ(0) - SK_2, \\ &\text{III.}\ \ \rho \neq 0, && \gamma_z \neq 0, && SJ(q) = SJ(0) - SK_2 \end{aligned} \tag{18.18}$$

we select the first. It corresponds to the ground-state energy

$$\mathscr{E}_0 = -\frac{1}{2} S^2 J(q). \tag{18.19}$$

The two other solutions are rejected, because they correspond to higher values of the ground-state energy $\mathscr{E}_0$ (the second solution represents pure ferromagnetic ordering with the spins oriented parallel to the symmetry axis; the third solution is possible for some fixed value of q and does not necessarily correspond to the energy minimum). Solution I of Eq. (18.18) describes the configuration of a simple helix (Fig. 6c) if the function

$$J(q) = \sum I(f) e^{i(f, q)}$$

has a maximum at some value $q \neq 0$. The angle by which the spins are rotated when translated from a given basal plane to the neighboring plane is equal to qa, where a is the distance between neighboring planes. For the helix to exist, the expression for $\mathscr{H}$ must include the interactions of the first and further neighbors.

For example, if the interaction is only between the nearest and second-nearest neighbors, then

$$J(q) = I_1 \cos(qa) + I_2 \cos(2qa).$$

Depending on the values and signs of the exchange integrals I_1, I_2, the following configurations are possible for which $\mathscr{E}_0$ is minimal:

$I_1 > 0,\ I_2 > 0$: ferromagnetic ordering [$\cos(qa) = 1$];

$I_1 < 0,\ I_2 > 0$: antiferromagnetic ordering [$\cos(qa) = -1$];

$I_1 > 0,\ I_2 < 0$: when $|I_1/4I_2| < 1$ – simple helix [$\cos(qa) = I_1/4I_2$];
when $|I_1/4I_2| > 1$ – ferromagnetic ordering;

$I_1 < 0,\ I_2 < 0$: when $|I_1/4I_2| < 1$ – simple helix [$\cos(qa) = -I_1/4I_2$];
when $|I_1/4I_2| > 1$ – antiferromagnetic ordering [$\cos(qa) = -1$].

We shall now determine the spectrum of elementary excitations.

For the operator of Eq. (18.13), the quadratic form of Eq. (18.12) is

$$\mathscr{H}_2 = \sum \left\{-2\lambda + \frac{1}{2} SK_2\rho^2\right\} n_f - \sum SI(f_1 - f_2)\left\{(A_{f_1}, A_{f_2})\, b_{f_1} b_{f_2} + \right. \tag{18.20}$$

$$\left. + 2(A^*_{f_1}, A_{f_2})\, b^+_{f_1} b_{f_2} + (A^*_{f_1}, A^*_{f_2})\, b^+_{f_1} b^+_{f_2}\right\} + \sum SK_2\left\{(A^z_f)^2 b^2_f + (A^{*z}_f)^2 b^{+2}_f\right\}.$$

The vectors A_f are defined in terms of the vectors γ_f using Eq. (5.10). According to Eq. (18.15), we have, in the case of solution I of Eq. (18.18),

$$\gamma_z = 0, \quad \gamma^x_f = \cos(q, f), \quad \gamma^y_f = \sin(q, f),$$

and formulas (5.10) assume the form

$$A^x_f = \frac{1}{4}\left(e^{-i(f, q)} - e^{i(f, q)}\right), \quad A^y_f = \frac{i}{4}\left(e^{-i(f, q)} + e^{i(f, q)}\right),$$

$$A^z_f = \frac{1}{2} \quad \left(\varphi_f = (f, q) = \arctan\frac{\gamma^y_f}{\gamma^x_f}\right). \tag{18.21}$$

Using these expressions, we easily find that

$$(A_{f_1}, A_{f_2}) = \frac{1}{4}\left(1 - \frac{e^{i(q, f_1-f_2)} + e^{-i(q, f_1-f_2)}}{2}\right),$$
$$(A^*_{f_1}, A_{f_2}) = \frac{1}{4}\left(1 + \frac{e^{i(q, f_1-f_2)} + e^{-i(q, f_1-f_2)}}{2}\right), \quad (A^z_f)^2 = \frac{1}{4}. \tag{18.22}$$

Finally, substituting λ_f of Eqs. (18.18) and Eq. (18.22) into Eq. (18.20), we obtain

$$\mathscr{H}_2 = \frac{1}{2}\sum R^*_{ff'} b_f b_{f'} + \sum S_{ff'} b^+_f b_{f'} + \\ + \frac{1}{2}\sum R_{ff'} b^+_f b^+_{f'}, \tag{18.23}$$

where the following notation is used:

$$S_{ff'} = \left\{SJ(q) + \frac{1}{2}SK_2\right\}\delta_{ff'} - \\ - \frac{S}{2} I(f - f')\left\{1 + \frac{e^{i(q, f-f')} + e^{-i(q, f-f')}}{2}\right\},$$
$$R_{ff'} = \frac{1}{2}SK_2\delta_{ff'} - \\ - \frac{S}{2} I(f - f')\left\{1 - \frac{e^{i(q, f-f')} + e^{-i(q, f-f')}}{2}\right\}, \tag{18.24}$$
$$R^*_{ff'} = R_{ff'}.$$

The quadratic form (18.23) is diagonalized to

$$\mathscr{H}_2 = \Delta\mathscr{E}_0 + \sum E_\nu \xi^+_\nu \xi_\nu,$$
$$\Delta\mathscr{E}_0 = -\sum E_\nu |v_{f\nu}|^2 \tag{18.25}$$

using Eq. (13.6). The transformation coefficients ($u_{f\nu}$, $v_{f\nu}$) are found from the solution of the system of equations (13.3):

$$E_\nu u_{f\nu} = \sum_{f'} S_{ff'} u_{f'\nu} + \sum_{f'} R_{ff'} v_{f'\nu},$$
$$-E_\nu v_{f\nu} = \sum_{f'} S_{ff'} v_{f'\nu} + \sum_{f'} R_{ff'} u_{f'\nu}. \tag{18.26}$$

We shall seek the solution of the system (18.26) in the form

$$u_{f\nu} = \frac{u_\nu}{\sqrt{N}} e^{-i(f, \nu)}, \quad v_{f\nu} = \frac{v_\nu}{\sqrt{N}} e^{-i(f, \nu)}. \tag{18.27}$$

Substituting Eq. (18.27) into Eq. (18.26), we obtain

$$\begin{aligned} \{E_\nu - S(\nu)\}\, u_\nu - R(\nu)\, v_\nu &= 0, \\ -R(\nu)\, u_\nu + \{-E_\nu - S(\nu)\}\, v_\nu &= 0, \end{aligned} \tag{18.28}$$

where

$$S(\nu) = \sum_{f'} S_{ff'} e^{i(\nu, f-f')}, \quad R(\nu) = \sum_{f'} R_{ff'} e^{i(\nu, f-f')}. \tag{18.29}$$

From the condition of the solubility of the system of equations (18.28), we find the spectrum of elementary excitations:

$$\begin{aligned} E_\nu &= \sqrt{S^2(\nu) - R^2(\nu)} = \\ &= \{SJ(q) + SK_2 - SJ(\nu)\}^{1/2} \left\{ SJ(q) - \frac{SJ(q+\nu) + SJ(q-\nu)}{2} \right\}^{1/2}. \end{aligned} \tag{18.30}$$

For small values of the wave vector ($\nu \rightarrow 0$), the expressions under the root signs can be expanded as series in powers of ν. Since $J(\nu)$ reaches its maximum value at $\nu = q$, we shall assume

$$J(q \pm \nu) \approx J(q) - \beta\nu^2, \quad \beta > 0. \tag{18.31}$$

Substituting Eq. (18.31) into Eq. (18.30), we obtain

$$\begin{aligned} E_\nu &\approx w\nu, \\ w &= S\sqrt{\beta\{J(q) + K_2 - J(0)\}}. \end{aligned} \tag{18.32}$$

Finally, using Eqs. (18.28) and the normalization condition of the functions (u,v), which, in this case, has the form

$$u_\nu^2 - v_\nu^2 = 1,$$

and bearing in mind that the quantities $S(\nu)$, $R(\nu)$ of Eq. (18.29) are real, we can easily show that

$$u_\nu^2 = \frac{S(\nu) + E_\nu}{2E_\nu}, \quad v_\nu^2 = \frac{S(\nu) - E_\nu}{2E_\nu}, \quad u_\nu v_\nu = \frac{R(\nu)}{2E_\nu}. \tag{18.33}$$

Thus, we have obtained the complete solution of the problem of the spin-wave spectrum of a simple helix.

Ferromagnetic Helix. Substances with helical structures of the rare-earth type have hexagonal packing. Instead of a real lattice, we shall consider a simpler lattice which will much facilitate the calculations. In this way, we lose the optical modes of the spectrum; however, since, in general, we shall consider only the weakly excited states, this will be of little importance. Moreover, we shall require the Hamiltonian of the system to have the necessary symmetry and spin ordering of the ferromagnetic-helix type (Fig. 6d). We shall assume that the hexagonal axis is the z-axis. The Hamiltonian of the system can then be written in the form [Kaplan (1961), Cooper et al. (1962)]

$$\begin{aligned}\mathscr{H} = &-\frac{1}{2}\sum I(f_1 - f_2)(S_{f_1}, S_{f_2}) - \\ &-\frac{1}{2}\sum I_a(f_1 - f_2) S^z_{f_1} S^z_{f_2} + \\ &+\sum \frac{K_4}{4!}(S^z_f)^4 + \sum \frac{K_6}{6!}(S^z_f)^6. \end{aligned} \tag{18.34}$$

The last term in the Hamiltonian allows for the anisotropy in the basal plane. To shorten the calculations, we shall neglect this term, assuming that $K_6 = 0$.

The classical form of $\mathscr{E}_0$, corresponding to the operator form of Eq. (18.34), is

$$\begin{aligned}\mathscr{E}_0 = &-\frac{1}{2}\sum S^2 I(f_1 - f_2)(\gamma_{f_1}, \gamma_{f_2}) - \\ &-\frac{1}{2}\sum S^2 I_a(f_1 - f_2)\gamma^z_{f_1}\gamma^z_{f_2} + \frac{1}{4!} S^4 K_4 \sum (\gamma^z_f)^4. \end{aligned} \tag{18.35}$$

The conditions for the minimum of the form (18.35), given by Eq. (18.11), will be written as follows:

$$\begin{aligned} -\sum_{f'} SI(f - f')\gamma^z_{f'} - \sum_{f'} SI_a(f - f')\gamma^z_{f'} + \\ + \frac{1}{3!} S^3 K_4 (\gamma^z_f)^3 - 2\lambda_f \gamma^z_f = 0, \end{aligned} \tag{18.36}$$

$$-\sum_{f'} SI(f - f')\gamma^{\pm}_{f'} - 2\lambda_f \gamma^{\pm}_f = 0,$$

where

$$\gamma^{\pm}_f = \gamma^x_f \pm i\gamma^y_f, \quad (\gamma^z_f)^2 + \gamma^-_f\gamma^+_f = 1. \tag{18.37}$$

We shall seek the solution of the system (18.36) in the form

$$\gamma^{\pm}_f = \rho e^{\pm i(q, f)}, \quad \gamma^z_f = \gamma_z = \text{const} \quad (\rho = \sqrt{1 - (\gamma_z)^2}); \tag{18.38}$$

in this solution, the parameter λ_f is independent of the number of the site:

$$-2\lambda = SJ(q) = SJ(0) + SJ_a(0) - \frac{1}{3!} S^3 K_4 \gamma_z^2. \tag{18.39}$$

This solution corresponds to spin ordering of the ferromagnetic-helix type.

Further solutions of the system (18.36) are

$$\gamma_z = 0, \quad \rho = 1, \quad -2\lambda = SJ(q)$$

and

$$\gamma_z = 1, \quad \rho = 0, \quad -2\lambda = SJ(0) + SJ_a(0) - \frac{1}{3!} S^3 K_4,$$

corresponding to ordering of the simple- and ferromagnetic-helix types, respectively. We shall now assume that $\mathscr{E}_0$ takes the lowest value given by the solution (18.38).

Substituting Eq. (18.38) into Eq. (18.35), we can easily show that the minimum of $\mathscr{E}_0$ is reached when

$$J(q) = \max. \tag{18.40}$$

If the maximum of J(q) is reached at $q \neq 0$, then the simple helical spin configuration is stable; if this maximum occurs at $q = 0$, the ferromagnetic configuration is stable. The pitch d of the helix is found from the condition $\gamma^{\pm}_{f+d} = \gamma^{\pm}_{f}$, and is equal to $2\pi/q$.

We shall now determine the spectrum of elementary excitations of the system. The quadratic form (18.12) for the operator (18.34) is

$$\begin{aligned}
\mathscr{H}_2 = &\sum\left\{-2\lambda + \frac{1}{4} SK_4 (S\gamma_z)^2 \rho^2\right\} n_f - \\
&-\frac{1}{2}\sum\left\{2SI(f_1 - f_2)(A_{f_1}, A_{f_2}) + \frac{1}{2}\rho^2 SI_a(f_1 - f_2)\right\} b_{f_1} b_{f_2} - \\
&-\sum\left\{2SI(f_1 - f_2)(A^*_{f_1}, A_{f_2}) + \frac{1}{2}\rho^2 SI_a(f_1 - f_2)\right\} b^+_{f_1} b_{f_2} - \\
&-\frac{1}{2}\sum\left\{2SI(f_1 - f_2)(A^*_{f_1}, A^*_{f_2}) + \frac{1}{2}\rho^2 SI_a(f_1 - f_2)\right\} b^+_{f_1} b^+_{f_2} + \\
&+\frac{1}{2}\sum \frac{SK_4}{4}(S\gamma_z)^2 \rho^2 \{b_f^2 + b_f^{+2}\}.
\end{aligned} \tag{18.41}$$

The scalar products (A_{f_1}, A_{f_2}), $(A^*_{f_1}, A_{f_2})$ in Eq. (18.41) are easily calculated using Eq. (5.10). Using Eqs. (5.10), (18.37), and (18.38), with

$$\varphi_f = (q, f) = \tan^{-1}\frac{\gamma_f^y}{\gamma_f^x}, \tag{18.42}$$

we find that

$$\begin{aligned} (A_{f_1}, A_{f_2}) &= \frac{\rho^2}{4}[1-\cos(q, f_1-f_2)], \\ (A^*_{f_1}, A_{f_2}) &= \frac{\rho^2}{4}[1-\cos(q, f_1-f_2)] + \\ &+ \frac{1}{2}[\cos(q, f_1-f_2) - i\gamma_z \sin(q, f_1-f_2)]. \end{aligned} \tag{18.43}$$

Substituting Eqs. (18.39) and (18.43) into Eq. (18.41), and using the standard notation, we obtain:

$$\mathcal{H}_2 = \frac{1}{2}\sum R^*_{f_1 f_2} b_{f_1} b_{f_2} + \sum S_{f_1 f_2} b^+_{f_1} b_{f_2} + \frac{1}{2}\sum R_{f_1 f_2} b^+_{f_1} b_{f_2}, \tag{18.44}$$

where

$$\begin{aligned} R_{f_1 f_2} &= \frac{1}{4} S K_4 (S\gamma_z)^2 \rho^2 \delta_{f_1, f_2} - \\ &\quad - \frac{1}{2}\rho^2 S\{I(f_1-f_2)[1-\cos(q, f_1-f_2)] + I_a(f_1-f_2)\}, \\ S_{f_1 f_2} &= \left\{SJ(q) + \frac{1}{4} S K_4 (S\gamma_z)^2 \rho^2\right\}\delta_{f_1, f_2} - \\ &\quad - \frac{1}{2}\rho^2 S\{I(f_1-f_2)[1-\cos(q, f_1-f_2)] + I_a(f_1-f_2)\} - \\ &\quad - SI(f_1-f_2)[\cos(q, f_1-f_2) - i\gamma_z \sin(q, f_1-f_2)]. \end{aligned} \tag{18.45}$$

The quadratic form (18.44) is diagonalized to the form of Eqs. (13.7)-(13.8),

$$\begin{aligned} \mathcal{H}_2 &= \Delta\mathcal{E}_0 + \sum E_\nu \xi^+_\nu \xi_\nu, \\ \Delta\mathcal{E}_0 &= -\sum E_\nu |v_{f\nu}|^2, \end{aligned} \tag{18.46}$$

by the canonical transformation of the variables given by Eq. (13.6):

$$b_f = \sum (u_{f\nu}\xi_\nu + v_{f\nu}\xi^+_\nu).$$

where ξ^+, ξ are the new Bose operators, E_ν and $u_{f\nu}$, $v_{f\nu}$ are the eigenvalues and eigenvectors of the system of equations

$$E_\nu u_{f\nu} = \sum_{f'} S_{ff'} u_{f'\nu} + \sum_{f'} R_{ff'} v_{f'\nu},$$
$$-E_\nu v_{f\nu} = \sum_{f'} S^*_{ff'} v_{f'\nu} + \sum_{f'} R^*_{ff'} u_{f'\nu}.$$

Since the quantities $S_{ff'}$, $R_{ff'}$ depend only on the difference between the coordinates of two sites, we shall seek the solution in the form

$$u_{f\nu} = \frac{u_\nu}{\sqrt{N}} e^{-i(f,\nu)}, \quad v_{f\nu} = \frac{v_\nu}{\sqrt{N}} e^{-i(f,\nu)}. \tag{18.47}$$

We have the following system of equations to determine u_ν, v_ν:

$$\begin{gathered}[E_\nu - S(\nu)]\, u_\nu - R(\nu)\, v_\nu = 0, \\ -R(-\nu)\, u_\nu - [E_\nu + S(-\nu)]\, v_\nu = 0,\end{gathered} \tag{18.48}$$

where

$$S(\nu) = \sum_{f'} S_{f-f'} e^{i(\nu, f-f')}, \quad R(\nu) = \sum_{f'} R_{f-f'} e^{i(\nu, f-f')}. \tag{18.49}$$

The normalization condition (13.4) for the functions u_ν, v_ν is

$$|u_\nu|^2 - |v_\nu|^2 = 1. \tag{18.50}$$

From Eqs. (18.48) and (18.50), we easily find

$$E_\nu = \frac{1}{2}\left\{S(\nu) - S(-\nu) \pm \sqrt{[S(\nu) + S(-\nu)]^2 - 4R(\nu)R(-\nu)}\right\}, \tag{18.51}$$

$$\begin{gathered}|u_\nu|^2 = \frac{[E_\nu + S(-\nu)]^2}{[E_\nu + S(-\nu)]^2 - R(\nu)R(-\nu)} = \\ = \frac{S(\nu) + S(-\nu) + \sqrt{[S(\nu) + S(-\nu)]^2 - 4R(\nu)R(-\nu)}}{2\sqrt{[S(\nu) + S(-\nu)]^2 - 4R(\nu)R(-\nu)}}, \\ |v_\nu|^2 = 1 - |u_\nu|^2, \\ |u_\nu v_\nu| = \frac{|R(\nu)|}{\sqrt{[S(\nu) + S(-\nu)]^2 - 4R(\nu)R(-\nu)}}.\end{gathered} \tag{18.52}$$

where, according to Eqs. (18.45) and (18.49),

$$\begin{aligned}R(\nu) &= \frac{1}{4} SK_4 (S\gamma_z)^2 \rho^2 - \\ &\quad - \frac{1}{2}\rho^2 S\left[J(\nu) + J_a(\nu) - \frac{J(\nu+q)+J(\nu-q)}{2}\right],\\ S(\nu) &= SJ(q) + \frac{1}{4} SK_4 (S\gamma_z)^2 \rho^2 - \\ &\quad - \frac{1}{2}\rho^2 S\left[J(\nu) + J_a(\nu) - \frac{J(\nu+q)+J(\nu-q)}{2}\right] - \\ &\quad - S\left[\frac{J(\nu+q)+J(\nu-q)}{2} - \gamma_z \frac{J(\nu+q)-J(\nu-q)}{2}\right].\end{aligned} \tag{18.53}$$

Substituting Eq. (18.53) into Eq. (18.51), we obtain the final expression for the spectrum of elementary excitations of a spin system having the ferromagnetic-helix ordering:

$$\begin{aligned}E_\nu &= \frac{1}{2} S\gamma_z [J(q+\nu) - J(q-\nu)] \pm \\ &\quad \pm \left\{SJ(q) + 2R(\nu) - S\frac{J(q+\nu)+J(q-\nu)}{2}\right\}^{1/2} \times \\ &\quad \times \left\{SJ(q) - S\frac{J(q+\nu)+J(q-\nu)}{2}\right\}^{1/2}.\end{aligned} \tag{18.54}$$

An approximate expression for the spectrum is obtained by expanding the quantities on the right-hand side of Eq. (18.54) as a series in powers of ν, using terms up to ν^2, inclusive. Since $J(\nu)$ has a maximum at $\nu = q$, we obtain

$$\begin{aligned}E_\nu &\approx w\nu,\\ w &= S^2\gamma_z\rho \sqrt{\frac{1}{3}\beta K_4} \quad (\rho = \sqrt{1-\gamma_z^2}).\end{aligned} \tag{18.55}$$

where β is defined in Eq. (18.31). In accordance with the remarks made in Sec. 13 about the selection of the sign of E_ν, we must take the plus sign in front of the root in Eqs. (18.51) and (18.54).

The calculation method described here can be applied to find elementary excitations for other types of spin ordering, and to allow for the influence of an external magnetic field.

Sec. 19. Ferromagnetic Resonance and Other Applications of the ASQ Method

Elementary Theory of Ferromagnetic Resonance.* Under certain conditions, an alternating magnetic field can excite normal-mode oscillations of a spin system (spin waves); this is known as ferromagnetic resonance. The resonance is accompanied by the absorption of the alternating-field energy by the spin system. Since, usually, only some of the normal modes are excited, the absorption has a more or less sharp maximum at the alternating-field frequency equal to the energy of the excited oscillations.† This frequency is known as the resonance frequency.

Since the normal modes interact, the energy of the oscillations excited by the external field is redistributed between all the other oscillations. Consequently, the resonance maxima are broadened to a lesser or greater extent.

To observe ferromagnetic resonance, a sample is placed in crossed steady (H) and alternating [h(t)] magnetic fields. The classical analog of the excitation of spin waves is the excitation of the forced precession of spins about the direction of a steady field. The resonance is observed when the frequency of the alternating field becomes equal to the free-precession frequency. If the sample is isotropic, the resonance is possible only if $H \neq 0$. If the sample is anisotropic, we can also have resonance when $H = 0$. In the latter case, we have "intrinsic" ferromagnetic resonance; the magnetocrystalline anisotropy takes on the role of the steady field. We shall consider which oscillation modes can be excited in spin systems. The alternating field will be assumed to be uniform over the whole sample.

We shall write the operator of the energy of interaction between the spin system and the alternating field in the form

$$\mathcal{V}(t) = -\sum_f \mu_f (S_f, h(t)). \tag{19.1}$$

*For a more general treatment, see Secs. 31, 38, and 39.

†We shall use a system of units in which $\hbar = 1$, so that the frequency and energy have the same dimensions.

Following the general procedure of the ASQ method, we shall use Eq. (5.8) to obtain new spin operators:

$$\mathscr{V}(t) = -\sum_f \mu_f (\gamma_f, h(t)) S_f^z - \\ -\sum \mu_f \{(A_f, h(t)) S_f^+ + (A_f^*, h(t)) S_f^-\}, \qquad (19.2)$$

and then we shall use Eq. (12.19) to go over from the spin operators to the Bose operators. Consequently, we shall obtain

$$\mathscr{V}(t) = -\sum_f \mu_f S_f (\gamma_f, h(t)) + \sum_f \mu_f (\gamma_f, h(t)) n_f - \\ -\sum_f \mu_f \sqrt{2S_f} \{(A_f, h(t)) b_f + (A_f^*, h(t)) b_f^+\}. \qquad (19.3)$$

We shall use Eq. (13.6) to transform from the variables b_f^+, b_f to the variables ξ_ν^+, ξ_ν in terms of which the main part of the Hamiltonian (the quadratic form) is diagonal. Consequently, we obtain an expression for $\mathscr{V}$ (t) in normal coordinates:

$$\mathscr{V}(t) = -\sum_f \mu_f (\gamma_f, h(t)) \Big(S_f - \sum_\nu |v_{f\nu}|^2\Big) + \\ + \sum \mu_f (\gamma_f, h(t)) \{(u_{f\nu_1}^* u_{f\nu_2} + v_{f\nu_1}^* v_{f\nu_2}) \xi_{\nu_1}^+ \xi_{\nu_2} + \\ + v_{f\nu_1} u_{f\nu_2} \xi_{\nu_1} \xi_{\nu_2} + u_{f\nu_1}^* v_{f\nu_2}^* \xi_{\nu_1}^+ \xi_{\nu_2}^+\} - \\ - \sum \mu_f \sqrt{2S_f} \{(A_f, h(t)) + (A_f^*, h(t))\} \times \\ \times \{(u_{f\nu} + v_{f\nu}) \xi_\nu + (u_{f\nu}^* + v_{f\nu}^*) \xi_\nu^+\}. \qquad (19.4)$$

It is obvious that the alternating field h(t) will first of all excite the normal modes with those values of ν which are present in $\mathscr{V}$ (t).

The first sum in Eq. (18.34) describes the change in the ground-state energy. It does not lead directly to the excitation of oscillations in the system, and it is usually ignored in considering ferromagnetic resonance. We shall omit this term in $\mathscr{V}$ (t).

The second sum represents processes in which two spin waves take part. The first term in this sum describes inelastic scattering, while the other two terms describe processes involving the annihilation or creation of two spin waves.

The third sum represents processes involving the creation or annihilation of one spin wave. Usually, only these processes are considered in the theory of ferromagnetic resonance.

We shall now consider two examples of ferromagnetic resonance.

For simplicity, we shall assume that a lattice consists of N equivalent sites, occupied by atoms of the same kind. Consequently, we shall further assume that

$$S_f = S, \quad \mu_f = \mu.$$

The field h(t) will be taken to be periodic:

$$h(t) = he^{i\omega t} + h^* e^{-i\omega t}.$$

We shall see later that this restriction is unimportant.

Isotropic Ferromagnet. Let the steady field be directed along the z-axis. Then, according to Eqs. (14.7) and (15.4), we have

$$\begin{gathered}\gamma_f^z = 1, \ \gamma_f^x = \gamma_f^y = 0; \ A_f^x = \frac{1}{2}, \quad A_f^y = -\frac{i}{2}, \quad A_f^z = 0; \\ u_{f\nu} = N^{-1/2} \exp i(f, \nu), \ v_{f\nu} = 0.\end{gathered} \tag{19.5}$$

If the field h(t) is directed along the x-axis, then

$$\begin{aligned}\mathscr{V}(t) &= -\mu \frac{\sqrt{2S}}{\sqrt{N}} \sum_{f,\nu} (e^{i(f,\nu)}\xi_\nu + e^{-i(f,\nu)}\xi_\nu^+)\, h(t) = \\ &= -\mu \sqrt{2SN}\,(\xi_0 + \xi_0^+)\, h(t).\end{aligned} \tag{19.6}$$

Hence, we see that only the spin waves with $\nu = 0$ can be excited. The probability of a transition in the spin system under the action of the perturbation given by Eq. (19.6) contains a δ-function of the energies,

$$\delta(\omega - E_0),$$

where ω is the frequency of the external field, E_0 is the energy of the spin wave with $\nu = 0$. Consequently, the resonance is possible if the external-field frequency is

$$\omega = E_0 = \mu H. \tag{19.7}$$

We can easily see that this is the frequency of the uniform precession of spin moments in the field H. When H = 0, the resonance frequency also vanishes, i.e., resonance is impossible.

If the field h(t) is directed along the z-axis, then

$$\mathcal{V}(t) = \frac{\mu}{N} \sum_{f,\nu} e^{-i(f,\nu-\nu')} \xi_\nu^+ \xi_{\nu'} h(t) = \mu \sum_\nu n_\nu h(t). \qquad (19.8)$$

Since the perturbation energy operator (19.8) is, in this approximation, diagonal with respect to the occupation numbers of the spin waves, it does not give rise to any transitions involving a change in the number of spin waves, i.e., resonance is impossible.

Simple Helix. We shall now consider intrinsic resonance. According to Eqs. (18.18), (18.21), and (18.27), we have, in this case

$$\begin{aligned} &\gamma_f^x = \cos(q, f), \qquad \gamma_f^y = \sin(q, f), \quad \gamma_f^z = 0; \\ &A_f^x = -\frac{i}{2}\sin(q, f), \qquad A_f^y = \frac{i}{2}\cos(q, f), \quad A_f^z = \frac{1}{2}; \\ &u_{f\nu} = \frac{u_\nu}{\sqrt{N}} \exp(-i(f, \nu)), \quad v_{f\nu} = \frac{v_\nu}{\sqrt{N}} \exp(-i(f, \nu)). \end{aligned} \qquad (19.19)$$

Let the field h(t) be directed along the anisotropy axis:

$$h_\alpha(t) = h(t)\,\delta_{\alpha, z}. \qquad (19.10)$$

Substituting Eqs. (19.9) and (19.10) into Eq. (19.4), we obtain, after making some simple calculations,

$$\mathcal{V}(t) = -\frac{\mu}{2}\sqrt{2SN}\,(u_0 + v_0)(\xi_0 + \xi_0^+)\,h(t). \qquad (19.11)$$

Consequently, spin waves with $\nu = 0$ can be excited. However, since, according to Eq. (18.32), $E_0 = 0$, it follows from the law of conservation of energy,

$$\omega - E_0 = 0$$

that the resonance frequency is zero, i.e., that resonance is impossible in a longitudinal alternating field.

Let h(t) be directed along the x-axis:

$$h_\alpha(t) = h(t)\,\delta_{\alpha, x}. \qquad (19.12)$$

In this case, the substitution of Eqs. (19.9) and (19.12) into Eq. (19.4) gives

$$\begin{aligned}\mathcal{V}(t) = {} & \mu h(t) \sum_{\nu_1, \nu_2} \{\Delta(q + \nu_1 - \nu_2) + \Delta(-q + \nu_1 - \nu_2)\} \times \\ & \times \left(u^*_{\nu_1} u_{\nu_2} + v^*_{\nu_1} v_{\nu_2}\right) \xi^+_{\nu_1} \xi_{\nu_2} + \\ & + \mu h(t) \sum_{\nu_1, \nu_2} \{\Delta(q + \nu_1 + \nu_2) + \Delta(-q + \nu_1 + \nu_2)\} \times \\ & \times \left(u^*_{\nu_1} v^*_{\nu_2} \xi^+_{\nu_1} \xi^+_{\nu_2} + u_{\nu_1} v_{\nu_2} \xi_{\nu_1} \xi_{\nu_2}\right) - \\ & - \frac{1}{4} \mu h(t) \sqrt{2SN} \left\{(u_q - v_q)(\xi_{-q} - \xi_q) + \left(u^*_q - v^*_q\right)\left(\xi^+_{-q} - \xi^+_q\right)\right\}.\end{aligned} \tag{19.13}$$

The first term describes the inelastic scattering of spin waves. Such scattering is possible if the laws of conservation of momentum and energy are satisfied simultaneously:

$$\pm q + \nu_1 - \nu_2 = 0, \quad \omega - E_{\nu_1} + E_{\nu_2} = 0. \tag{19.14}$$

The second term describes processes involving the simultaneous creation or annihilation of two spin waves. In this case, the laws of conservation are

$$\pm q + \nu_1 + \nu_2 = 0, \quad \pm \omega + E_{\nu_1} + E_{\nu_2} = 0. \tag{19.15}$$

Finally, the third term describes processes involving the creation or annihilation of one spin wave with the wave vector ν = q. These processes are possible provided the following condition is satisfied:

$$\pm \omega + E_{\pm q} = 0. \tag{19.16}$$

The processes described by Eqs. (19.14) and (19.15) do not have a sharp maximum at any definite frequency ω. On the other hand, the absorption in the processes described by Eq. (19.16) occurs at a fixed frequency, corresponding to the excitation of spin waves having a wave vector equal to the reciprocal of the pitch of the helix.

The case q = 0 corresponds to resonance when the spins have ferromagnetic ordering and the case $q = \pi/a$ corresponds to resonance when the spins have antiferromagnetic ordering.

Further Applications of the ASQ Method. We shall now consider some of the applications of the approximate second quantization (ASQ) method in order to get some idea of its possibilities. Such a review of applications is useful also because, in some cases, the solution in the spin-wave approximation can be considered as a sufficiently satisfactory zeroth approximation in investigations of the problem by other methods – for example, by the Green's function method (cf., Chap. VIII). The problems of transport processes in strongly magnetic substances will not be considered here. They are sufficiently fully dealt with in the reviews of Van Kranendonk and Van Vleck (1958), Akhiezer, Bar'yakhtar, and Kaganov (1960a, b), Borovik–Romanov (1962), Pakhomov and Smol'kov (1962), in the collection "Ferromagnetic Resonance" (1961), and in the monographs of Gurevich (1960) and Smit and Wijn (1962).

Uniaxial Ferromagnets. Tyablikov (1950) and Tyablikov and Siklos (1959) calculated the temperature and field dependences of the magnetization of a uniaxial ferromagnet for those cases when the external field was applied along or at right angles to the anisotropy axis, and they determined the temperature and field dependences of the first anisotropy constant. They considered the case of spin $S = 1/2$.

Turov and Irkhin (1958) have shown that in calculations of the effects of temperature in the vicinity of the critical values of the field, the equilibrium magnetization vector should be determined from the conditions for the free-energy minimum, and not for the minimum of the ground-state energy. This is because, in the latter case, the error due to an inexact determination of the critical field is greater than the temperature effects being calculated.

Potapkov (1958) determined the temperature and field dependences of the first anisotropy constant for arbitrary spin, using Dyson's spin-wave model.

In strong external fields $H \gg H_a$ (H_a is the effective anisotropy field), the approximate expression for the first anisotropy constant has the form

$$\begin{gathered} K_1 = \frac{1}{2} I_a S^2 \left\{ 1 - 3vS^{-1} Z_{3/2}\left(\frac{\mu H}{\vartheta}\right)\left(\frac{\vartheta}{4\pi\alpha}\right)^{3/2} \right\}, \\ I_a \ll \vartheta \ll I, \quad I_a = \frac{1}{N} \sum_{f_1, f_2} I_a(f_1, f_2), \end{gathered} \tag{19.17}$$

where I_a is determined in accordance with Eq. (7.9), and α represents the coefficient of ν^2 in Eq. (15.13).

Anisotropy of Cubic Crystals. Calculations of the first anisotropy constant for the Hamiltonian of Eq. (7.10) have been carried out, in the usual spin-wave approximation, by Tyablikov and Gusev (1956) for $S = \frac{1}{2}$ and by Pal (1954) for $S \geq \frac{1}{2}$ and for complete saturation. Potapkov (1957) calculated the first constant for spin $S \geq \frac{1}{2}$ in Dyson's spin-wave model. The results of these calculations give the following expression for the first anisotropy constant:

$$K_1 = \frac{2}{4!} I_4 S^4 \left\{ 1 - 10 v S^{-1} Z_{3/2}\left(\frac{\mu H}{\vartheta}\right)\left(\frac{\vartheta}{4\pi\alpha}\right)^{3/2} \right\},$$
$$I_4 = \frac{1}{N} \sum_{f_1, \ldots, f_4} I_4(f_1, f_2, f_3, f_4), \qquad (19.18)$$

where I_4 is determined in accordance with Eq. (7.11).

We must remember that the expansions of the anisotropy constants in powers of temperature, of the type given by Eqs. (19.17) and (19.18), are, in general, meaningful only if $\vartheta \gg I_a, I_4$.

The temperature and field dependences of the anisotropy constants, of the type given by Eqs. (19.17) and (19.18), are in agreement with the experimental data of Puzei (1957, 1963).

Ferromagnets with Complex Lattices (ordered binary alloys) were considered by Kondorskii and Pakhomov (1953). They determined the spectrum of elementary excitations and the temperature dependence of the magnetization.

Ferrimagnets. Isotropic ferrimagnets have been treated using the ASQ method by Vonsovskii and Seidov (1954), Kondorskii et al. (1956), Siklos (1957), and Tyablikov (1956b, 1959a). Depending on the assumptions made, temperature dependences of the $\vartheta^{3/2}$ type and ϑ^2 type were obtained for the spontaneous magnetization. *

The expression for the magnetization of a two-sublattice ferrimagnet has the following form in the zeroth approximation:

*Vonsovskii and Seidov's treatment (1954) should not be interpreted literally but as formulated in Tyablikov's paper (1959a).

$$M_0=\begin{cases} M_2-M_1, & 0\leq H\leq H_1; \\ \sqrt{M_1M_2}\,\frac{H}{H_e}, & H_1\leq H\leq H_2; \\ M_2+M_1, & H_2\leq H, \end{cases} \tag{19.19}$$

where M_1, M_2 are the total moments of the first and second sublattices:

$$H_1=H_e\frac{M_2-M_1}{\sqrt{M_1M_2}},\quad H_2=H_e\frac{M_2+M_1}{\sqrt{M_1M_2}},\quad H_e=\left(\frac{J_{12}(0)\,J_{21}(0)}{4\mu_1\mu_2}\right)^{1/2},$$
$$J_{ij}(0)=\sum_{h_j}I(h_i-h_j) \tag{19.20}$$

or

$$H_e=\left(\frac{K^2}{4M_1M_2}\right)^{1/2},\quad K=\sum_{f,g}I(f-g).$$

The relative distributions of the sublattice moments and the field dependence of the total magnetization are shown schematically in Figs. 4 and 5.

Yakovlev (1957, 1958) considered the ground state of a uniaxial anisotropic ferrimagnet with two sublattices, the dependence of the magnetization on an external field, and temperature corrections to this dependence.

In general, the magnetic structure of ferrimagnets is complex, and only sometimes can we consider it as an assembly of two sublattices superimposed on each other. Yafet and Kittel (1952) considered the possible types of structure using the molecular field approximation* and neglecting the magnetic anisotropy. They showed that, depending on the assumptions about the molecular field constants, each of the sublattices f and g may in turn be split into two or more sublattices.

At 0°K, the sublattice magnetization is maximal and the molecular field approximation reduces to the quasi-classical approximation. Therefore, the results obtained in the molecular field approximation and extrapolated to zero temperature, can be

*For the molecular field approximation, see Secs. 1 and 21.

used as the zeroth approximation in solving the problem by the ASQ method.

The ground state of a three-sublattice ferrimagnet was considered by Gusev and Pakhomov (1961). They showed that, when certain assumptions are made about the signs of the exchange integrals, the collinear spin configurations should be stable. They obtained an expression for the total magnetic moment which gave better agreement with the experimental results than did the empirical formula of Néel [see Belov and Zaitseva (1958)].

The spin-wave spectrum was investigated in spinel ferrites by Kaplan (1958) and in yttrium ferrite garnets by Meyer and Harris (1962). As expected, one of the modes of the spectrum represented the "acoustical" vibrations, and the other modes were "optical."

Anisotropic Antiferromagnets. The magnetic and thermodynamic properties of antiferromagnets depend strongly on their magnetic anisotropy. The anisotropy has the strongest effect when the energy of a crystal in an external field is less than, or of the order of, the anisotropy energy. From general considerations, it follows that two critical fields should exist for uniaxial antiferromagnets: an external field which suppresses the anisotropy, and another field in which the antiferromagnetic ordering is transformed into the ferromagnetic configuration.

The low-temperature properties of antiferromagnets have been considered many times in the spin-wave approximation. The magnetic anisotropy was allowed for either as some effective anisotropy field of the type given by Eq. (7.13), or by adding a term of the "exchange" anisotropy type (7.9) or of the one-ion anisotropy type (7.11) to a pure exchange Hamiltonian.

Anderson (1950b, 1952) and Tessman (1952) used a semiclassical method and allowed for the one-ion anisotropy; Nakamura (1952) used the Holstein–Primakoff representation of Eq. (5.16) and allowed for the one-ion anisotropy of the (7.11) type; Ziman (1952a, b) and Kubo (1953) also used the Holstein–Primakoff representation and introduced the anisotropy as an effective field of the type given by Eq. (7.13). A detailed analysis of the main assumptions in the spin-wave approximation was carried out by

Marshall (1955a, b), using an antiferromagnet model with the one-ion uniaxial anisotropy. Marshall was the first to formulate clearly the difficulties of the spin-wave theory due to the appearance of nonphysical states. A shortcoming of these investigations was the consideration of only weak fields ($H < H_a$), in which no spin rearrangement (reversal) occurs.

The case of arbitrary external fields was considered by Tyablikov and Amatuni (1956), and Amatuni (1956, 1957), the anisotropy being introduced as an effective field. The energy of the ground state, the magnetization, and the susceptibility were calculated, and temperature corrections found. The dependence of the magnetization on an external field, for various orientations of the latter with respect to the anisotropy axis, was, in general, found to agree with the experimental data of van den Handel et al. (1952). However, in arbitrary fields, a transition of the second kind was predicted for spin reversal (in the first critical field), which was not in agreement with the experimental data. The introduction of the exchange-type anisotropy allowed Amatuni (1958) to remove this difficulty. The spectrum of elementary excitations and the thermodynamic properties of a system with a Hamiltonian of this type were considered by Rudoi (1963a, b).

A spin-wave treatment of cubic antiferromagnets with fcc and bcc lattices was given by ter Haar and Lines (1962), and by Lines (1963). They allowed for the rhombohedral anisotropy and considered the ground-state energy, the spectrum of elementary excitations, and the sublattice magnetization for the case $H = 0$.

Magnetostriction of Ferromagnets. The magnetostriction of hexagonal crystals with the rhombohedral anisotropy was considered in the spin-wave approximation by Gusev (1955), who was thus able to determine the magnetostriction constants as functions of temperature and an external magnetic field. Such dependences were predicted earlier by Akulov (1939).

The magnetostriction constants of ferromagnets can be positive or negative. This feature of the quantum theory was first pointed out by Vonsovskii (1940b).

Magnetic Substances with Helical Structures. Kaplan (1961) and Cooper et al. (1962) showed that for such substances one can find spin Hamiltonians having the required symmetry, and that one can investigate the possible types of ordering observed experimentally. They also considered ferromagnetic resonance in such structures and showed that the resonance conditions were quite different from those for the collinear structures.

Nagamiya (1962) and Nagamiya et al. (1962a) investigated changes in the spin distribution in a simple helix under the action of an external field applied in the basal plane. It was found that in weak fields the spin orientation altered slightly and the resultant magnetic moment was directed along the external field. At the first critical field H_c, a transition of the first kind occurred with the spin configuration transforming into a fan-shaped structure; a large resultant moment was produced by such a transition. A further increase in the field led to a continuous decrease in the angle of the fan. At the second critical field H_f ($\approx 2H_c$), the angle of the fan became zero; at the same time a transition of the second kind took place.

Lyons (1963) investigated helical structures in the molecular-field approximation [see also Kaplan's paper (1961)]; his results can be used partly in calculating the ground state by the ASQ method, if the magnetization at a given temperature is replaced with the magnetization at 0°K.

Cooper and Elliot (1963) and Cooper et al. (1962) investigated, in the spin-wave approximation, the conditions for ferromagnetic resonance in simple and ferromagnetic helices; they allowed for the influence of an external field. Bar'yakhtar et al. (1964a) calculated, again in the spin-wave approximation, the susceptibility tensor for ferromagnetic resonance in the ferromagnetic helix.

Within the framework of the s – f exchange model, Miwa (1963) showed that a helical ordering of the f-electrons give rise to discontinuities in the spectrum of the conduction electrons. Savchenko and Bar'yakhtar (1963) used the same model to investigate relaxation processes.

Chapter VI

Molecular Field Method and Perturbation Theory

The molecular field approximation is used in the theory of magnetism at high temperatures. The exchange interaction Hamiltonian is replaced, on the basis of some obvious physical considerations, by a Hamiltonian representing the interaction of spins with some effective field (molecular field), which is proportional to the magnetization of the substance. We shall use the theorems relating to the minimum values of the free energy, and thereby we shall formulate a principle which will then be used to deduce the molecular field equations. In concluding this chapter, we shall consider the simplest variant of the perturbation theory for high temperatures.

Sec. 20. Principle of the Free Energy Minimum

Theorem 1.* If $\{\varphi_n\}$ is an arbitrary complete system of orthonormalized functions, which are not the eigenfunctions of the Hamiltonian $\mathscr{H}$, of a system then

$$F(\mathscr{H}) \leq F_{\mathrm{mod}}(\mathscr{H}), \tag{20.1}$$

where F $(\mathscr{H})$ is the intrinsic free energy of the system,

$$F(\mathscr{H}) = -\vartheta \ln Q, \quad Q = \sum_{\nu} e^{-\frac{\mathscr{E}_\nu}{\vartheta}}, \tag{20.2}$$

*See Peierls (1938).

$\mathscr{E}_\nu$ are eigenfunctions of the Hamiltonian $\mathscr{H}$, $F_{\text{mod}}(\mathscr{H})$ is the "model" free energy, which gives approximately the upper limit of the intrinsic free energy:

$$F_{\text{mod}}(\mathscr{H}) = -\vartheta \ln Q_{\text{mod}}, \quad Q_{\text{mod}} = \sum_n e^{-\frac{1}{\vartheta}\mathscr{H}_{nn}},$$
$$\mathscr{H}_{nn} = (\varphi_n^*, \mathscr{H}\varphi_n). \tag{20.3}$$

The inequality (20.1) may also be written in the following way:

$$Q \geq Q_{\text{mod}}. \tag{20.4}$$

The relationships represented by the equality sign in Eqs. (20.1) and (20.4) applies if φ_n are eigenfunctions of the Hamiltonian of the system.

Proof (by Contradiction). We shall consider a function satisfying the conditions

$$\frac{df}{d\xi} < 0, \quad \frac{d^2 f}{d\xi^2} > 0, \tag{20.5}$$

and we shall show that, if

$$Q = \sum_n f(\mathscr{E}_n), \quad Q_{\text{mod}} = \sum_n f(\mathscr{H}_{nn}), \tag{20.6}$$

then the inequality (20.4) applies. In particular, the conditions (20.5) are satisfied by the function $f(\xi) = \exp(-\xi/\vartheta)$.

We shall number the quantities $\mathscr{E}_n$ and $\mathscr{H}_{nn}$ in their increasing order:

$$\mathscr{E}_n \leq \mathscr{E}_m, \quad \mathscr{H}_{nn} \leq \mathscr{H}_{mm}, \quad \text{if} \quad n < m.$$

We shall introduce partial sums*

$$Q^{(N)} = \sum_{n=1}^{N} f(\mathscr{E}_n), \quad Q_{\text{mod}}^{(N)} = N \sum_{n=1}^{N} f(\mathscr{H}_{nn}) \tag{20.7}$$

---

*Since, on the strength of the conditions (20.5), the function $f(\xi)$ decreases monotonically as the argument increases, the sums in Eq. (20.7) have upper limits:

$$Q^{(N)} \leq Nf(\mathscr{E}_1), \quad Q_{\text{mod}}^{(N)} \leq fN(\mathscr{H}_{11}).$$

and we shall show first that the inequality (20.4) is satisfied by them, and then we shall go over to the limit $N \to \infty$.

We shall assume that the functions $\varphi_1, \ldots, \varphi_N$ are not the eigenfunctions of the Hamiltonian of the system, and that the sum $Q^{(N)}_{\text{mod}}$ of Eq. (20.7) has its maximum value when these functions are used. Then, we have at least one nondiagonal element which is not equal to zero:

$$\mathcal{H}_{n_0m_0} \neq 0, \quad m_0 \leq N, \quad n_0 \neq m_0. \tag{20.8}$$

We shall consider the three possible cases:

$$\begin{aligned} &\text{a)} \quad n_0 > N; \\ &\text{b)} \quad n_0 \leq N, \quad \mathcal{H}_{m_0m_0} \neq \mathcal{H}_{n_0n_0}; \\ &\text{c)} \quad n_0 \leq N, \quad \mathcal{H}_{m_0m_0} = \mathcal{H}_{n_0n_0}. \end{aligned} \tag{20.9}$$

a) We shall introduce new functions:

$$\varphi'_n = \varphi_n + \varepsilon\varphi_{n_0}\Delta(m_0 - n), \tag{20.10}$$

where ε is a small parameter ($|\varepsilon| \ll 1$). Then, to within terms of the order of ε, inclusive, we have

$$\mathcal{H}'_{m_0m_0} = (\varphi'_{m_0}, \mathcal{H}\varphi'_{m_0}) = \mathcal{H}_{m_0m_0} + (\varepsilon\mathcal{H}_{m_0n_0} + \varepsilon^*\mathcal{H}_{n_0m_0}). \tag{20.11}$$

Consequently, the increment in $Q^{(N)}_{\text{mod}}$ is

$$Q'^{(N)}_{\text{mod}} - Q^{(N)}_{\text{mod}} = \left(\frac{df}{d\xi}\right)_{\xi = \mathcal{H}_{m_0m_0}} (\varepsilon\mathcal{H}_{m_0n_0} + \varepsilon^*\mathcal{H}_{n_0m_0}). \tag{20.12}$$

By selecting the appropriate sign for ε, the right-hand side of Eq. (20.12) can be made positive and, therefore, $Q'^{(N)}_{\text{mod}} > Q^{(N)}_{\text{mod}}$. Consequently, the assumption that $Q^{(N)}_{\text{mod}}$ has its maximum value for the functions $\varphi_1, \ldots, \varphi_N$ is not true.

b) We shall introduce new functions:

$$\varphi'_n = \varphi_n + \varepsilon\varphi_{n_0}\Delta(m_0 - n) - \varepsilon\varphi_{m_0}\Delta(n_0 - n). \tag{20.13}$$

Then, to within terms of the order of ε, inclusive, we have

$$\mathcal{H}'_{m_0m_0} = \mathcal{H}_{m_0m_0} + (\varepsilon\mathcal{H}_{m_0n_0} + \varepsilon^*\mathcal{H}_{n_0m_0}),$$
$$\mathcal{H}'_{n_0n_0} = \mathcal{H}_{n_0n_0} - (\varepsilon\mathcal{H}_{m_0n_0} + \varepsilon^*\mathcal{H}_{n_0m_0}), \tag{20.14}$$

and the increment in $Q^{(N)}_{\text{mod}}$ is

$$Q'^{(N)}_{\text{mod}} - Q^{(N)}_{\text{mod}} = $$
$$= \left\{\left(\frac{df}{d\xi}\right)_{\xi=\mathcal{H}_{m_0m_0}} - \left(\frac{df}{d\xi}\right)_{\xi=\mathcal{H}_{n_0n_0}}\right\}(\varepsilon\mathcal{H}_{m_0n_0} + \varepsilon^*\mathcal{H}_{n_0m_0}). \tag{20.15}$$

Since, by definition, the function f is monotonic and $\mathcal{H}_{m_0m_0} \neq \mathcal{H}_{n_0n_0}$, the expression in braces (curly brackets) in Eq. (20.15) is not equal to zero, and the right-hand side of Eq. (20.15) can be made positive by a suitable selection of ε; this again contradicts the initial assumption.

c) In this case, $\mathcal{H}_{m_0m_0} = \mathcal{H}_{n_0n_0}$ and we must include terms up to ε^2, inclusive. Bearing in mind the normalization condition for trial functions, we shall assume

$$\varphi'_n = \begin{cases} \varphi_n, & n \neq m_0,\ n_0; \\ \dfrac{\varphi_{m_0} + \varepsilon\varphi_{n_0}}{\sqrt{1+|\varepsilon|^2}}, & n = m_0; \\ \dfrac{\varphi_{n_0} - \overset{*}{\varepsilon}\varphi_{m_0}}{\sqrt{1+|\varepsilon|^2}}, & n = n_0. \end{cases} \tag{20.16}$$

We can easily see that the increment in $Q^{(N)}_{\text{mod}}$,

$$Q'^{(N)}_{\text{mod}} - Q^{(N)}_{\text{mod}} = \left(\frac{d^2f}{d\xi^2}\right)_{\xi=\mathcal{H}_{m_0m_0}} (\varepsilon\mathcal{H}_{m_0n_0} + \varepsilon^*\mathcal{H}_{n_0m_0})^2, \tag{20.17}$$

is always positive, contradicting the initial assumption.

Consequently, for finite values of N, the quantity $Q^{(N)}_{\text{mod}}$ does not reach its maximum for any system of functions $\varphi_1, \ldots, \varphi_N$ which are not the eigenfunctions of the Hamiltonian of the system. Since the partial sums are bounded from above, we always have

$$Q^{(N)} > Q^{(N)}_{\text{mod}}. \tag{20.18}$$

Let us assume that, as $N \to \infty$, the sums in Eq. (20.7) converge uniformly to the values of Eq. (20.6). The inequality (20.4) applies at the limiting values given by Eq. (20.6). In fact, if Q_{mod} were to exceed Q, one could always find a number N such that $Q^{(N)}_{mod} > Q^{(N)}$, but this is impossible, as already proved. This completes the proof of Theorem 1.

Using Theorem 1, we can formulate a variational principle for the approximate determination of the free energy of a system.

Let the functions $\{\varphi_n\}$ depend on some arbitrary parameter σ. Then, the model free energy will also depend on σ. Since, in accordance with Theorem 1,

$$F \leq F_{mod} = -\vartheta \ln \sum_n e^{-\frac{1}{\vartheta}\mathcal{H}_{nn}(\sigma)}, \qquad (20.19)$$

the best approximation for the upper limit of the free energy F is obtained by selecting the values of the parameter σ in accordance with the condition for the minimum of the model free energy F_{mod}.

Let the Hamiltonian of the system, $\mathcal{H}$, be written in the form

$$\mathcal{H} = \mathcal{H}_0(\sigma) + \mathcal{H}'(\sigma) \equiv \mathcal{H}_0(\sigma) + [\mathcal{H} - \mathcal{H}_0(\sigma)], \qquad (20.20)$$

where $\mathcal{H}_0(\sigma)$ is some operator depending on the parameter σ. The form of the operator $\mathcal{H}_0(\sigma)$ will be selected on the basis of convenience in calculations. We shall use $\mathscr{E}_n^0$ and φ_n to denote the eigenvalues and the eigenfunctions of the operator $\mathcal{H}_0$ and we shall use $\mathcal{H}'_{nn}$ to denote the diagonal matrix elements of the operator $\mathcal{H}'$ in terms of the functions φ_n. We shall assume that φ_n are not the eigenfunctions of the total Hamiltonian $\mathcal{H}$. Clearly, $\mathscr{E}_n^0$ and $\mathcal{H}'_{nn}$ are also some functions of the parameter σ. We shall use the system of functions $\{\varphi_n\}$ as a trial system of functions for Theorem 1. Then,

$$\mathcal{H}_{nn} = \mathscr{E}_n^0 + \mathcal{H}'_{nn} \equiv \mathscr{E}_n^0 + [\mathcal{H}_{nn} - \mathscr{E}_n^0]. \qquad (20.21)$$

According to Eqs. (20.1) and (20.3), the free energy satisfies the inequality

$$F(\mathcal{H}) \leq -\vartheta \ln \sum_n e^{-\frac{1}{\vartheta}\left(\mathscr{E}_n^0 + \mathcal{H}'_{nn}\right)}.$$

We shall now assume that the operator $\mathscr{H}'$ can be considered as a "small perturbation" compared with the operator $\mathscr{H}$. To within quantities of the first order of smallness with respect to $\mathscr{H}'$ we have

$$Q_{\text{mod}} = \sum_n e^{-\frac{\mathscr{E}_n^0}{\vartheta}} - \frac{1}{\vartheta} \sum_n \mathscr{H}'_{nn} e^{-\frac{\mathscr{E}_n^0}{\vartheta}}. \tag{20.22}$$

Substituting Eq. (20.22) into Eq. (20.1), we obtain, in this approximation,

$$F(\mathscr{H}) \leq F(\mathscr{H}_0) + \frac{\mathrm{Sp}\left(\mathscr{H}' e^{-\frac{\mathscr{H}_0}{\vartheta}}\right)}{\mathrm{Sp}\left(e^{-\frac{\mathscr{H}_0}{\vartheta}}\right)}. \tag{20.23}$$

In this case, the best approximation to the upper limit of the free energy is obtained by selecting the value of the parameter σ from the condition for the minimum of the right-hand side of Eq. (20.23). The formulation of the variational principle of Eq. (20.23) is more restricted than the initial formulation of Eq. (20.1).

The variational principle in the form of Eq. (20.23) can be strengthened by removing the limitation of the smallness of the operator $\mathscr{H}'$. For this purpose, we shall use the following theorem.

Theorem 2.* Let the Hamiltonian $\mathscr{H}$ of a system be represented in the form

$$\mathscr{H} = \mathscr{H}_0 + \mathscr{H}'. \tag{20.24}$$

Then we have the inequality

$$F(\mathscr{H}) \leq F_{\text{mod}}(\mathscr{H}), \tag{20.25}$$

where $F(\mathscr{H})$ is the intrinsic free energy and $F_{\text{mod}}(\mathscr{H})$ is the

*See Bogolyubov (1956).

model free energy of the system:

$$F_{\text{mod}}(\mathscr{H}) = F(\mathscr{H}_0) + \frac{\operatorname{Sp}\left(\mathscr{H}' e^{-\frac{\mathscr{H}_0}{\vartheta}}\right)}{\operatorname{Sp}\left(e^{-\frac{\mathscr{H}_0}{\vartheta}}\right)},$$

$$F(\mathscr{H}_0) = -\vartheta \ln \operatorname{Sp}\left(e^{-\frac{\mathscr{H}_0}{\vartheta}}\right). \tag{20.26}$$

Proof. Let $\mathscr{H}$ depend on some parameter ξ : $\mathscr{H} = \mathscr{H}(\xi)$. We shall consider the quantity $\exp[\mathscr{H}(\xi)t]$ where t is also some parameter. We can easily see that

$$\frac{d}{dt} e^{\mathscr{H}(\xi)t} = \mathscr{H}(\xi) e^{\mathscr{H}(\xi)t}. \tag{20.27}$$

Hence, we obtain an equation for the determination of the derivative of $\exp[\mathscr{H}(\xi)t]$ with respect to ξ:

$$\frac{d}{dt}\left\{\frac{d}{d\xi} e^{\mathscr{H}(\xi)t}\right\} = \mathscr{H}(\xi)\frac{d}{d\xi} e^{\mathscr{H}(\xi)t} + \frac{d\mathscr{H}(\xi)}{d\xi} e^{\mathscr{H}(\xi)t},$$

$$\frac{d}{d\xi} e^{\mathscr{H}(\xi)t} = 0 \text{ when } t = 0. \tag{20.28}$$

We shall assume that

$$\frac{d}{d\xi} e^{\mathscr{H}(\xi)t} = e^{\mathscr{H}(\xi)t} V(\xi). \tag{20.29}$$

Then, we have the following equation for V:

$$\frac{dV(\xi)}{dt} = e^{-\mathscr{H}(\xi)t}\frac{d\mathscr{H}(\xi)}{d\xi} e^{\mathscr{H}(\xi)t}, \tag{20.30}$$

$$V(\xi) = 0 \text{ when } t = 0.$$

This means that

$$\frac{d}{d\xi} e^{\mathscr{H}(\xi)} = e^{\mathscr{H}(\xi)} V(\xi) = e^{\mathscr{H}(\xi)} \int_0^1 e^{-\mathscr{H}(\xi)t}\frac{d\mathscr{H}(\xi)}{d\xi} e^{\mathscr{H}(\xi)t}\, dt. \tag{20.31}$$

We shall consider the special case

$$\mathscr{H}(\xi) = \mathscr{A} + \mathscr{B}\xi, \tag{20.32}$$

where $\mathscr{A}$, $\mathscr{B}$ are Hermitian operators, and we shall show that there is a convexity condition,

$$\frac{d^2}{d\xi^2} \operatorname{Sp} e^{\mathscr{A} + \mathscr{B}\xi} \geq 0. \tag{20.33}$$

In fact, in accordance with Eq. (20.31), we have

$$\frac{d}{d\xi}\,\mathrm{Sp}\, e^{\mathcal{A}+\mathcal{B}\xi} = \mathrm{Sp}\,\frac{d}{d\xi}\, e^{\mathcal{A}+\mathcal{B}\xi} =$$

$$= \mathrm{Sp}\, e^{\mathcal{A}+\mathcal{B}\xi}\int_0^1 e^{-(\mathcal{A}+\mathcal{B}\xi)t}\,\mathcal{B}\,e^{(\mathcal{A}+\mathcal{B}\xi)t}\,dt = \mathrm{Sp}\,(e^{\mathcal{A}+\mathcal{B}\xi}\mathcal{B}).$$

Consequently,

$$\frac{d^2}{d\xi^2}\,\mathrm{Sp}\,(e^{\mathcal{A}+\mathcal{B}\xi}) = \frac{d}{d\xi}\,\mathrm{Sp}\,(e^{\mathcal{A}+\mathcal{B}\xi}\mathcal{B}) =$$

$$= \mathrm{Sp}\left(e^{\mathcal{A}+\mathcal{B}\xi}\int_0^1 e^{-(\mathcal{A}+\mathcal{B}\xi)t}\,\mathcal{B}\,e^{(\mathcal{A}+\mathcal{B}\xi)t}\,dt\,\mathcal{B}\right).$$

We shall use the representation in which the operator (20.32) is diagonal, and we shall denote its eigenvalues by $\mathcal{E}_n$. Then,

$$\frac{d^2}{d\xi^2}\,\mathrm{Sp}\,(e^{\mathcal{A}+\mathcal{B}\xi}) = \sum_{m,\,n} e^{\mathcal{E}_m}\int_0^1 e^{-\mathcal{E}_m t}\,\mathcal{B}_{mn}e^{\mathcal{E}_n t}\,dt\,\mathcal{B}_{nm} =$$

$$= \sum_{m,\,n} |\,\mathcal{B}_{mn}|^2\,\frac{e^{\mathcal{E}_n} - e^{\mathcal{E}_m}}{\mathcal{E}_n - \mathcal{E}_m} \geq 0,$$

which proves the inequality (20.33).

However, if

$$\frac{d^2 f}{d\xi^2} \geq 0,$$

where $f(\xi)$ is some continuous differentiable function, then

$$f(\xi) \geq f(0) + \xi f'(0).$$

Consequently,

$$\mathrm{Sp}\,(e^{\mathcal{A}+\mathcal{B}\xi}) \geq \mathrm{Sp}\,(e^{\mathcal{A}}) + \xi\,\mathrm{Sp}\,(\mathcal{B}e^{\mathcal{A}}). \tag{20.34}$$

Assuming that

$$\mathcal{A} = -\frac{\mathcal{H}_0}{\vartheta},\quad \mathcal{B} = -\frac{\mathcal{H}' - \overline{\mathcal{H}}'}{\vartheta},\quad \xi = 1,$$

$$\overline{\mathcal{H}}' = \frac{\mathrm{Sp}\left(\mathcal{H}' e^{-\frac{\mathcal{H}_0}{\vartheta}}\right)}{\mathrm{Sp}\left(e^{-\frac{\mathcal{H}_0}{\vartheta}}\right)}, \tag{20.35}$$

we obtain from the preceding inequality

$$\mathrm{Sp}\left(e^{-\frac{1}{\vartheta}(\mathcal{H}_0+\mathcal{H}')}\right)\geq e^{-\frac{\overline{\mathcal{H}}'}{\vartheta}}\,\mathrm{Sp}\left(e^{-\frac{\mathcal{H}_0}{\vartheta}}\right). \tag{20.36}$$

Using the definitions of the free energy given by Eqs. (8.9) and (8.13), we now obtain

$$F(\mathcal{H})\leq F(\mathcal{H}_0)+\overline{\mathcal{H}}'=F(\mathcal{H}_0)+\frac{\mathrm{Sp}\left(\mathcal{H}'e^{-\frac{\mathcal{H}_0}{\vartheta}}\right)}{\mathrm{Sp}\left(e^{-\frac{\mathcal{H}_0}{\vartheta}}\right)}. \tag{20.37}$$

The inequality (20.25) of Theorem 2 follows directly from the above equation.

On the basis of the inequality (20.37), we can – as in the preceding case – formulate an approximate method for the determination of the free energy of a system. For this purpose, we shall represent the Hamiltonian of the system in the form of Eqs. (20.20)-(20.21). Then, the right-hand side of the inequality (20.37) gives the best approximation to the upper limit of the free energy, if the value of the parameter σ is determined from the condition for the minimum of the model free energy. This variational principle allows us always to obtain an estimate of the upper limit of the free energy, but such an estimate might be quite rough. The success of the application of this principle in the solution of actual problems depends to a considerable degree on how successfully the Hamiltonian is separated into the parts $\mathcal{H}_0$ and $\mathcal{H}'$.

In conclusion, we note that the minimum principle (20.37) can be considered to be a generalization of Fock's well-known minimum principle (1930) for the ground-state energy. Let $\mathcal{E}$ be the first eigenvalue of the total Hamiltonian $\mathcal{H}=\mathcal{H}_0+\mathcal{H}'$, and $\mathcal{E}_0$, φ_0 be the first eigenvalue and the first eigenfunction of the operator $\mathcal{H}_0$. According to this principle,

$$\mathcal{E}\leq\frac{(\varphi_0^*,\,\mathcal{H}\varphi_0)}{(\varphi_0^*,\,\varphi_0)},$$

and, consequently,

$$\mathcal{E}\leq\mathcal{E}_0+\frac{(\varphi_0^*,\,\mathcal{H}'\varphi_0)}{(\varphi_0^*,\,\varphi_0)}.$$

We shall now consider the application of the variational principle for the free energy in problems occurring in the theory of magnetism.

Sec. 21. Isotropic Ferromagnets

We shall consider first an isotropic ferromagnet. We shall assume that its lattice consists of N equivalent sites and the spin at each site is S = ½. The Hamiltonian for an isotropic ferromagnet, given by Eqs. (14.14)-(14.16), will be rewritten in the form

$$\tilde{\mathcal{H}} = \mathcal{E}_0 + \mathcal{H}_0 + \mathcal{H}', \tag{21.1}$$

where

$$\mathcal{E}_0 = -\frac{1}{2} N\mu H - \frac{1}{8} NJ(0) \quad \left(J(0) = \sum_f I(f)\right). \tag{21.2}$$

$$\mathcal{H}_0 = \left(\mu H + \frac{1}{2}\sigma_{1/2} J(0)\right) \sum_f n_f. \tag{21.3}$$

$$\mathcal{H}' = \frac{1}{2}(1 - \sigma_{1/2}) J(0) \sum_f n_f - \frac{1}{2} \sum_{f_1, f_2} I(f_1 - f_2) n_{f_1} n_{f_2} - \\ - \frac{1}{2} \sum_{f_1, f_2} I(f_1 - f_2) b_{f_1}^+ b_{f_2}. \tag{21.4}$$

Here, b_f^+, b_f are the Pauli operators; $\sigma_{1/2}$ is a parameter which has to be determined from the condition for the minimum of the model free energy (20.37). Obviously, $\mathcal{H} = \mathcal{H}_0 + \mathcal{H}'$ is independent of $\sigma_{1/2}$.

We shall bear in mind that the occupation numbers n_f assume the values of 0 and 1, and that the operators b_f^+, b_f are not diagonal with respect to the occupation numbers. We shall use the notation

$$\alpha = \mu H + \frac{1}{2}\sigma_{1/2} J(0) \tag{21.5}$$

and we shall calculate the spurs (traces) in Eq. (20.37) for the operators $\mathcal{H}_0$, $\mathcal{H}'$, given by Eqs. (21.3) and (21.4).

We can easily see that

$$\mathrm{Sp}\left(e^{-\frac{\mathcal{H}_0}{\vartheta}}\right) = \underset{(\ldots,\, n_f,\, \ldots)}{\mathrm{Sp}} \left(e^{-\frac{\alpha}{\vartheta}\sum n_f}\right) = \\ = \prod_f \sum_{n_f} e^{-\frac{\alpha}{\vartheta} n_f} = \left(1 + e^{-\frac{\alpha}{\vartheta}}\right)^N. \qquad (21.6)$$

We note that

$$\mathrm{Sp}\left(n_f e^{-\frac{\mathcal{H}_0}{\vartheta}}\right) = e^{-\frac{\alpha}{\vartheta}}\left(1 + e^{-\frac{\alpha}{\vartheta}}\right)^{N-1},$$

$$\mathrm{Sp}\left(b_f e^{-\frac{\mathcal{H}_0}{\vartheta}}\right) = 0,$$

and

$$\frac{\mathrm{Sp}\left(\mathcal{H}' e^{-\frac{\mathcal{H}_0}{\vartheta}}\right)}{\mathrm{Sp}\left(e^{-\frac{\mathcal{H}_0}{\vartheta}}\right)} = \frac{N}{2}\left(1 - \sigma_{1/2}\right)\bar{n}J(0) - \frac{N}{2}\bar{n}^2 J(0), \qquad (21.7)$$

$$\bar{n} = \frac{\mathrm{Sp}\left(n_f e^{-\frac{\mathcal{H}_0}{\vartheta}}\right)}{\mathrm{Sp}\left(e^{-\frac{\mathcal{H}_0}{\vartheta}}\right)} = \left(e^{\frac{\alpha}{\vartheta}} + 1\right)^{-1}. \qquad (21.8)$$

We shall substitute the expressions (21.6) and (21.7) into Eq. (20.37), and we shall take into account the contribution of the ground-state energy F ($\mathcal{H}_0$) to the free energy $\mathcal{E}_0$. Consequently, we shall obtain the following expression for the model free energy

$$F_{\mathrm{mod}} = \mathcal{E}_0 - N\vartheta \ln\left(1 + e^{-\frac{\alpha}{\vartheta}}\right) + \\ + \frac{N}{2}\left(1 - \sigma_{1/2}\right)\bar{n}J(0) - \frac{N}{2}\bar{n}^2 J(0), \qquad (21.9)$$

where α and $\bar{n}$ are defined in accordance with Eqs. (21.5) and (21.8).

From the condition of minimum of the form (21.9) with respect to $\sigma_{1/2}$, we obtain the following equation for the determination of $\sigma_{1/2}$:

$$\sigma_{1/2} = 1 - 2\bar{n} \qquad (21.10)$$

or, in the more usual form,

$$\sigma_{1/2} = \tanh \frac{\mu H + \frac{1}{2}\sigma_{1/2} J(0)}{2\vartheta}. \tag{21.11}$$

Since $S_f^z = ½ - n_f$, $\sigma_{1/2}$ is the relative magnetization per site.

The total magnetization of the system is $(N/2)\mu\sigma_{1/2}$. The same value is obtained by differentiating the model free energy (21.9) with respect to the external field. Bearing in mind the condition for the minimum of F_{mod} with respect to $\sigma_{1/2}$, we obtain:

$$M = -\frac{\partial F_{mod}}{\partial H} = \frac{1}{2} N\mu - \frac{\partial F_{mod}}{\partial \alpha}\frac{\partial \alpha}{\partial H} = M_0 \sigma_{1/2}, \tag{21.12}$$

where $M_0 = N\mu/2$ is the saturation magnetization of the ferromagnet.

An approximate value of the free energy is

$$F = \min F_{mod} = \mathcal{E}_0 - N\vartheta \ln\left(1 + e^{-\frac{\alpha}{\vartheta}}\right) + \frac{1}{2} N J(0)\bar{n}^2, \tag{21.13}$$

where $\mathcal{E}_0$, α, $\bar{n}$ are defined by Eqs. (21.2), (21.5), and (21.8).

Equation (21.11) for the relative magnetization σ is usually called the molecular field equation. This name is due to the fact that the expression under the hyperbolic tangent can be interpreted as the magnetic moment energy in some effective field,

$$H_{eff} = H + H_W, \quad H_W = \frac{\sigma_{1/2}}{2\mu} J(0), \tag{21.14}$$

which consists of the external field H and an internal or molecular Weiss field H_W. Usually, the dependence of H_W on the magnetization is separated out in the form:

$$H_W = \frac{1}{2}\sigma_{1/2} H'_W, \quad H'_W = \frac{1}{\mu} J(0), \tag{21.15}$$

where H'_W is known as the molecular field constant.

Using the notation of Eqs. (21.14) and (21.15), we shall write

Eq. (21.11) in the standard form:

$$m = m_0 \tanh \frac{\mu H + mH'_W}{2\vartheta}, \tag{21.16}$$

where $m = (\mu/2)\sigma_{1/2}$ is the magnetization at temperature ϑ, and $m_0 = \mu/2$ is the saturation magnetization per site.

It is evident from Eq. (21.14) that the molecular field has clear physical meaning: it is a quantity which has the dimensions of the magnetic field and is proportional to the average energy of the exchange interaction per site J(0) and to the relative magnetization $\sigma_{1/2}$. From the definition in Eq. (21.14) or Eq. (21.15), it is also clear that the molecular field is only a very rough approximation. Therefore, when we use it, we should expect to obtain only a general qualitative picture.

The analysis of the molecular field equations is well known and can be found in many textbooks [see, for example, the book of Vonsovskii and Shur (1948)]. We shall simply state here the main results.

We shall consider the case H = 0. Then, Eq. (21.11) assumes the form

$$\sigma_{1/2} = \tanh \frac{\sigma_{1/2}}{\tau} \quad \left(\tau = \frac{4\vartheta}{J(0)}\right). \tag{21.17}$$

Equation (21.17) has a trivial solution: $\sigma_{1/2} = 0$. In this case, the spontaneous magnetization is absent.

Equation (21.17) also has a nontrivial solution $\sigma_{1/2} \neq 0$ if $\tau < \tau_C$, where τ_C is the critical or Curie temperature.

We shall also consider the behavior of the magnetization near the point τ_C. Since

$$\left(\frac{d}{d\sigma_{1/2}} \tanh \frac{\sigma_{1/2}}{\tau}\right)_{\sigma_{1/2}=0} = \tau^{-1}; \quad \tanh \frac{\sigma_{1/2}}{\tau} \leq \frac{\sigma_{1/2}}{\tau},$$

it follows that when $\tau \geq 1$ there is only the trivial solution. The critical temperature τ_C, at which the nontrivial solution disappears, is

$$\tau_C = 1 \quad \text{or} \quad \vartheta_C = \frac{1}{4} J(0). \tag{21.18}$$

In the vicinity of the Curie point, the right-hand side of Eq. (21.17) may be expanded as a series in powers of the argument, and only the first terms of the expansion need be used:

$$\tanh y = y - \frac{y^3}{3} + O(y^5).$$

This gives the following approximate expression for the temperature dependence of the magnetization near the Curie point:

$$\sigma_{1/2} \approx \sqrt{3\tau^2(1-\tau)} \approx \sqrt{3(1-\tau)}. \tag{21.19}$$

Hence, it follows that at the Curie point there is a phase transition of the second kind. This is in agreement with the thermodynamic theory of phase transitions in ferromagnets [Landau and Lifshits (1959)].

If $\tau \to 0$, then we can easily find the following asymptotic expression for $\sigma_{1/2}$:

$$\sigma_{1/2} \sim 1 - 2e^{-\frac{2}{\tau}}. \tag{21.20}$$

In the preceding chapter, we mentioned that, at low temperatures, the magnetization of an isotropic ferromagnet varies with temperature as $\tau^{3/2}$ (cf., Sec. 15). Therefore, at low temperatures, the molecular field approximation, which leads to an exponential dependence, is far too rough; it only shows the general tendency of the approach of the magnetization to saturation.

We shall now estimate the order of magnitude of the molecular field. For typical ferromagnets, the Curie temperatures are $T_C \approx 10^3$°K (cf., Table 7), and $\mu \approx 10^{-20}$ cgs emu. Substituting these values into Eq. (21.15), we obtain, using Eq. (21.18),

$$H'_W = 4\frac{kT_C}{\mu} \approx 10^7 \text{ Oe}. \tag{21.21}$$

These are extremely high fields. Therefore, below the Curie point the magnetization of ferromagnets is close to their saturation magnetization.

In conclusion, it should be mentioned that the quantities H_W and H'_W behave only formally as fields: in fact, we are dealing with the electrostatic interaction between electrons.

Sec. 22. Ferrimagnets

Ferrimagnets are usually modeled by an assembly of several ferromagnetic sublattices between which there is an exchange interaction of a given sign. We shall consider here the extension of the molecular field approximation to a two-sublattice system. As before, we shall use the minimum principle for the free energy (cf., Sec. 20) to deduct the equations.

We shall consider only the case of a magnetically isotropic crystal. We shall use f and g to denote the sites in the first and second sublattices, and we shall assume that the internal ordering in the sublattices is ferromagnetic and the ordering between the sublattices is antiferromagnetic. Accordingly, we shall assume

$$I(f_1-f_2)\geq 0, \quad I(g_1-g_2)\geq 0, \quad I(f-g)\leq 0. \tag{22.1}$$

The quantities referring to the first and second sublattices will be denoted by the subscripts "1" and "2." We shall use N_i, S_i, μ_i to denote, respectively, the number of atoms in a sublattice, the value of the spin, and the magnetic moment of atoms of type i. In general, $N_1 \neq N_2$, $S_1 \neq S_2$, $\mu_1 \neq \mu_2$, so that the magnetic moments of the two sublattices taken as a whole are not equal. Without affecting the generality of our treatment, we may assume that

$$\begin{gathered} M_{10} > M_{20}\,(M_{10}=N_1S_1\mu_1, \quad M_{20}=N_2S_2\mu_2), \\ H^{\alpha}=H\delta_{\alpha,z} \quad (\alpha=x,\ y,\ z) \end{gathered} \tag{22.2}$$

(H is an external magnetic field).

We shall write the Hamiltonian of our system in the form

$$\begin{aligned} \tilde{\mathcal{H}} = &-\mu_1\sum(H,\ S_f)-\mu_2\sum(H,\ S_g)- \\ &-\tfrac{1}{2}\sum I(f_1-f_2)(S_{f_1},\ S_{f_2})-\tfrac{1}{2}\sum I(g_1-g_2)(S_{g_1},\ S_{g_2})- \\ &-\sum I(f-g)(S_f,\ S_g), \end{aligned} \tag{22.3}$$

where S_f, S_g are the spin operators of atoms in the first and second sublattices.

We shall now use Eqs. (5.8)-(5.10) and (5.16) to go over to the operators S_f^z and $S_f^{\pm}$, and we shall assume that the spins are oriented parallel to one another within each sublattice. Then, we can make the following assumptions about the classical vectors γ, which occur in the transformation coefficients of Eq. (5.8):

$$\gamma_f^\alpha = \gamma_1^\alpha, \quad \gamma_g^\alpha = \gamma_2^\alpha \quad (\alpha = x, y, z),$$
$$\sum_\alpha (\gamma_1^\alpha)^2 = 1, \quad \sum_\alpha (\gamma_2^\alpha)^2 = 1. \tag{22.4}$$

Consequently, we obtain the following expression for $\widetilde{\mathscr{H}}$:

$$\widetilde{\mathscr{H}} = \mathscr{E}_0 + \sum a_1 n_f + \sum a_2 n_g -$$
$$- \frac{1}{2} \sum \gamma_1^2 I(f_1 - f_2) n_{f_1} n_{f_2} - \frac{1}{2} \sum \gamma_2^2 I(g_1 - g_2) n_{g_1} n_{g_2}$$
$$- \sum (\gamma_1, \gamma_2) I(f - g) n_f n_g + \mathscr{H}' .$$
$$[n_h = 0, \ldots, 2S_h \quad (h = f, g)], \tag{22.5}$$

where

$$\mathscr{E}_0 = - N_1 \mu_1 S_1 (\gamma_1, H) - N_2 \mu_2 S_2 (\gamma_2, H) -$$
$$- \frac{1}{2} N_1 J_{11}(0) S_1^2 \gamma_1^2 - \frac{1}{2} N_2 J_{22}(0) S_2^2 \gamma_2^2 - N_1 J_{12}(0) S_1 S_2 (\gamma_1, \gamma_2). \tag{22.6}$$

$$N_1 J_{11}(0) = \sum_{f_1, f_2} I(f_1 - f_2),$$
$$N_1 J_{12}(0) = N_2 J_{21}(0) = \sum_{f, g} I(f - g), \ldots,$$
$$a_1 = \mu_1 (\gamma_1, H) + J_{11}(0) S_1 \gamma_1^2 + J_{12}(0) S_2 (\gamma_1, \gamma_2),$$
$$a_2 = \mu_2 (\gamma_2, H) + J_{22}(0) S_2 \gamma_2^2 + J_{21}(0) S_1 (\gamma_1, \gamma_2) \tag{22.7}$$

and $\mathscr{H}'$ is the part of the operator $\widetilde{\mathscr{H}}$ which is not diagonal with respect to the occupation numbers n_h.

We shall now derive the molecular field equations for our system. For this purpose, we shall use the minimum principle for the free energy in the form of Eq. (20.37).

We shall divide the Hamiltonian of Eq. (22.5) into two parts:

$$\widetilde{\mathscr{H}} = \mathscr{H}_0 + \mathscr{H}_1, \tag{22.8}$$

where

$$\mathscr{H}_0 = \mathscr{E}_0 + \sum \alpha_1 n_f + \sum \alpha_2 n_g,$$
$$\mathscr{H}_1 = \widetilde{\mathscr{H}} - \mathscr{H}_0. \tag{22.9}$$

The undetermined parameters α_1, α_2 will be found from the condition for the minimum of the model energy (20.37). First, we shall calculate the model free energy.

Since the Hamiltonian $\mathscr{H}_0$ of Eq. (22.9) splits into a system of individual Hamiltonians for each site, we can easily see that

$$Q_0 = \mathrm{Sp}\left(e^{-\frac{\mathscr{H}_0}{\vartheta}}\right) = e^{-\frac{\mathscr{E}_0}{\vartheta}}\left(\sum_{n=0}^{2S_1} e^{-\frac{\alpha_1 n}{\vartheta}}\right)^{N_1}\left(\sum_{n=0}^{2S_2} e^{-\frac{\alpha_2 n}{\vartheta}}\right)^{N_2}.$$

Hence,

$$F(\mathscr{H}_0) = \mathscr{E}_0 - N_1\vartheta \ln\left(\sum_{n=0}^{2S_1} e^{-\frac{\alpha_1}{\vartheta}n}\right) - N_2\vartheta \ln\left(\sum_{n=0}^{2S_2} e^{-\frac{\alpha_2}{\vartheta}n}\right). \tag{22.10}$$

Bearing in mind the equivalence of the sites in each of the sublattices, we now have

$$\langle\mathscr{H}_1\rangle = \frac{\mathrm{Sp}\left(\mathscr{H}_1 e^{-\frac{\mathscr{H}_0}{\vartheta}}\right)}{\mathrm{Sp}\left(e^{-\frac{\mathscr{H}_0}{\vartheta}}\right)} =$$
$$= N_1(a_1-\alpha_1)\bar{n}_1 + N_2(a_2-\alpha_2)\bar{n}_2 - \frac{N_1}{2}J_{11}(0)\gamma_1^2\bar{n}_1^2 -$$
$$- \frac{N_2}{2}J_{22}(0)\gamma_2^2\bar{n}_2^2 - N_1 J_{12}(0)(\gamma_1, \gamma_2)\bar{n}_1\bar{n}_2, \tag{22.11}$$

where

$$\bar{n}_i = B_{S_i}\left(\frac{\alpha_i}{\vartheta}\right) \quad (i = 1, 2), \tag{22.12}$$

$$B_S(x) = \frac{\sum_{n=0}^{2S} n e^{-nx}}{\sum_{n=0}^{2S} e^{-nx}} \tag{22.13}$$

(B_S is a Brillouin function).

The model free energy is equal to the sum of the expressions (22.10) and (22.11):

$$F_{\mathrm{mod}}(\widetilde{\mathscr{H}}) = F(\mathscr{H}_0) + \langle\mathscr{H}_1\rangle. \tag{22.14}$$

From the condition for the minimum of the expression (22.14) we find equations for the determination of the values of α_i, γ_i:

$$\frac{\partial F_{\text{mod}}}{\partial \alpha_i} = \left(\frac{\partial F_{\text{mod}}}{\partial \alpha_i}\right)_{\bar{n}_i} + \sum_j \left(\frac{\partial F_{\text{mod}}}{\partial \bar{n}_j}\right)_{\alpha_i} \frac{\partial \bar{n}_j}{\partial \alpha_i} = 0,$$

$$\frac{\partial}{\partial \gamma_i^\alpha}\left(F_{\text{mod}} - \frac{1}{2}\sum_{j,\beta} \lambda_j N_j (\gamma_j^\beta)^2\right) =$$

$$= \left(\frac{\partial F_{\text{mod}}}{\partial \gamma_i^\alpha}\right)_{\bar{n}_j} + \sum_j \left(\frac{\partial F_{\text{mod}}}{\partial \bar{n}_j}\right)_{\gamma_i^\alpha} \frac{\partial \bar{n}_j}{\partial \gamma_i^\alpha} - \lambda_i N_i \gamma_i^\alpha = 0$$

(λ_i are the indeterminate Lagrange multipliers).

Noting that $(\partial F/\partial \alpha_i)_{\bar{n}_i} = 0$, and using the explicit form of the quantities a_i given by Eq. (22.7), we now obtain the following system of equations:

$$\left.\begin{aligned} \alpha_1 &= \mu_1(\gamma_1, H) + J_{11}(0) S_1\sigma_1\gamma_1^2 + J_{12}(0) S_2\sigma_2(\gamma_1, \gamma_2), \\ \alpha_2 &= \mu_2(\gamma_2, H) + J_{21}(0) S_1\sigma_1(\gamma_2, \gamma_1) + J_{22}(0) S_2\sigma_2\gamma_2^2, \end{aligned}\right\} \quad (22.15)$$

$$\left.\begin{aligned} &(J_{11}(0) S_1^2\sigma_1^2 + \lambda_1)\gamma_1^\alpha + J_{12}(0) S_1 S_2\sigma_1\sigma_2\gamma_2^\alpha = \\ &\qquad = -S_1\sigma_1\mu_1 H\delta_{\alpha, z}, \\ &J_{21}(0) S_2 S_1\sigma_2\sigma_1\gamma_1^\alpha + (J_{22}(0) S_2^2\sigma_2^2 + \lambda_2)\gamma_2^\alpha = \\ &\qquad = -S_2\sigma_2\mu_2 H\delta_{\alpha, z}, \\ &\sum_\alpha (\gamma_1^\alpha)^2 = 1, \quad \sum_\alpha (\gamma_2^\alpha)^2 = 1, \end{aligned}\right\} \quad (22.16)$$

where $\sigma_i = S_i^{-1}\langle S_{h_i}\rangle = 1 - (\bar{n}_i/S_i)$ is the relative magnetization per site in the sublattice i; according to Eq. (22.12) this magnetization is equal to

$$\sigma_i = 1 - S_i^{-1} B_{S_i}\left(\frac{\alpha_i}{\vartheta}\right) \quad (i = 1, 2), \quad (22.17)$$

where B_S is the Brillouin function of Eq. (22.13).

The system of equations (22.15)-(22.17) represents the required molecular field equations for a two-sublattice isotropic ferromagnet.

Gusev (1959) used this method to obtain equations of the type of (22.15)-(22.17) for the special case $S_1 = S_2 = 1/2$. In this case ($S = 1/2$), we have $1 - 2B_{1/2}(\alpha/\vartheta) = \tanh(\alpha/2\vartheta)$ and Eq. (22.17) assumes the standard form:

$$\sigma_1 = \tanh\frac{\alpha_1}{2\vartheta}, \quad \sigma_2 = \tanh\frac{\alpha_2}{2\vartheta}. \quad (22.18)$$

The equilibrium configurations of the spin moments are found by solving the system of equations (22.16). When $\sigma_i = 1$, this system is identical with the system of equations defining the equilibrium configurations of spins at zero temperature in the approximate second quantization method (cf., Sec. 16). Therefore, we can use the known results, considering the partial magnetizations σ_i as parameters.

For convenience, we shall introduce the following notation for the sublattice magnetizations:

$$M_i^\alpha = M_i \gamma_i^\alpha, \quad M_i = M_{i0}\sigma_i = N_i S_i \mu_i \sigma_i, \tag{22.19}$$

so that the resultant magnetization is obviously

$$M^z = M_1^z + M_2^z. \tag{22.20}$$

Using the condition for the minimum of the free energy, we can easily show that, depending on the value of the external field, one of the following solutions is obtained.

<u>**Weak Fields:** $0 \leq H \leq H_1'$.</u> In this case,

$$\gamma_1^z = 1, \quad \gamma_2^z = -1; \tag{22.21}$$

the magnetic moments of the sublattices are antiparallel and the resultant moment is oriented along the field, its value being given by

$$M = M_1 - M_2 = M_{10}\sigma_1 - M_{20}\sigma_2. \tag{22.22}$$

The system of Eqs. (22.15)-(22.17) reduces to the following:

$$\left.\begin{aligned} \alpha_1 &= \mu_1 H + J_{11}(0) S_1\sigma_1 + |J_{12}(0)| S_2\sigma_2, \\ \alpha_2 &= -\mu_2 H + |J_{21}(0)| S_1\sigma_1 + J_{22}(0) S_2\sigma_2, \end{aligned}\right\} \tag{22.23}$$

$$\sigma_i = 1 - S_i^{-1} B_{S_i}\left(\frac{\alpha_i}{\vartheta}\right) \quad (i = 1, 2). \tag{22.24}$$

Usually, the molecular field approximation is applied to "weak" fields and, consequently, the molecular field equations are understood to be of the type of Eqs. (22.23)-(22.24). In the phenomenological derivation of these equations, the quantities corresponding to the average values of the exchange integrals I are introduced as some constants.

Moderate Fields: $H_1' \leq H \leq H_2'$. In this case,

$$\gamma_1^x = -\frac{J_{12}(0)\, S_2\sigma_2\mu_2}{J_{21}(0)\, S_1\sigma_1\mu_1}\gamma_2^x, \quad \gamma_1^y = -\frac{J_{12}(0)\, S_2\sigma_2\mu_2}{J_{21}(0)\, S_1\sigma_1\mu_1}\gamma_2^y,$$

$$\gamma_1^z = -\frac{\mu_2 H}{J_{21}(0)\, S_1\sigma_1} - \frac{J_{12}(0)\, S_2\sigma_2\mu_2}{J_{21}(0)\, S_1\sigma_1\mu_1}\gamma_2^z \tag{22.25}$$

or, in terms of components of the sublattice magnetization vectors,

$$M_1^x + M_2^x = 0, \quad M_1^y + M_2^y = 0,$$

$$M^z = M_1^z + M_2^z = \frac{1}{S_1 S_2}\sqrt{M_{10}M_{20}}\,\frac{H}{H_e}. \tag{22.26}$$

Strong Fields: $H \geq H_2'$. In this case, $\gamma_1^z = \gamma_2^z = 1$, and the resultant magnetization is

$$M^z = M_1^z + M_2^z = M_{10}\sigma_1 + M_{20}\sigma_2. \tag{22.27}$$

The critical fields H_1' and H_2', at which transitions take place from one configuration to another, have approximately the following form:

$$H_1' \approx \frac{M_1 - M_2}{\sqrt{M_{10}M_{20}}} H_e, \quad H_2' \approx \frac{M_1 + M_2}{\sqrt{M_{10}M_{20}}} H_e, \tag{22.28}$$

$$H_e = \sqrt{\frac{N_1 J_{12}(0)\, N_2 J_{21}(0)}{M_{10}M_{20}}} = \sqrt{\frac{J_{12}(0)\, J_{21}(0)}{S_1\mu_1 S_2\mu_2}}, \tag{22.29}$$

where M_i, M_{i0} are, respectively, the magnetization at a given temperature and the saturation magnetization of the sublattice i [cf., Eqs. (22.2) and (22.19)]. At zero temperature, the values of H_1' and H_2' are identical with the corresponding values obtained in the determination of the stable spin configurations in the approximate second quantization method.

Sec. 23. Applications of the Molecular Field Method

A theoretical analysis of the magnetic properties of ferrimagnets on the basis of the model of several coupled ferromagnetic sublattices was first carried out by Néel (1932, 1936, 1948) and Van Vleck (1941, 1951) in the molecular field approximation, which was introduced phenomenologically. Later, these results were refined by Yafet and Kittel (1952). Vlasov and Ishmukhametov (1954) gave a quantum-mechanical derivation of the molecular field equations for ferrimagnets using the method of the "energy centers of gravity." The variational principle in the formulation described here was extended by Kvasnikov (1957) to antiferromagnets, by Gusev (1959, 1960) to the two-sublattice model of a ferrimagnet, and by Gusev and Pakhomov (1961) to the three-sublattice model.

The molecular field method is not limited to the applications just referred to. Numerous applications of the method to the theory of antiferromagnetism and ferrimagnetism are described in the collective works "Ferromagnetic Resonance" (1952), "Antiferromagnetism" (1956), "Ferromagnetic Resonance" (1961), in the reviews by Borovik–Romanov (1962), Pakhomov and Smol'kov (1962), and in a monograph by Smit and Wijn (1962).

Sec. 24. Perturbation Theory for High Temperatures

Elementary Theory. The molecular field approximation is applicable to the range of temperatures close to the Curie point. Near this point, the method gives results identical with the thermodynamic treatment [cf., Landau and Lifshits (1959)].

We may assume that above the Curie point the average energy of the exchange interaction is small compared with the thermal energy: $J(0)/\vartheta \ll 1$. This ratio can be regarded as a small parameter in the theory and we can use the perturbation method to calculate the free energy of a system.

We shall write the Hamiltonian of a system in the form

$$\tilde{\mathscr{H}} = \mathscr{E}_0 + \mathscr{H}_0 + \varepsilon\mathscr{H}', \tag{24.1}$$

where $\varepsilon\mathscr{H}'$ is a "small" perturbation.

The free energy of Eq. (8.13) will be written as follows:

$$F = \mathscr{E}_0 - \vartheta \ln \mathrm{Sp}\, U_\beta \quad (\beta = \vartheta^{-1}), \tag{24.2}$$

where

$$U_\beta = e^{-\beta(\mathscr{H}_0 + \varepsilon\mathscr{H}')}. \tag{24.3}$$

We shall now represent U_β in the form of a series in powers of ε, using the thermodynamic perturbation theory. We shall assume that the free energy and the average values of various operators are known for the unperturbed system.

The quantity U_β of Eq. (24.3) may be regarded as the solution of the equation

$$\frac{dU_\beta}{d\beta} = -(\mathcal{H}_0 + \varepsilon\mathcal{H}')\,U_\beta \tag{24.4}$$

for the initial condition $U_0 = 1$. We shall substitute into Eq. (24.4) the expression

$$U_\beta = e^{-\beta\mathcal{H}_0}\,\mathcal{U}_\beta. \tag{24.5}$$

For the new unknown function $\mathcal{U}_\beta$, we obtain the equation

$$\frac{d\mathcal{U}_\beta}{d\beta} = -\,\varepsilon\mathcal{H}'(\beta)\,\mathcal{U}_\beta, \qquad \mathcal{U}_{\beta=0} = 1, \tag{24.6}$$

where

$$\mathcal{H}'(\beta) = e^{\beta\mathcal{H}_0}\mathcal{H}'e^{-\beta\mathcal{H}_0}. \tag{24.6a}$$

Integrating Eq. (24.6) by the method of successive approximations, we obtain the solution in the form of a series in powers of ε:

$$\mathcal{U}_\beta = \sum_{p=0}^{\infty}(-1)^p\,\varepsilon^p \int_0^\beta \cdots \int_0^{\beta_{p-1}} d\beta_1 \ldots d\beta_p \mathcal{H}'(\beta_1)\,\mathcal{H}'(\beta_2)\ldots\mathcal{H}'(\beta_p). \tag{24.7}$$

The partition function is given by

$$Q = \mathrm{Sp}\,U_\beta = Q_0\,\langle\mathcal{U}_\beta\rangle, \tag{24.8}$$

where the symbol < ... > denotes, in contrast to other cases, averaging over the density matrix for the unperturbed Hamiltonian:

$$\langle\mathcal{U}_\beta\rangle = \frac{\mathrm{Sp}\,(e^{-\beta\mathcal{H}_0}\mathcal{U}_\beta)}{\mathrm{Sp}\,(e^{-\beta\mathcal{H}_0})}, \qquad Q_0 = \mathrm{Sp}\,(e^{-\beta\mathcal{H}_0}). \tag{24.9}$$

The expression for the free energy is written in the following way:

$$F = \mathcal{E}_0 - \vartheta \ln Q_0 - \vartheta \ln \langle\mathcal{U}_\beta\rangle. \tag{24.10}$$

Thus, the problem of calculating the free energy in the form of a series in powers of the perturbation reduces to the calculation of the average value of the operator $\mathcal{U}_\beta$ (24.7).

We shall calculate the spur (trace) of Eq. (24.9) in a basis system of eigenfunctions of the unperturbed Hamiltonian $\mathcal{H}_0$. We shall use $\mathcal{E}_\nu$, C_ν to denote the eigenvalues and the eigenfunctions of the operator $\mathcal{H}_0$. Then, integration with respect to β_i in Eq. (24.7) is carried out easily, and we obtain the following expression:

$$\langle \mathcal{U}_\beta \rangle = \sum_{p=0}^{\infty} \varepsilon^p x_p, \tag{24.11}$$

where

$$x_0 = 1, \quad x_1 = -\beta Q_0^{-1} \sum_\nu e^{-\beta \mathcal{E}_\nu} \mathcal{H}'_{\nu\nu},$$
$$x_2 = Q_0^{-1} \sum_{\nu,\mu} e^{-\beta \mathcal{E}_\nu} \mathcal{H}'_{\nu\mu} \mathcal{H}'_{\mu\nu} \frac{e^{\beta(\mathcal{E}_\nu - \mathcal{E}_\mu)} - 1 - \beta(\mathcal{E}_\nu - \mathcal{E}_\mu)}{(\mathcal{E}_\nu - \mathcal{E}_\mu)^2}, \ \ldots,$$
$$\mathcal{H}'_{\nu\mu} = (C_\nu^*, \mathcal{H}' C_\mu). \tag{24.12}$$

At high temperatures $\beta(\mathcal{E}_\nu - \mathcal{E}_\mu)$ in the exponentials of Eq. (24.12) may be regarded as small quantities. Therefore, approximately,

$$x_2 = \frac{\beta^2}{2} Q_0^{-1} \sum_{\nu,\mu} e^{-\beta \mathcal{E}_\nu} \mathcal{H}'_{\nu\mu} \mathcal{H}'_{\mu\nu} + Q(\beta^3). \tag{24.13}$$

We shall substitute Eq. (24.11) into Eq. (24.10) and expand the right-hand part of the resultant expression as a series in powers of ε. Consequently, we obtain the following expression for the free energy:

$$F = \mathcal{E}_0 - \vartheta \ln Q_0 - \varepsilon \vartheta x_1 - \varepsilon^2 \vartheta \left(x_2 - \frac{x_1^2}{2} \right) - O(\varepsilon^3). \tag{24.14}$$

We shall apply the results obtained to calculate the magnetization of an isotropic ferromagnet with spin $S = 1/2$. We shall write the Hamiltonian in the form of Eq. (24.1), assuming that

$$\mathcal{E}_0 = -\frac{1}{2} N\mu H - \frac{1}{8} N J(0), \tag{24.15}$$

$$\mathcal{H}_0 = \left(\mu H + \frac{1}{2} J(0) \right) \sum_f n_f, \tag{24.16}$$

$$\mathscr{H}' = -\frac{1}{2}\sum_{f_1, f_2} I(f_1-f_2)\, b^+_{f_1} b_{f_2} - \frac{1}{2}\sum_{f_1, f_2} I(f_1-f_2)\, n_{f_1} n_{f_2}. \qquad (24.17)$$

We shall carry out calculations including terms up to ε^2. We shall confine ourselves to the nearest-neighbor approximation. Then, we can easily find x_1, x_2:

$$x_1 = N\frac{Iz}{2\vartheta}\bar{n}^2,$$

$$x_2 = N\frac{I^2 z}{4\vartheta^2}\bar{n}(1-\bar{n})\left\{\frac{1}{2}+\bar{n}(1-\bar{n})+2z\bar{n}^2\right\}+N^2\frac{I^2 z^2}{8\vartheta^2}\bar{n}^4,$$

$$\bar{n} = \left(e^{\frac{a}{\vartheta}}+1\right)^{-1}, \quad a = \mu H + \frac{1}{2}Iz, \quad I = \frac{1}{z}J(0), \qquad (24.18)$$

where z is the number of nearest neighbors, I is the exchange interaction energy for two neighboring spins.

Substituting Eqs. (24.15) and (24.18) into Eq. (24.14), we obtain the following expression for the free energy in the second approximation of the thermodynamic perturbation theory:

$$F = -\frac{1}{2}N\mu H - \frac{1}{8}NIz - N\vartheta\ln\left(1+e^{-\frac{a}{\vartheta}}\right) -$$
$$-\frac{\varepsilon}{2}NIz\bar{n}^2 - \frac{\varepsilon^2}{4\vartheta}NI^2 z\bar{n}(1-\bar{n})\left\{\frac{1}{2}+\bar{n}(1-\bar{n})+2z\bar{n}^2\right\}. \qquad (24.19)$$

Nonphysical terms, containing powers of the number of particles N (or of the volume of the system) higher than the first, appear in the calculation of F by means of Eqs. (24.7) and (24.9).

These nonphysical terms disappear only in the final result, due to complete mutual compensation in a given order of the perturbation theory. In our case, the term N^2 in x_2 is compensated by the term $x_1^2/2$ in Eq. (24.14), etc.

At high temperatures, the small quantity is the difference

$$t = 1 - 2\bar{n} = \tanh\frac{a}{2\vartheta}. \qquad (24.20)$$

Therefore, the expression for F can be conveniently rewritten in the following form:

$$F = -\frac{1}{2}N\mu H - \frac{1}{8}NIz - N\vartheta\ln\left(1+e^{-\frac{a}{\vartheta}}\right) -$$

$$-\frac{\varepsilon}{8} NIz(1-t)^2 - \frac{\varepsilon^2}{16\vartheta} NI^2 z(1-t^2)\left[\frac{1}{2} + \frac{1-t^2}{4} + \frac{z}{2}(1-t)^2\right]. \quad (24.21)$$

Hence, by differentiation with respect to H, we obtain the following expression for the relative magnetization $\sigma_{1/2}$:

$$\sigma_{1/2} = -\frac{2}{N\mu}\frac{\partial F}{\partial H} = t - \varepsilon(1-t^2)(1-t)(4\tau)^{-1} -$$
$$-\varepsilon^2(1-t^2)\left(1 + \frac{2t}{z} - 3t^2 + \frac{2z-1}{2z}t^3\right)(4\tau)^{-2} + O(\varepsilon^3), \quad (24.22)$$

where the following notation is used:

$$\tau = \frac{\vartheta}{Iz}, \quad h = \frac{\mu H}{Iz}, \quad t = \tanh\frac{2h+\varepsilon}{4\tau}. \quad (24.23)$$

We can easily see that the expansion for the magnetization in terms of powers of the small parameter, as given by Eq. (24.22), is essentially an expansion in powers of the reciprocal of temperature: $\tau = \vartheta/Iz$.

Usually, the Zeeman energy operator $-\mu H \Sigma n_f$ is used as $\mathscr{H}_0$. In going over to this case, we assume that $t = \tanh[(2h+\varepsilon)/4\tau]$ and expand the coefficients of various powers of ε in Eq. (24.22). We thus obtain

$$\sigma_{1/2} = t_0 + \varepsilon(1-t_0^2)\frac{t_0}{4\tau} +$$
$$+ \varepsilon^2(1-t_0^2)\left(\frac{z-2}{z} - \frac{2z-1}{2z}t_0^2\right)\frac{t_0}{(4\tau)^2} + O(\varepsilon^3), \quad (24.24)$$

where

$$t_0 = \tanh\frac{h}{2\tau}. \quad (24.25)$$

The expansion (24.24) was obtained first by Opekhowski (1937) in precisely the same way. A common deficiency of the elementary perturbation theory is the appearance of nonphysical terms proportional to powers of N higher than the first. The perturbation theory for large systems was improved by Van Hove (1955, 1956), Goldstone (1957), Hugenholtz (1957a,b), Dyson (1956a), and Bloch and de Dominicis (1958) in such a way that the nonphysical terms do not appear at all in calculations. This variant of the perturbation theory was used by Rudoi (1963a,b) to investigate high-temperature expansions in the theory of magnetism. Stinchcombe et al. (1962) investigated both high- and low-temperature states within the perturbation theory framework.

The expression for the susceptibility can be found easily from Eq. (24.24) as a series in powers of the reciprocal of temperature:

$$\chi_{1/2} = \mu\left(\frac{\partial\sigma_{1/2}}{\partial H}\right)_{H=0} = \frac{\mu^2}{4\vartheta}\sum_{n=0}^{\infty}\frac{a_n}{(4\tau)^n}; \qquad (24.26)$$

the first three coefficients are

$$a_0 = 1, \quad a_1 = 1, \quad a_2 = \frac{z-1}{z}. \qquad (24.27)$$

In the case of arbitrary spin ($S \geq 1/2$),

$$\chi_S = \frac{\mu^2 S(S+1)}{3\vartheta}\sum_{n=0}^{\infty} a_n\left(\frac{S(S+1)}{3\tau}\right)^n; \qquad (24.28)$$

the coefficients a_0, a_1, a_2 are defined in accordance with Eq. (24.27).

We shall bear in mind that above the Curie point the susceptibility obeys the Curie–Weiss law of Eq. (1.1). Therefore, we can define the Curie temperature as the zero value of the reciprocal of the susceptibility: $\chi^{-1}(\vartheta_C) = 0$. The values of the Curie temperature were determined by this method by Brown and Luttinger (1955) and Brown (1956) in the fourth and fifth order of the perturbation theory. Table 9 lists the values of the Curie temperature in units of Iz for the simple cubic, bcc, and fcc lattices.

Behavior of the Susceptibility in the Vicinity of the Curie Point. The perturbation theory series for χ given by Eq. (24.28) can be very roughly represented by the Curie–Weiss law of Eq. (1.1). In fact, if we restrict ourselves to the first three terms in Eq. (24.28), and if we neglect the difference between the coefficient a_2 and unity, we shall have

$$\chi_S \approx \frac{\mu^2 S(S+1)}{3\vartheta}\left\{1 + \left(\frac{3\tau}{S(S+1)}\right)^{-1} + \left(\frac{3\tau}{S(S+1)}\right)^{-2} + O(\tau^{-3})\right\} = \\ = \frac{\mu^2 S(S+1)}{3\vartheta}\frac{1}{1-\vartheta/\vartheta_{Cp}} + O(\vartheta^{-4}). \qquad (24.29)$$

Table 9. Values of the Temperature ϑ_C in Units of Iz (Values of ϑ_C calculated using the statistical perturbation theory; I represents the exchange integral for the nearest neighbors, and z is the number of such neighbors)*

Type of lattice	Spin					
	1/2	1	3/2	2	5/2	3
Simple cubic	0.161	0.447	0.886	1.455	2.151	2.978
BCC	0.149	0.488	0.964	1.573	2.319	3.199
FCC	0.177	0.531	1.028	1.667	2.447	3.370

*The values are taken from the paper by Brown and Luttinger (1955). Brown (1956) reported only certain improvements in ϑ_C, obtained from the fifth approximation.

where

$$\vartheta_{Cp} = \frac{S(S+1)}{3} Iz \tag{24.30}$$

is the paramagnetic Curie temperature.

A more detailed analysis of the perturbation theory series* leads to the conclusion that the susceptibility at the Curie point has a singularity different from that given by Eq. (24.29). Investigations of this problem have dealt with the behavior of the susceptibility of an isotropic ferromagnet allowing only for the nearest-neighbor interaction. To analyze the temperature dependence of the susceptibility, Domb and Sykes (1962) calculated the coefficients a_n of Eq. (24.28), selected asymptotic expressions for these coefficients, and used them to reproduce the behavior of χ in the vicinity of the Curie point. Gammal et al. (1963) determined the temperature dependence of χ by deriving an approximate expression by means of Padé approximants. We shall

*See the papers of Domb and Sykes (1962), Gammal et al. (1963), and the references cited in these papers.

give only some general idea about the results of these investigations, because a more complete presentation would involve cumbersome calculations. For simplicity, we shall use the Padé approximant method.

We shall consider some function $f(x)$ represented as a series:

$$f(x)=\sum_{n=0}^{\infty} C_n x^n, \quad C_0=1. \tag{24.31}$$

The Padé approximant $P_n^m(x)$ for a function $f(x)$ is the rational-fraction function

$$P_n^m(x)=\frac{A_m(x)}{B_n(x)}=\frac{1+a_1x+\ldots+a_mx^m}{1+b_1x+\ldots+b_nx^n}, \tag{24.32}$$

whose coefficients a_p, b_q are found from the condition that the coefficients of the powers of x, up to x^{m+n}, inclusive, vanish in the function

$$f(x)B_n(x)-A_m(x). \tag{24.33}$$

If $f(x)$ is a polynomial of the degree L, then $m+n < L$.

Let us assume that the first L coefficients of the series are known. Then, we can find the explicit form of the approximants $P_n^m(x)$ and from them reproduce the behavior of the function $f(x)$.*

Gammal et al. (1963) used the functions $\ln \chi$ and $\chi^{3/4}$ for interpolation. The numerical values of the first coefficients C_n ($n = 1, 2, \ldots, 5$), which were necessary in the calculations, were found from the coefficients of the expansion of χ in terms of ϑ^{-1}, given by Rushbrooke and Wood (1958). Consequently, it was possible to establish that the temperature dependence of the initial susceptibility (when $H \to 0$) was quite well described by the formula

$$\chi\sim(\vartheta-\vartheta_c)^{-4/3}. \tag{24.34}$$

A similar result has been obtained by other methods (with the same degree of rigor); at present, it is assumed that this result can be regarded as reliable.

*For details of this method, see Baker and Gammal (1961) and Baker et al. (1961).

Baker et al. (1964) developed the method of obtaining high-temperature expansions using the irreducible representations of symmetry groups. Wojtowicz and Joseph (1964) obtained an expansion including the interactions of the nearest and second-nearest neighbors.

Chapter VII

Green's Function Method

In the present chapter, we shall present the method of double-time Green's temperature-dependent functions (retarded and advanced).* This method is widely used in statistical mechanics, because it is an effective means of calculating the observable macroscopic properties of a system as well as the microscopic quantities (for example, the energy and lifetime of elementary excitations). The Green's function method is also convenient in the calculation of transport coefficients, such as the electrical conductivity, susceptibility, etc. The results obtained are applicable to a wide range of temperatures. We shall give below the basic information about Green's functions, and we shall develop the mathematical apparatus necessary for the solving of actual problems.

Sec. 25. Time Correlation Functions and Green's Functions

Let $\mathcal{A}(t)$ and $\mathcal{B}(t')$ be some operators in the Heisenberg representation:

$$\mathcal{A}(t) = e^{i\mathcal{H}t}\mathcal{A}(0)\,e^{-i\mathcal{H}t}, \quad \mathcal{B}(t') = e^{i\mathcal{H}t'}\,\mathcal{B}(0)\,e^{-i\mathcal{H}t'}, \tag{25.1}$$

where $\mathcal{H}$ is the Hamiltonian of the system. We shall assume that $\mathcal{H}$ includes the term $-\lambda\mathcal{N}$ (λ is the chemical potential, $\mathcal{N}$

*Green's temperature-dependent functions, proposed by Matsubara (1955), are also used extensively. The different variants of the Green's function method and its very numerous applications in other branches of statistical mechanics are described in the reviews by Zubarev (1960), Alekseev (1961), and in the books by Bonch-Bruevich and Tyablikov (1961) and Abrikosov, Gor'kov, and Dzyaloshinskii (1962).

is the operator for the total number of particles in the system). In general, the operators $\mathcal{A}$ and $\mathcal{B}$ are products of quantized wave functions or of particle creation and annihilation operators.

The equations of motion for these operators have the form

$$i\frac{dA(t)}{dt} = [\mathcal{A}(t), \mathcal{H}] = \mathcal{A}(t)\mathcal{H} - \mathcal{H}\mathcal{A}(t). \tag{25.2}$$

In general, the commutator on the right-hand side of Eq. (25.2) includes products of a larger number of the second quantization operators than the operator $\mathcal{A}$.

We shall define the time correlation functions by

$$F_{\mathcal{A}\mathcal{B}}(t, t') = \langle \mathcal{A}(t)\mathcal{B}(t')\rangle, \tag{25.3}$$

where the symbol < ... > represents, as usual, statistical averaging and the use of the Hamiltonian $\mathcal{H}$ [cf., Eq. (8.4)].

We can easily see that the functions (25.3) depend only on the difference of the time arguments. In fact, because of the invariance of the spur (trace) under cyclic permutation of the cofactors, we have

$$\begin{aligned}\mathrm{Sp}\left\{\mathcal{A}(t)\mathcal{B}(t')e^{-\frac{\mathcal{H}}{\vartheta}}\right\} &= \\ &= \mathrm{Sp}\left\{e^{i\mathcal{H}(t-t')}\mathcal{A}(0)e^{-i\mathcal{H}(t-t')}\mathcal{B}(0)e^{-\frac{\mathcal{H}}{\vartheta}}\right\} = \\ &= \mathrm{Sp}\left\{\mathcal{A}(0)e^{-i\mathcal{H}(t-t')}\mathcal{B}(0)e^{i\mathcal{H}(t-t')}e^{-\frac{\mathcal{H}}{\vartheta}}\right\}.\end{aligned}$$

Consequently,

$$F_{\mathcal{A}\mathcal{B}}(t, t') = F_{\mathcal{A}\mathcal{B}}(t - t'). \tag{25.4}$$

When the time arguments are equal, t = t', the time averages (25.3) reduce to the usual statistical averages:

$$F_{\mathcal{A}\mathcal{B}}(t, t) = F_{\mathcal{A}\mathcal{B}}(0) = \langle \mathcal{A}(0)\mathcal{B}(0)\rangle. \tag{25.5}$$

Differentiating a correlation function with respect to one of the arguments, for example, the first argument, we can obtain an equation which gives the development of this function with time. Since the right-hand side contains, in general, products of a larger number of second quantization operators than the same side in the original expression, we obtain an infinite chain (set) of coupled equations for the correlation functions of increasing order:

$$i\frac{d}{dt}\langle\mathcal{A}(t)\,\mathcal{B}(t')\rangle=\langle[\mathcal{A}(t),\ \mathcal{H}]\,\mathcal{B}(t')\rangle;$$
$$i\frac{d}{dt}\langle[\mathcal{A}(t),\ \mathcal{H}]\,\mathcal{B}(t')\rangle=\langle[[\mathcal{A}(t),\ \mathcal{H}],\ \mathcal{H}]\,\mathcal{B}(t')\rangle;\ \ldots \quad (25.6)$$

The structure of this chain is similar to that of the chains of equations which give the classical correlation functions and which were introduced by Bogolyubov (1946). This system of equations can be used to determine the correlation functions of interest to us by truncating it in some way. It is more convenient to define the time correlation functions in terms of Green's functions.

We shall define the double-time retarded (r), advanced (a), and causal (c) Green's functions by the relationships*:

$$G^{(r)}_{\mathcal{A}\mathcal{B}}(t,t')=\langle\langle\mathcal{A}(t)\,|\,\mathcal{B}(t')\rangle\rangle^{(r)}=\theta(t-t')\langle[\mathcal{A}(t),\ \mathcal{B}(t')]_{\eta}\rangle, \quad (25.7)$$
$$G^{(a)}_{\mathcal{A}\mathcal{B}}(t,t')=\langle\langle\mathcal{A}(t)\,|\,\mathcal{B}(t')\rangle\rangle^{(a)}=-\theta(t'-t)\langle[\mathcal{A}(t),\ \mathcal{B}(t')]_{\eta}\rangle, \quad (25.8)$$
$$G^{(c)}_{\mathcal{A}\mathcal{B}}(t,t')=\langle\langle\mathcal{A}(t)\,|\,\mathcal{B}(t')\rangle\rangle^{(c)}=\langle T_{\eta}\,\mathcal{A}(t)\,\mathcal{B}(t')\rangle, \quad (25.9)$$

where

$$[\mathcal{A}(t),\ \mathcal{B}(t')]_{\eta}=\mathcal{A}(t)\,\mathcal{B}(t')-\eta\mathcal{B}(t')\,\mathcal{A}(t), \quad (25.10)$$
$$T_{\eta}\,\mathcal{A}(t)\,\mathcal{B}(t')=\theta(t-t')\,\mathcal{A}(t)\,\mathcal{B}(t')+\eta\theta(t'-t)\,\mathcal{B}(t')\,\mathcal{A}(t), \quad (25.11)$$
$$\theta(x)=\begin{cases}1, & x>0,\\ 0, & x<0.\end{cases} \quad (25.12)$$

---

*The various definitions of the Green's functions of the type given by Eqs. (25.7)-(25.9) differ by a constant factor on the right-hand side. Thus, Zubarev (1960) gives $G^{(r)}_{\mathcal{A}\mathcal{B}}(t,t')=-i\theta(t-t')\langle\ldots\rangle$; Bonch-Bruevich and Tyablikov (1961) have $G^{(r)}_{\mathcal{A}\mathcal{B}}(t,t')=i\theta(t-t')\ \langle\ldots\rangle$ etc.

It is also possible to introduce noncommutator Green's functions of the type

$$G^{(r)}_{\mathcal{A}\mathcal{B}}(t,t')=\theta(t-t')\langle\mathcal{A}(t)\,\mathcal{B}(t')\rangle;\ldots$$

and many-time Green's functions [cf., for example, Bonch-Bruevich (1959), Tyablikov and P'u Fu-ch'o (1961)].

Either $\eta = 1$ or $\eta = -1$ is selected on the basis of convenience, irrespective of the form of the commutation relationships for the operators $\mathcal{A}$, $\mathcal{B}$. It is usually assumed that $\eta = 1$ if these operators are represented in terms of the Bose operators, and $\eta = -1$ if they are represented in terms of the Fermi operators.

Since the Green's functions are defined in terms of the time correlation functions of the form given in Eq. (25.3), the Green's functions also depend on the difference between the time arguments:

$$G^{(j)}_{\mathcal{A}\mathcal{B}}(t, t') = G^{(j)}_{\mathcal{A}\mathcal{B}}(t - t') \quad (j = r, a, c). \tag{25.13}$$

By definition, a Green's function $\langle\langle \mathcal{A} | \mathcal{B} \rangle\rangle^{(j)}$ depends linearly on each of the operator arguments, $\mathcal{A}$ or $\mathcal{B}$, and, therefore,

$$\langle\langle \alpha_1 \mathcal{A}_1 + \alpha_2 \mathcal{A}_2 | \mathcal{B} \rangle\rangle^{(j)} = \alpha_1 \langle\langle \mathcal{A}_1 | \mathcal{B} \rangle\rangle^{(j)} + \alpha_2 \langle\langle \mathcal{A}_2 | \mathcal{B} \rangle\rangle^{(j)}, \tag{25.14}$$

where α_1 and α_2 are arbitrary numbers.

We shall now derive the chain of coupled equations for the Green's functions. For this purpose, we shall differentiate Eqs. (25.7)-(25.9) with respect to t. We shall write the decoupling function θ (t) in the following way:

$$\theta(t) = \int_{-\infty}^{t} e^{\varepsilon t'} \delta(t')\, dt' \quad (\varepsilon \to +0), \tag{25.15}$$

where δ (t) is Dirac's δ-function. Since

$$\frac{d\theta(t)}{dt} = \delta(t)$$

and the equations of motion for the operators in the Heisenberg representation are given by Eq. (25.2), we obtain

$$i \frac{d}{dt} \langle\langle \mathcal{A}(t) | \mathcal{B}(t') \rangle\rangle^{(j)} = i\delta(t - t') \langle [\mathcal{A}(t), \mathcal{B}(t)]_\eta \rangle + \\ + \langle\langle [\mathcal{A}(t), \mathcal{H}] | \mathcal{B}(t') \rangle\rangle^{(j)} \quad (j = r, a, c). \tag{25.16}$$

The equations (25.16) differ from the equations (25.6) for the correlation functions by the presence of an inhomogeneous term with δ-type factors. Consequently, we are dealing with equations similar to the usual equations for the influence functions (Green's functions). This is why the expressions of the type (25.7)-(25.9) are called Green's functions.

As in the equations for the time correlation functions, the right-hand side of Eq. (25.16) includes a Green's function of products of a larger number of operators than the initial function. Differentiating the function $\langle\langle[\mathcal{A}(t), \mathcal{H}]|\mathcal{B}(t')\rangle\rangle^{(j)}$, with respect to t, we obtain an equation of the (25.16) type, the right-hand side of which includes the function $\langle\langle[[\mathcal{A}(t), \mathcal{H}], \mathcal{H}]|\mathcal{B}(t')\rangle\rangle^{(j)}$. Consequently, we obtain an infinite chain (set) of coupled equations for the series of functions

$$\langle\langle\mathcal{A}|\mathcal{B}\rangle\rangle^{(j)};\ \langle\langle[\mathcal{A}, \mathcal{H}]|\mathcal{B}\rangle\rangle^{(j)};\ \langle\langle[[\mathcal{A}, \mathcal{H}], \mathcal{H}]|\mathcal{B}\rangle\rangle^{(j)};\ \ldots$$

We note that the chain of equations is the same for the retarded, advanced, and causal Green's functions.

We shall now consider the Fourier time transforms of the Green's functions:

$$G^{(j)}_{\mathcal{A}\mathcal{B}}(t-t') = \int_{-\infty}^{\infty} G^{(j)}_{\mathcal{A}\mathcal{B}}(E)\, e^{-iE(t-t')}\, dE,$$

$$G^{(j)}_{\mathcal{A}\mathcal{B}}(E) = \frac{1}{2\pi}\int_{-\infty}^{\infty} G^{(j)}_{\mathcal{A}\mathcal{B}}(t)\, e^{iEt}\, dt \qquad (j = r,\ a,\ c). \tag{25.17}$$

The chain of equations given by Eq. (25.16) now assumes the form

$$E\langle\langle\mathcal{A}|\mathcal{B}\rangle\rangle^{(j)}_{E} = \frac{i}{2\pi}\langle[\mathcal{A}, \mathcal{B}]_{\eta}\rangle + \langle\langle[\mathcal{A}, \mathcal{H}]|\mathcal{B}\rangle\rangle^{(j)}_{E};\ \ldots \tag{25.18}$$

We shall call the Fourier time transforms of the Green's functions the E-representation of the Green's functions or, if it does not cause any misunderstanding, we shall simply refer to these transforms as the Green's functions.

To solve the chain of equations (25.16), we must stipulate the boundary conditions for t (which are different for the retarded, advanced, and causal functions). The form of these conditions may be established from the definitions (25.7)-(25.9) of the Green's functions themselves. However, it is more convenient to use the Fourier time transforms of the Green's functions. Then, the boundary conditions are in the form of spectral representations of the Green's functions or dispersion relationships which give the rules for bypassing the poles of the Fourier transforms of the Green's functions and, consequently, define the boundary conditions for the Green's functions [cf., Bonch-Bruevich and Tyablikov (1961)].

We note that the appearance of infinite chains of coupled equations of the (25.6) or (25.16) type is unavoidable for a system of interacting particles, and it is due to the fact that one cannot consider a single particle or a group of particles independently of the rest of the system. In the approach considered here, the main problem is to develop approximate methods for the solution of the chain of equations for the Green's functions. The solution reduces to the decoupling of the chain in some way or another.

Sec. 26. Spectral Representations*

a. Spectral Representations of the Correlation Functions. We shall use $\mathcal{E}_\nu$ and C_ν to denote, respectively, the eigenvalues and eigenfunctions of the Hamiltonian of a system:

$$(\mathcal{H} - \mathcal{E}_\nu) C_\nu = 0. \tag{26.1}$$

Using the completeness of the system of functions $\{C_\nu\}$, we shall write the average value of the product of two operators in the following form:

$$\langle \mathcal{B}(t') \mathcal{A}(t) \rangle = Q^{-1} \sum_\nu (C^*_\nu, \mathcal{B}(t') \mathcal{A}(t) C_\nu) e^{-\frac{\mathcal{E}_\nu}{\vartheta}} =$$
$$= Q^{-1} \sum_{\mu, \nu} (C^*_\nu, \mathcal{B}(t') C_\mu)(C^*_\mu, \mathcal{A}(t) C_\nu) e^{-\frac{\mathcal{E}_\nu}{\vartheta}},$$

where Q is the partition function of Eq. (8.12).

Bearing in mind that the time dependence of the operators $\mathcal{A}$, $\mathcal{B}$ is given by Eq. (25.1), and that

$$(\exp i\mathcal{H}t) C_\nu = (\exp i\mathcal{E}_\nu t) C_\nu,$$

*We shall follow the work of Bogolyubov and Tyablikov (1959) and Bogolyubov (1961).

we have

$$\langle \mathcal{B}(t')\mathcal{A}(t)\rangle = \\ = Q^{-1}\sum_{\mu,\nu}(C_\nu^*, \mathcal{B}(0)C_\mu)(C_\mu^*, \mathcal{A}(0)C_\nu)e^{-\frac{\mathcal{E}_\nu}{\vartheta}}e^{-i(\mathcal{E}_\nu-\mathcal{E}_\mu)(t-t')}$$

and, similarly,

$$\langle \mathcal{A}(t)\mathcal{B}(t')\rangle = Q^{-1}\sum_{\mu,\nu}(C_\nu^*, \mathcal{B}(0)C_\mu)\times \\ \times(C_\mu^*, \mathcal{A}(0)C_\nu)e^{-\frac{\mathcal{E}_\nu}{\vartheta}}e^{-\frac{1}{\vartheta}(\mathcal{E}_\mu-\mathcal{E}_\nu)}e^{-i(\mathcal{E}_\nu-\mathcal{E}_\mu)(t-t')}.$$

These expressions will be rewritten in the following form:

$$\langle \mathcal{B}(t')\mathcal{A}(t)\rangle = \int_{-\infty}^{\infty} I_{\mathcal{A}\mathcal{B}}(\omega)e^{-i\omega(t-t')}\,d\omega, \tag{26.2}$$

$$\langle \mathcal{A}(t)\mathcal{B}(t')\rangle = \int_{-\infty}^{\infty} I_{\mathcal{A}\mathcal{B}}(\omega)e^{\frac{\omega}{\vartheta}}e^{-i\omega(t-t')}\,d\omega, \tag{26.3}$$

where the following notation is used:

$$I_{\mathcal{A}\mathcal{B}}(\omega) = \\ = Q^{-1}\sum_{\mu,\nu}(C_\nu^*, \mathcal{B}(0)C_\mu)(C_\mu^*, \mathcal{A}(0)C_\nu)e^{-\frac{\mathcal{E}_\nu}{\vartheta}}\delta(\mathcal{E}_\nu-\mathcal{E}_\mu-\omega). \tag{26.4}$$

The formulas (26.2) and (26.3) are known as the spectral representations of the time correlation functions. The quantity $I_{\mathcal{A}\mathcal{B}}(\omega)$ is called the spectral density or the spectral function. Equations (26.2) and (26.3) are essentially formulas for the expansion in terms of the Fourier integrals of the argument difference $(t - t')$.

When $t = t'$, the formulas (26.2) and (26.3) assume the form

$$\langle \mathcal{B}(0)\mathcal{A}(0)\rangle = \int_{-\infty}^{\infty} I_{\mathcal{A}\mathcal{B}}(\omega)\,d\omega, \tag{26.5}$$

$$\langle \mathcal{A}(0)\mathcal{B}(0)\rangle = \int_{-\infty}^{\infty} I_{\mathcal{A}\mathcal{B}}(\omega)e^{\frac{\omega}{\vartheta}}\,d\omega. \tag{26.6}$$

In formulas (26.2), (26.3) and (26.5), (26.6) we give the time and conventional correlation functions in terms of the spectral density. We shall show that the density is given in terms of the Green's functions. Thus, the problem of calculating the correlation functions reduces to finding the appropriate Green's functions.

We shall multiply Eqs. (26.2) and (26.5) by η and subtract the results from Eqs. (26.3) and (26.6), respectively. This gives

$$\langle [\mathcal{A}(t), \mathcal{B}(t')]_\eta \rangle = \int_{-\infty}^{\infty} I_{\mathcal{A}\mathcal{B}}(\omega)\left(e^{\frac{\omega}{\vartheta}} - \eta\right) e^{-i\omega(t-t')}\, d\omega, \tag{26.7}$$

$$\langle [\mathcal{A}(0), \mathcal{B}(0)]_\eta \rangle = \int_{-\infty}^{\infty} I_{\mathcal{A}\mathcal{B}}(\omega)\left(e^{\frac{\omega}{\vartheta}} - \eta\right) d\omega. \tag{26.8}$$

These exact relationships can be used to check the calculations.

b. Spectral Representations of the Green's Functions. We shall now find the spectral representations of the Green's functions.

For the retarded Green's functions, we have, in accordance with Eqs. (25.7), (25.10), and (25.17),

$$G^{(r)}_{\mathcal{A}\mathcal{B}}(E) = \frac{1}{2\pi}\int_{-\infty}^{\infty} dt e^{iE(t-t')}\theta(t-t')\langle \mathcal{A}(t)\mathcal{B}(t') - \eta\mathcal{B}(t')\mathcal{A}(t)\rangle$$

or, using Eqs. (26.7) and (26.3),

$$G^{(r)}_{\mathcal{A}\mathcal{B}}(E) = \frac{1}{2\pi}\int_{-\infty}^{\infty} d\omega I_{\mathcal{A}\mathcal{B}}(\omega)\left(e^{\frac{\omega}{\vartheta}} - \eta\right)\int_{-\infty}^{\infty} dt e^{i(E-\omega)t}\theta(t).$$

We shall allow for the fact that the decoupling function $\theta(t)$ can be represented in the form of a contour integral*:

$$\theta(t) = \frac{i}{2\pi}\int_{-\infty}^{\infty} dE \frac{e^{-iEt}}{E + i\varepsilon} \qquad (\varepsilon > 0). \tag{26.9}$$

---

*Formula (26.9) is obtained from Eq. (25.15) by the substitution

$$\delta(t) = \frac{1}{2\pi}\int_{-\infty}^{\infty} dE e^{-iEt}.$$

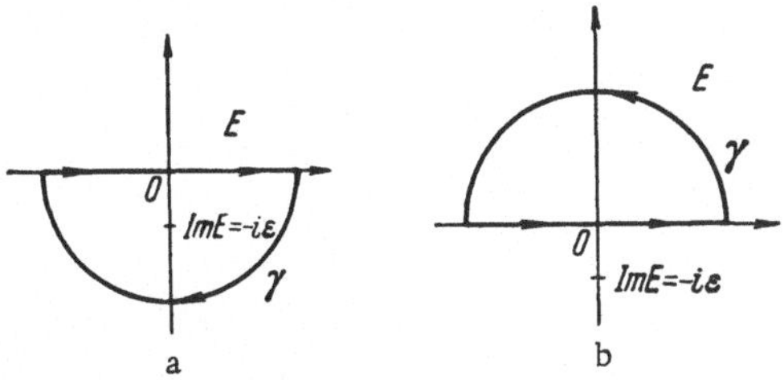

Fig. 10. Integration contour γ in the formula for θ (t) (26.9): a) t > 0; b) t < 0.

We shall assume that the variable E is complex. Let t > 0; then the contour γ in the complex E-plane should be closed in the lower half-plane. Since the integrand has a pole in that half-plane, E = $-i\varepsilon$, the theorem on residues gives

$$\theta(t) = -\frac{i}{2\pi}\int_{\gamma} dE \frac{e^{-iEt}}{E+i\varepsilon} = 1.$$

If t < 0, the contour should be closed in the upper half-plane, where the integrand has no singularities, and θ (t) = 0. Figure 10 shows the integration contours for the formula (26.9) when t > 0 and t < 0.

Using Eq. (26.9), we obtain the following expression for the integral with respect to t:

$$\int_{-\infty}^{\infty} dt e^{i(E-\omega)t}\theta(t) = \frac{i}{E-\omega+i\varepsilon}. \tag{26.10}$$

Consequently, we obtain the following expression for the retarded function:

$$G^{(r)}_{\mathcal{AB}}(E) = \frac{i}{2\pi}\int_{-\infty}^{\infty}\left(e^{\frac{\omega}{\vartheta}} - \eta\right) I_{\mathcal{AB}}(\omega)\frac{d\omega}{E-\omega+i\varepsilon}; \tag{26.11}$$

similarly, for the advanced function, we find

$$G^{(a)}_{\mathcal{AB}}(E) = \frac{i}{2\pi}\int_{-\infty}^{\infty}\left(e^{\frac{\omega}{\vartheta}} - \eta\right) I_{\mathcal{AB}}(\omega)\frac{d\omega}{E-\omega-i\varepsilon}. \tag{26.12}$$

We shall next consider E as a complex variable. Taking into consideration the fact that the spectral density $I_{\mathcal{AB}}(\omega)$ is, by definition, the same in the formulas (26.11) and (26.12), we can combine these two formulas:

$$G_{\mathcal{A}\mathcal{B}}(E) = \frac{i}{2\pi}\int_{-\infty}^{\infty}\left(e^{\frac{\omega}{\vartheta}} - \eta\right) I_{\mathcal{A}\mathcal{B}}(\omega)\frac{d\omega}{E-\omega} = \tag{26.13}$$

$$= \begin{cases} G^{(r)}_{\mathcal{A}\mathcal{B}}(E), & \operatorname{Im} E > 0, \\ G^{(a)}_{\mathcal{A}\mathcal{B}}(E), & \operatorname{Im} E < 0. \end{cases}$$

The formulas (26.11)-(26.13) are known as the spectral representations of the Green's functions. The functions $G^{(r)}_{\mathcal{A}\mathcal{B}}(E)$ and $G^{(a)}_{\mathcal{A}\mathcal{B}}(E)$ are analytic in the upper and lower half-planes, respectively. They can be regarded as a single analytic function $G_{\mathcal{A}\mathcal{B}}(E)$, having singularities along its real axis (poles, or cut lines). Therefore, we shall now omit the indices (r) and (a) whenever this will not lead to confusion.

The analyticity of the function $G_{\mathcal{A}\mathcal{B}}(E)$ follows from the theorem of Bogolyubov and Parasyuk (1956) that the complex function G (E), generalized in the Sobolev–Shvarts sense, has analytic continuation in the upper (lower) half-plane of the complex variable E if its Fourier transformation G(t) vanishes at $t < 0$ ($t > 0$). The latter condition is necessary and sufficient.

In general, a Green's function is a many-valued function of a complex variable. Its singularities lie on the real axis on the first sheet of a Riemann surface; on other sheets, the singularities may shift to the complex plane. This may lead to the appearance of complex poles.

Similarly, we can find the spectral representations for the Green's causal function (25.9). They have the form:

$$G^{(c)}_{\mathcal{A}\mathcal{B}}(E) = \frac{i}{2\pi}\int_{-\infty}^{\infty} I_{\mathcal{A}\mathcal{B}}(\omega)\left\{\frac{e^{\frac{\omega}{\vartheta}}}{E-\omega+i\varepsilon} - \frac{\eta}{E-\omega-i\varepsilon}\right\}d\omega =$$

$$= \frac{i}{2\pi}\int_{-\infty}^{\infty}\left(e^{\frac{\omega}{\vartheta}} - \eta\right) I_{\mathcal{A}\mathcal{B}}(\omega)\left\{P\frac{1}{E-\omega} - i\pi\frac{e^{\frac{\omega}{\vartheta}}+\eta}{e^{\frac{\omega}{\vartheta}}-\eta}\,\delta(E-\omega)\right\}d\omega. \tag{26.14}$$

The Green's causal functions are defined only along the real axis and cannot be continued into the complex plane if $\vartheta \neq 0$. This complicates somewhat their applications. Therefore, we shall use only the retarded and advanced Green's functions.

If the function $G_{\mathcal{A}\mathcal{B}}(E)$ is known, we can find the spectral density $I_{\mathcal{A}\mathcal{B}}(\omega)$ from the relationship

$$\left(e^{\frac{\omega}{\vartheta}} - \eta\right) I_{\mathcal{A}\mathcal{B}}(\omega) = G_{\mathcal{A}\mathcal{B}}(\omega + i\varepsilon) - G_{\mathcal{A}\mathcal{B}}(\omega - i\varepsilon). \tag{26.15}$$

We then use the well-known relationship for the δ – function [Dirac (1960)]:

$$\frac{1}{x \pm i\varepsilon} = P\frac{1}{x} \mp i\pi\delta(x) \qquad (\varepsilon \to +0), \tag{26.16}$$

where P represents the principal value. Next, using the formulas (26.13), we find that Eq. (26.15) is satisfied identically:

$$G_{\mathcal{A}\mathcal{B}}(\omega + i\varepsilon) - G_{\mathcal{A}\mathcal{B}}(\omega - i\varepsilon) =$$

$$= \frac{i}{2\pi} \int_{-\infty}^{\infty} \left(e^{\frac{\omega'}{\vartheta}} - \eta\right) I_{\mathcal{A}\mathcal{B}}(\omega') \left(\frac{1}{\omega - \omega' + i\varepsilon} - \frac{1}{\omega - \omega' - i\varepsilon}\right) d\omega' =$$

$$= \int_{-\infty}^{\infty} \left(e^{\frac{\omega'}{\vartheta}} - \eta\right) I_{\mathcal{A}\mathcal{B}}(\omega')\, \delta(\omega - \omega')\, d\omega' = \left(e^{\frac{\omega}{\vartheta}} - \eta\right) I_{\mathcal{A}\mathcal{B}}(\omega).$$

After finding the spectral density $I_{\mathcal{A}\mathcal{B}}(\omega)$, we can calculate the statistical averages of the operator products of the type (25.3), (25.5). These quantities are of fundamental interest in statistical mechanics problems because they give the conventional correlation functions if $t = t'$ and the time correlation functions if $t \neq t'$. The latter are frequently used in investigations of nonstationary processes.

We shall now show that the singularities of the Green's functions determine the energy of elementary excitations in a system [cf., Bonch-Bruevich and Tyablikov (1961)].

Let a Green's function have poles only at the points ω_j on the real axis. In accordance with Eqs. (26.15), (26.16), the spectral density will have δ-type singularities at the points ω_j, and the correlation function (26.2) will oscillate at the frequencies ω_j. From the definition of the spectral density given by Eq. (26.4), it follows that when $\vartheta = 0$, the quantities ω_j are the exact eigenvalues of the Hamiltonian of the system. When $\vartheta \neq 0$, the quantities ω_j are functions of the temperature and the chemical potential, and they cannot be given a purely mechanical meaning. However, in this case also they represent undamped oscillations of the system.

In general, singularities of a more complicated kind may appear on the real axis, but then we can find a sheet of the analytic function G(E) on which the singularities shift to the complex plane and have, at least approximately, the form of poles. The imaginary parts of poles then represent the damping of the corresponding oscillations in the system. If the damping is sufficiently weak, we can introduce the concept of quasi-stationary states. The correlation functions will have the form $f(t)\exp(-i\omega_j t)$, where $f(t)$ describes the damping of the states [$f(t) \to 0$ when $t \to \infty$]. The constant representing the damping should be small compared with the frequencies ω_j measured from the ground state, since, otherwise, the concept of elementary excitations loses its meaning. In the future, when speaking of a spectrum of a system when $\vartheta \geq 0$, we shall mean these quasi-stationary states.

The problem of the relationship between elementary excitations and poles of the Green's functions was considered, using model representations, by ter Haar and Parry (1962) and Bolsterli (1960). It was considered in greater detail by Pike (1964), who showed that the problem could be reduced to an investigation of poles of a Green's function, obtained by averaging over the ground state of the system.

c. Spectral Representations at Zero Temperature. We shall now consider the spectral representations of the time correlation functions and the Green's functions at zero temperature ($\vartheta = 0$).

The eigenfunctions of the ground state will be denoted by C_ν^0 and its energy will be assumed to be zero:

$$\mathcal{H} C_\nu^0 = 0. \tag{26.17}$$

Using the definition (26.4) of the spectral density, we shall calculate the limits of the functions $I_{\mathcal{A}\mathcal{B}}(\omega)$, $I_{\mathcal{A}\mathcal{B}}(\omega)e^{\frac{\omega}{\vartheta}}$ as $\vartheta \to 0$. For example,

$$\lim_{\vartheta\to 0} I_{\mathcal{A}\mathcal{B}}(\omega) = \lim_{\vartheta\to 0} Q^{-1} \sum_{\mu,\nu} (C_\nu^*, \mathcal{B}(0) C_\mu)(C_\mu^*, \mathcal{A}(0) C_\nu) \times$$
$$\times e^{-\frac{\mathcal{E}_\nu}{\vartheta}} \delta(\mathcal{E}_\nu - \mathcal{E}_\mu - \omega) =$$
$$= Q_0^{-1} \sum_{\mu,\nu} (C_\nu^{*0}, \mathcal{B}(0) C_\mu)(C_\mu^*, \mathcal{A}(0) C_\nu^0) \delta(\mathcal{E}_\mu^0 + \omega).$$

where $\mathcal{E}_\mu^0$ are the eigenvalues of $\mathscr{H}$ when $\vartheta = 0$ ($\mathscr{H}$ includes the chemical potential), Q_0 is the value of Q when $\vartheta = 0$. Since $\mathcal{E}_\mu^0 \geq 0$, we can easily show that

$$\lim_{\vartheta \to 0} I_{\mathcal{A}\mathcal{B}}(\omega) = \theta(-\omega)\, I^0_{\mathcal{A}\mathcal{B}}(\omega) \tag{26.18}$$

and, similarly,

$$\lim_{\vartheta \to 0} I_{\mathcal{A}\mathcal{B}}(\omega)\, e^{\frac{\omega}{\vartheta}} = \theta(\omega)\, I^0_{\mathcal{A}\mathcal{B}}(\omega), \tag{26.19}$$

where

$$I^0_{\mathcal{A}\mathcal{B}}(\omega) = \begin{cases} Q_0^{-1} \sum\limits_{\mu,\nu} \left(C_\nu^{*0}, \mathcal{B}(0)\, C_\mu\right)\left(C_\mu^*, \mathcal{A}(0)\, C_\nu^0\right) \delta\left(\mathcal{E}_\mu^0 + \omega\right), & \omega < 0, \\ Q_0^{-1} \sum\limits_{\mu,\nu} \left(C_\nu^{*0}, \mathcal{A}(0)\, C_\mu\right)\left(C_\mu^*, \mathcal{B}(0)\, C_\nu^0\right) \delta\left(\mathcal{E}_\mu^0 - \omega\right), & \omega > 0. \end{cases} \tag{26.20}$$

We note that the function $I^0_{\mathcal{A}\mathcal{B}}(\omega)$ is continuous at zero, because the limits

$$\lim_{\omega \to -0} I^0_{\mathcal{A}\mathcal{B}}(\omega) = Q_0^{-1} \sum_{\mu,\nu} \left(C_\nu^{*0}, \mathcal{B}(0)\, C_\mu^0\right)\left(C_\mu^{*0}, \mathcal{A}(0)\, C_\nu^0\right),$$
$$\lim_{\omega \to +0} I^0_{\mathcal{A}\mathcal{B}}(\omega) = Q_0^{-1} \sum_{\mu,\nu} \left(C_\nu^{*0}, \mathcal{A}(0)\, C_\mu^0\right)\left(C_\mu^{*0}, \mathcal{B}(0)\, C_\nu^0\right)$$

coincide.

The spectral representations of the time correlation functions have, in accordance with Eqs. (26.2), (26.3), and (26.18), (26.19), the following form at zero temperature:

$$\langle \mathcal{B}(t')\, \mathcal{A}(t) \rangle = \int_{-\infty}^{0} I^0_{\mathcal{A}\mathcal{B}}(\omega)\, e^{-i\omega(t-t')}\, d\omega, \tag{26.21}$$

$$\langle \mathcal{A}(t)\, \mathcal{B}(t') \rangle = \int_{0}^{\infty} I^0_{\mathcal{A}\mathcal{B}}(\omega)\, e^{-i\omega(t-t')}\, d\omega. \tag{26.22}$$

We shall now go to the limit $\vartheta \to 0$ in the spectral representations (26.11), (26.12), (26.14) of the retarded, advanced, and causal Green's functions. This gives [Bogolyubov (1961), Bonch-Bruevich (1956)]

$$G^{(r)}_{\mathcal{AB}}(E)=\frac{i}{2\pi}\int\limits_{-\infty}^{\infty}\frac{\theta(\omega)-\eta\theta(-\omega)}{E-\omega+i\varepsilon}I^{0}_{\mathcal{AB}}(\omega)\,d\omega, \tag{26.23}$$

$$G^{(a)}_{\mathcal{AB}}(E)=\frac{i}{2\pi}\int\limits_{-\infty}^{\infty}\frac{\theta(\omega)-\eta\theta(-\omega)}{E-\omega-i\varepsilon}I^{0}_{\mathcal{AB}}(\omega)\,d\omega, \tag{26.24}$$

$$G^{(c)}_{\mathcal{AB}}(E)=\frac{i}{2\pi}\int\limits_{-\infty}^{\infty}\frac{\theta(\omega)-\eta\theta(-\omega)}{E-\omega+i\varepsilon\,[\theta(\omega)-\theta(-\omega)]}I^{0}_{\mathcal{AB}}(\omega)\,d\omega, \tag{26.25}$$

Hence, we see that the retarded and advanced Green's functions $G^{(r)}_{\mathcal{AB}}(E)$, $G^{(a)}_{\mathcal{AB}}(E)$ are always the limits ($\varepsilon\rightarrow+0$) of the following function of the complex variable E:

$$G_{\mathcal{AB}}(E)=\frac{i}{2\pi}\int\limits_{-\infty}^{\infty}\left(e^{\frac{\omega}{\vartheta}}-\eta\right)I_{\mathcal{AB}}(\omega)\frac{d\omega}{E-\omega},$$

while the causal Green's function has this property only at zero temperature. For finite temperatures, the function

$$G^{(c)}_{\mathcal{AB}}(E)=\frac{i}{2\pi}\int\limits_{-\infty}^{\infty}I_{\mathcal{AB}}(\omega)\left\{\frac{e^{\frac{\omega}{\vartheta}}}{E-\omega+i\varepsilon}-\frac{\eta}{E-\omega-i\varepsilon}\right\}d\omega$$

cannot be extended into the complex plane.

The effectiveness of the Green's function method is to a considerable extent due to the use of the spectral representations. Initially, these representations were introduced into quantum field theory by Cällen (1952) and Lehmann (1954). The spectral representations of the Green's functions in statistical mechanics were introduced by Bonch-Bruevich (1956), Landau (1958), Gor'kov (1958), Martin and Schwinger (1958), Bogolyubov and Tyablikov (1959), Kogan (1959), and Bonch-Bruevich and Kogan (1960). The spectral representations of this type were used first in statistical mechanics (theory of fluctuations and statistical mechanics of irreversible processes) by Callen and Welton (1951).

Sec. 27. Dispersion Relations*

The formulas which relate the real and imaginary components of the Green's functions for real values of E (E = ω) are known as the dispersion relations.

Using the relationship (26.16), we shall transform the formulas (26.11) and (26.12) into

$$G^{(j)}_{\mathcal{A}\mathcal{B}}(\omega) = \pm \frac{1}{2}\left(e^{\frac{\omega}{\vartheta}} - \eta\right) I_{\mathcal{A}\mathcal{B}}(\omega) + \\ + \frac{i}{2\pi} P \int_{-\infty}^{\infty} \left(e^{\frac{\omega'}{\vartheta}} - \eta\right) I_{\mathcal{A}\mathcal{B}}(\omega') \frac{d\omega'}{\omega - \omega'} \quad (j = r,\ a). \tag{27.1}$$

P represents the principal value and the plus sign is used for the retarded function, and the minus for the advanced function. We shall also assume that the spectral density is real.

The real and imaginary components of the functions (27.1) then become

$$\mathrm{Re}\, G^{(j)}_{\mathcal{A}\mathcal{B}}(\omega) = \pm \frac{1}{2}\left(e^{\frac{\omega}{\vartheta}} - \eta\right) I_{\mathcal{A}\mathcal{B}}(\omega),$$
$$\mathrm{Im}\, G^{(j)}_{\mathcal{A}\mathcal{B}}(\omega) = \frac{1}{2\pi} P \int_{-\infty}^{\infty} \left(e^{\frac{\omega'}{\vartheta}} - \eta\right) I_{\mathcal{A}\mathcal{B}}(\omega') \frac{d\omega'}{\omega - \omega'}. \tag{27.2}$$

Hence, we obtain the dispersion relations for the real and imaginary components of the Green's functions

$$\mathrm{Im}\, G^{(j)}_{\mathcal{A}\mathcal{B}}(\omega) = \pm \frac{1}{\pi} P \int_{-\infty}^{\infty} \frac{\mathrm{Re}\, G^{(j)}_{\mathcal{A}\mathcal{B}}(\omega')}{\omega - \omega'} d\omega', \tag{27.3}$$

*Dispersion relations have been considered by many workers. They represent, basically, the properties of limits of Cauchy-type integrals, which were established first by Yu. V. Sokhotskii in 1873 and later by K. Plemel in 1908. See, for example, Lavrent'ev and Shabat (1958).

where the plus sign is used for the retarded functions and the minus sign for the advanced functions.

The formula (26.14) for the causal Green's function may be written in the following form:

$$G^{(c)}_{\mathcal{AB}}(\omega)=\frac{1}{2}\left(e^{\frac{\omega}{\vartheta}}+\eta\right)I_{\mathcal{AB}}(\omega)+ \\ +\frac{i}{2\pi}P\int_{-\infty}^{\infty}\left(e^{\frac{\omega'}{\vartheta}}-\eta\right)I_{\mathcal{AB}}(\omega')\frac{d\omega'}{\omega-\omega'}. \qquad (27.4)$$

Noting that the real and imaginary components of the causal Green's function have the form

$$\operatorname{Re} G^{(c)}_{\mathcal{AB}}(\omega)=\frac{1}{2}\left(e^{\frac{\omega}{\vartheta}}+\eta\right)I_{\mathcal{AB}}(\omega),$$

$$\operatorname{Im} G^{(c)}_{\mathcal{AB}}(\omega)=\frac{1}{2\pi}P\int_{-\infty}^{\infty}\left(e^{\frac{\omega'}{\vartheta}}-\eta\right)I_{\mathcal{AB}}(\omega')\frac{d\omega'}{\omega-\omega'}, \qquad (27.5)$$

we obtain the dispersion relation

$$\operatorname{Im} G^{(c)}_{\mathcal{AB}}(\omega)=\frac{1}{\pi}P\int_{-\infty}^{\infty}\frac{e^{\frac{\omega'}{\vartheta}}-\eta}{e^{\frac{\omega'}{\vartheta}}+\eta}\,\frac{\operatorname{Re} G^{(c)}_{\mathcal{AB}}(\omega')}{\omega-\omega'}\,d\omega'. \qquad (27.6)$$

Comparing the formulas (27.2)-(27.3) and (27.5)-(27.6), we find that if we know one of the functions, for example, the causal function, the other two functions can be found through the spectral density. In this sense, the retarded, advanced, and causal Green's functions are equivalent.

Sec. 28. General Properties of the Green's Functions*

In addition to the dispersion relations, the Green's functions also obey a number of relationships which are independent of the actual form of the Hamiltonian of a system. We shall consider some of these in the present section.

In order to avoid references to the formulas given in earlier sections, we shall write once again the definitions of the retarded and advanced Green's functions,

$$\langle\langle \mathcal{A}(t) | \mathcal{B}(t')\rangle\rangle^{(r)} = \theta(t-t')\langle \mathcal{A}(t)\mathcal{B}(t') - \eta\mathcal{B}(t')\mathcal{A}(t)\rangle, \quad (28.1)$$

$$\langle\langle \mathcal{A}(t) | \mathcal{B}(t')\rangle\rangle^{(a)} = -\theta(t'-t)\langle \mathcal{A}(t)\mathcal{B}(t') - \eta\mathcal{B}(t')\mathcal{A}(t)\rangle \quad (28.2)$$

and their spectral representations

$$\langle\langle \mathcal{A} | \mathcal{B}\rangle\rangle_E^{(j)} = \frac{i}{2\pi}\int_{-\infty}^{\infty}\left(e^{\frac{\omega}{\vartheta}} - \eta\right) I_{\mathcal{A}\mathcal{B}}(\omega)\frac{d\omega}{E-\omega \pm i\varepsilon}$$
$$(j = r,\ a) \quad (28.3)$$

[in Eq. (28.3), the plus sign is used for the retarded function and the minus sign for the advanced function].

The time correlation functions are given by

$$\langle \mathcal{A}(t)\mathcal{B}(t')\rangle = \int_{-\infty}^{\infty} I_{\mathcal{A}\mathcal{B}}(\omega)\, e^{\frac{\omega}{\vartheta}} e^{-i\omega(t-t')}\, d\omega, \quad (28.4)$$

$$\langle \mathcal{B}^{+}(t')\mathcal{A}^{+}(t)\rangle = \int_{-\infty}^{\infty} I_{\mathcal{B}^{+}\mathcal{A}^{+}}(\omega)\, e^{\frac{\omega}{\vartheta}} e^{-i\omega(t'-t)}\, d\omega =$$
$$= \int_{-\infty}^{\infty} I_{\mathcal{A}^{+}\mathcal{B}^{+}}(\omega)\, e^{-i\omega(t-t')}\, d\omega. \quad (28.5)$$

The function (28.5) may be considered as the conjugate of (28.4) and, therefore,

$$\langle \mathcal{B}^{+}(t')\mathcal{A}^{+}(t)\rangle = \int_{-\infty}^{\infty} I^{*}_{\mathcal{A}\mathcal{B}}(\omega)\, e^{\frac{\omega}{\vartheta}} e^{i\omega(t-t')}\, d\omega.$$

Symmetry Properties of the Green's Functions.* Using the definition (28.2), we easily obtain:

$$\langle\langle \mathcal{A}(t) | \mathcal{B}(t')\rangle\rangle^{(a)} = -\theta(t'-t)\langle \mathcal{A}(t)\mathcal{B}(t') - \eta\mathcal{B}(t')\mathcal{A}(t)\rangle =$$
$$= \eta\theta(t'-t)\langle \mathcal{B}(t')\mathcal{A}(t) - \eta\mathcal{A}(t)\mathcal{B}(t')\rangle. \quad (28.6)$$

*See Bogolyubov (1961); for the symmetry properties of the correlation functions, see also Kubo (1957).

Employing the definition (28.1) of the retarded function, we see that

$$\langle\langle \mathcal{A}(t) \,|\, \mathcal{B}(t') \rangle\rangle^{(a)} = \eta \langle\langle \mathcal{B}(t') \,|\, \mathcal{A}(t) \rangle\rangle^{(r)}. \tag{28.7}$$

Using the Fourier transforms with respect to time, we obtain

$$\int_{-\infty}^{\infty} \langle\langle \mathcal{A} \,|\, \mathcal{B} \rangle\rangle_E^{(a)} e^{-iE(t-t')}\, dE = \eta \int_{-\infty}^{\infty} \langle\langle \mathcal{B} \,|\, \mathcal{A} \rangle\rangle_E^{(r)} e^{-iE(t'-t)}\, dE.$$

Consequently, for the real values of the argument (E = ω),

$$\langle\langle \mathcal{A} \,|\, \mathcal{B} \rangle\rangle_\omega^{(a)} = \eta \langle\langle \mathcal{B} \,|\, \mathcal{A} \rangle\rangle_{-\omega}^{(r)}. \tag{28.8}$$

To go over to the complex plane of E, we shall consider the spectral representations (28.3) of these functions:

$$\langle\langle \mathcal{A} \,|\, \mathcal{B} \rangle\rangle_E^{(a)} = \frac{i}{2\pi} \int_{-\infty}^{\infty} \left(e^{\frac{\omega}{\theta}} - \eta\right) I_{\mathcal{A}\mathcal{B}}(\omega) \frac{d\omega}{F - \omega - i\varepsilon},$$

$$\langle\langle \mathcal{B} \,|\, \mathcal{A} \rangle\rangle_{-E}^{(r)} = \frac{i}{2\pi} \int_{-\infty}^{\infty} \left(e^{\frac{\omega}{\theta}} - \eta\right) I_{\mathcal{A}\mathcal{B}}(\omega) \frac{d\omega}{-E - \omega + i\varepsilon}.$$

We see from these representations that both functions are analytic in the lower half-plane. Therefore, for the complex values of the argument, we have, instead of Eq. (28.8),

$$\langle\langle \mathcal{A} \,|\, \mathcal{B} \rangle\rangle_E = \eta \langle\langle \mathcal{B} \,|\, \mathcal{A} \rangle\rangle_{-E}. \tag{28.9}$$

We shall consider the expression which is the conjugate of (28.1):

$$\begin{aligned}
&\{\langle\langle \mathcal{A}(t) \,|\, \mathcal{B}(t') \rangle\rangle^{(r)}\}^* = \\
&\quad = -\eta\theta(t-t')\langle \mathcal{A}^+(t)\, \mathcal{B}^+(t') - \eta \mathcal{B}^+(t')\, \mathcal{A}^+(t) \rangle = \\
&\quad\quad = \theta(t-t')\langle \mathcal{B}^+(t')\, \mathcal{A}^+(t) - \eta \mathcal{A}^+(t)\, \mathcal{B}^+(t') \rangle.
\end{aligned}$$

According to the definitions of the retarded and advanced functions [cf., Eqs. (28.1) and (28.2)], the above expressions may be written as follows:

$$\{\langle\langle \mathcal{A}(t) | \mathcal{B}(t')\rangle\rangle^{(r)}\}^* = -\eta\langle\langle \mathcal{A}^+(t) | \mathcal{B}^+(t')\rangle\rangle^{(r)} = \tag{28.10}$$

$$= -\langle\langle \mathcal{B}^+(t') | \mathcal{A}^+(t)\rangle\rangle^{(a)}. \tag{28.11}$$

A comparison of the Fourier coefficients shows that, on the real axis,

$$\{\langle\langle \mathcal{A} | \mathcal{B}\rangle\rangle^{(r)}_{\omega}\}^* = -\eta\langle\langle \mathcal{A}^+ | \mathcal{B}^+\rangle\rangle^{(r)}_{-\omega} = \tag{28.12}$$

$$= -\langle\langle \mathcal{B}^+ | \mathcal{A}^+\rangle\rangle^{(a)}_{\omega}. \tag{28.13}$$

Comparison of the Fourier coefficients in the expressions (28.5), (28.6) gives, for the spectral densities,

$$I^*_{\mathcal{A}\mathcal{B}}(\omega) = I_{\mathcal{B}^+\mathcal{A}^+}(\omega), \tag{28.14}$$

$$I^*_{\mathcal{A}\mathcal{B}}(\omega)\, e^{\frac{\omega}{\theta}} = I_{\mathcal{A}^+\mathcal{B}^+}(-\omega). \tag{28.15}$$

Let $\mathcal{B}^+ = \mathcal{A}$. Then, it follows from Eq. (28.14) that $I_{\mathcal{A}\mathcal{A}^+}$ is real:

$$I^*_{\mathcal{A}\mathcal{A}^+}(\omega) = I_{\mathcal{A}\mathcal{A}^+}(\omega). \tag{28.16}$$

Then, the Green's function

$$\langle\langle \mathcal{A} | \mathcal{A}^+\rangle\rangle_E = \frac{i}{2\pi}\int_{-\infty}^{\infty}\left(e^{\frac{\omega}{\theta}} - \eta\right) I_{\mathcal{A}\mathcal{A}^+}(\omega)\,\frac{d\omega}{E-\omega} \tag{28.17}$$

is a function of a complex variable with purely imaginary coefficients. Consequently,

$$\{\langle\langle \mathcal{A} | \mathcal{A}^+\rangle\rangle_E\}^* = -\langle\langle \mathcal{A} | \mathcal{A}^+\rangle\rangle_{E^*}. \tag{28.18}$$

From the fact that $I_{\mathcal{A}\mathcal{A}^+}(\omega)$ of Eq. (28.16) is real, it follows that the usual averages are also real:

$$\langle \mathcal{A}\mathcal{A}^{+}\rangle = \langle \mathcal{A}\mathcal{A}^{+}\rangle^{*} = \int_{-\infty}^{\infty} I_{\mathcal{A}\mathcal{A}^{+}}(\omega)\, e^{\frac{\omega}{\vartheta}}\, d\omega,$$

$$\langle \mathcal{A}^{+}\mathcal{A}\rangle = \langle \mathcal{A}^{+}\mathcal{A}\rangle^{*} = \int_{-\infty}^{\infty} I_{\mathcal{A}\mathcal{A}^{+}}(\omega)\, d\omega. \qquad (28.19)$$

According to Eq. (26.8), we have

$$\int_{-\infty}^{\infty} \left(e^{\frac{\omega}{\vartheta}} - \eta\right) I_{\mathcal{A}\mathcal{B}}(\omega)\, d\omega = \langle \mathcal{A}\mathcal{B} - \eta\mathcal{B}\mathcal{A}\rangle. \qquad (28.20)$$

Let $\mathcal{B} = \mathcal{A}$, $\eta = 1$. Then we obtain

$$\int_{-\infty}^{\infty} I_{\mathcal{A}\mathcal{A}}(\omega) \left(e^{\frac{\omega}{\vartheta}} - 1\right) d\omega = 0. \qquad (28.21)$$

Let $\mathcal{A} = a_f$, $\mathcal{B} = a_f^{+}$, where a_f^{+}, a_f are operators with communications relationships of the Bose or Fermi type:

$$a_f a_{f'}^{+} - \eta a_{f'}^{+} a_f = \Delta(f - f').$$

Then, from Eq. (28.20), it follows that

$$\int_{-\infty}^{\infty} \left(e^{\frac{\omega}{\vartheta}} - \eta\right) I_{a_f a_{f'}^{+}}(\omega)\, d\omega = \Delta(f - f'). \qquad (28.22)$$

For the Pauli operators, we have

$$b_f b_{f'}^{+} - b_{f'}^{+} b_f = (1 - 2n_f)\, \Delta(f - f'),$$

and, therefore, in accordance with Eq. (28.20), we obtain

$$\int_{-\infty}^{\infty} \left(e^{\frac{\omega}{\vartheta}} - 1\right) I_{b_f b_{f'}^{+}}(\omega)\, d\omega = \langle 1 - 2n_f\rangle\, \Delta(f - f'). \qquad (28.23)$$

Symmetry Properties of the Green's Functions in the Case of Time Inversion Invariance. Let the equations of motion for the operators $\mathcal{A}$, $\mathcal{B}$ be invariant with respect to time inversion:

$$t \to -t, \quad i \to -i. \tag{28.24}$$

Then, the operators $\mathcal{A}(t)$, $\mathcal{B}(t)$ go into themselves under the transformation (28.24).

Let us now consider Eq. (28.4). We shall carry out the transformations: $t \to -t$, $t' \to -t'$, $i \to -i$. Then, $I_{\mathcal{A}\mathcal{B}}(\omega)$ on the right-hand side is, in general, replaced with $I^*_{\mathcal{A}\mathcal{B}}(\omega)$. However, since the left-hand side remains unchanged, it follows that the spectral density is real:

$$I^*_{\mathcal{A}\mathcal{B}}(\omega) = I_{\mathcal{A}\mathcal{B}}(\omega). \tag{28.25}$$

The corresponding Green's function

$$\langle\langle \mathcal{A} \mid \mathcal{B} \rangle\rangle_E = \frac{i}{2\pi} \int_{-\infty}^{\infty} \left(e^{\frac{\omega}{\vartheta}} - \eta\right) I_{\mathcal{A}\mathcal{B}}(\omega) \frac{d\omega}{E - \omega} \tag{28.26}$$

is a function of a complex variable with purely imaginary coefficients. Therefore,

$$\{\langle\langle \mathcal{A} \mid \mathcal{B} \rangle\rangle_E\}^* = -\langle\langle \mathcal{A} \mid \mathcal{B} \rangle\rangle_{E^*}. \tag{28.27}$$

From Eqs. (28.4), (28.6), and (28.25), it follows that

$$\begin{aligned} \langle \mathcal{A}(t)\, \mathcal{B}(t') \rangle &= \langle \mathcal{B}^+(t)\, \mathcal{A}^+(t') \rangle, \\ \langle \mathcal{B}(t')\, \mathcal{A}(t) \rangle &= \langle \mathcal{A}^+(t')\, \mathcal{B}^+(t) \rangle. \end{aligned} \tag{28.28}$$

We shall multiply the second of the expressions in Eq. (28.28) by η and subtract the product from the first expression; multiplying the results by $\theta\ (t - t')$ or $-\theta\ (t' - t)$, we obtain

$$\langle\langle \mathcal{A}(t) \mid \mathcal{B}(t') \rangle\rangle^{(j)} = \langle\langle \mathcal{B}^+(t) \mid \mathcal{A}^+(t') \rangle\rangle^{(j)} \quad (j = r,\ a). \tag{28.29}$$

Hence, we obtain the following expressions for the Fourier transforms with respect to time:

$$\langle\langle \mathcal{A} \mid \mathcal{B} \rangle\rangle^{(j)}_\omega = \langle\langle \mathcal{B}^+ \mid \mathcal{A}^+ \rangle\rangle^{(j)}_\omega, \tag{28.30}$$

$$\langle\langle \mathcal{A} \mid \mathcal{B} \rangle\rangle_E = \langle\langle \mathcal{B}^+ \mid \mathcal{A}^+ \rangle\rangle_E. \tag{28.31}$$

Let $\mathcal{A}$, $\mathcal{B}$ be self-adjoint operators. Then, if $t = t'$, it follows from Eq. (28.28) that $\langle \mathcal{A}\mathcal{B} - \mathcal{B}\mathcal{A} \rangle = 0$. Next, using Eq. (28.20), we find

$$\int_{-\infty}^{\infty}\left(e^{\frac{\omega}{\vartheta}}-\eta\right) I_{\mathcal{A}\mathcal{B}}(\omega)\,d\omega=(1-\eta)\langle\mathcal{A}\mathcal{B}\rangle. \tag{28.32}$$

We note that the formulas (28.25)-(28.32) apply only to those operators which are invariant under time inversion (28.24).

<u>Behavior of the Retarded Green's Functions on the Real Axis.</u> We shall introduce the matrix

$$G(\omega)=\begin{pmatrix} G_{11}(\omega) & G_{12}(\omega) \\ G_{21}(\omega) & G_{22}(\omega)\end{pmatrix}, \tag{28.33}$$

where

$$G_{11}(\omega)=\langle\langle\mathcal{B}\mid\mathcal{B}\rangle\rangle_{\omega}^{(r)},\quad G_{12}(\omega)=\langle\langle\mathcal{B}\mid\mathcal{B}^{+}\rangle\rangle_{\omega}^{(r)},\ \ldots, \tag{28.34}$$

and its inverse $F(\omega) = G^{-1}(\omega)$.

Using the formula (28.12), we find

$$G_{11}(\omega)=-\eta G_{22}^{*}(-\omega),\quad G_{12}(\omega)=-\eta G_{21}^{*}(-\omega), \tag{28.35}$$

and, if the equations of motion are invariant under time inversion, we obtain, using Eq. (28.30)

$$G_{11}(\omega)=G_{22}(\omega). \tag{28.36}$$

The same relationships apply also to the elements of the inverse matrix $F(\omega)$. For the determinants of the matrices G and F, we have the conditions

$$\begin{aligned} D(\omega)&=D^{*}(-\omega)=G_{11}G_{22}-G_{12}G_{21},\\ \Delta(\omega)&=\Delta^{*}(-\omega)=F_{11}F_{22}-F_{12}F_{21}. \end{aligned} \tag{28.37}$$

Let $G(\omega)$ have a pole at a point $\omega = \omega_R$. At the poles of the function $G(\omega)$, the function $F(\omega)$ is regular. Therefore, expressing $G_{ij}(\omega)$ in terms of $F_{ij}(\omega)$, we find that the poles of the function $G(\omega)$ correspond to the zeros of the equations $\Delta(\omega) = 0$, $\Delta(-\omega) = 0$. Consequently, in the region of a pole, we have

$$\begin{aligned} \Delta^{-1}(\omega)&=R(\omega)\left\{2\omega_R\Lambda(\omega)\right\}^{-1},\\ R(\omega)&=\frac{1}{\omega-\omega_R+i\varepsilon}-\frac{1}{\omega+\omega_R+i\varepsilon}, \end{aligned} \tag{28.38}$$

where $\Lambda(\omega)$ is some function which is regular in the vicinity of the point $\omega = \omega_R$.

An Inequality for the Spectral Densities and Green's Functions. According to Eqs. (26.4) and (28.14), the spectral densities obey the relationships

$$I_{\mathcal{A}\mathcal{A}^+}(\omega) \geq 0,$$
$$I^*_{\mathcal{A}\mathcal{B}}(\omega) = I_{\mathcal{B}^+\mathcal{A}^+}(\omega). \tag{28.39}$$

Next, it follows from the definition (26.4) that the spectral density is a bilinear form with respect to the operators $\mathcal{A}$, $\mathcal{B}$.

Using these relationships, we can show [Bogolyubov (1961)] that the spectral densities obey the inequality

$$I_{\mathcal{A}\mathcal{A}^+}(\omega)\, I^*_{\mathcal{B}\mathcal{B}^+}(\omega) \geq |I_{\mathcal{A}\mathcal{B}}(\omega)|^2, \tag{28.40}$$

and, if E = 0, the commutator Green's functions satisfy the inequality

$$G_{\mathcal{A}\mathcal{A}^+}(0)\, G^*_{\mathcal{B}\mathcal{B}^+}(0) \geq |G_{\mathcal{A}\mathcal{B}}(0)|^2. \tag{28.41}$$

To prove this, we shall consider an arbitrary bilinear form $\Phi(\mathcal{A}, \mathcal{B})$, having the properties

$$\Phi(\mathcal{A}, \mathcal{A}^+) \geq 0, \quad \Phi(\mathcal{B}, \mathcal{B}^+) \geq 0,$$
$$\{\Phi(\mathcal{A}, \mathcal{B})\}^* = \Phi(\mathcal{B}^+, \mathcal{A}^+), \tag{28.42}$$

and we shall show that Φ obeys the inequality

$$\Phi(\mathcal{A}, \mathcal{A}^+)\, \Phi^*(\mathcal{B}, \mathcal{B}^+) \geq |\Phi(\mathcal{A}, \mathcal{B})|^2. \tag{28.43}$$

In this proof, we shall bear in mind that, according to Eq. (28.42),

$$\Phi(x\mathcal{A} + y^*\mathcal{B}^+, x^*\mathcal{A}^+ + y\mathcal{B}) \geq 0$$

(x, y are arbitrary complex numbers), or

$$|x|^2 \Phi(\mathcal{A}, \mathcal{A}^+) + x^*y^*\Phi(\mathcal{B}^+, \mathcal{A}^+) + xy\Phi(\mathcal{A}, \mathcal{B}) + $$
$$+ |y|^2 \Phi(\mathcal{B}^+, \mathcal{B}) \geq 0. \tag{28.44}$$

We shall now assume

$$x^* = -\Phi(\mathcal{A}, \mathcal{B}), \quad x = -\Phi^*(\mathcal{A}, \mathcal{B}) = -\Phi(\mathcal{B}^+, \mathcal{A}^+),$$
$$y = y^* = \Phi(\mathcal{A}, \mathcal{A}^+).$$

Then, Eq. (28.44) transforms into

$$-|\Phi(\mathcal{A}, \mathcal{B})|^2 \Phi(\mathcal{A}, \mathcal{A}^+) + \{\Phi(\mathcal{A}, \mathcal{A}^+)\}^2 \Phi(\mathcal{B}^+, \mathcal{B}) \geq 0.$$

If $\Phi(\mathcal{A}, \mathcal{A}^+) \neq 0$, then the inequality (28.43) follows. We shall now show that if $\Phi(\mathcal{A}, \mathcal{A}^+) = 0$, then $\Phi(\mathcal{A}, \mathcal{B}) = 0$.

For this purpose, we shall substitute into Eq. (28.44)

$$x^* = -q\Phi(\mathcal{A}, \mathcal{B}), \quad x = -q\Phi(\mathcal{B}^+, \mathcal{A}^+),$$
$$\Phi(\mathcal{A}, \mathcal{A}^+) = 0, \quad y = y^* = 1,$$

where q is an arbitrary positive number.

As a result of this operation, we obtain

$$-2q|\Phi(\mathcal{A}, \mathcal{B})|^2 + \Phi(\mathcal{B}^+, \mathcal{B}) \geq 0.$$

When $\Phi(\mathcal{A}, \mathcal{B}) \neq 0$ and q is sufficiently large, the left-hand side of the above inequality can be made negative. Since this is impossible, it follows that $\Phi(\mathcal{A}, \mathcal{B}) = 0$ when $\Phi(\mathcal{A}, \mathcal{A}^+) = 0$.

We shall now allow for the fact that the conditions of Eq. (28.42) are satisfied by the bilinear forms

$$\Phi(\mathcal{A}, \mathcal{B}) = I_{\mathcal{A}\mathcal{B}}(\omega),$$
$$\Phi(\mathcal{A}\mathcal{B}) = \frac{1}{i} G_{\mathcal{A}\mathcal{B}}(0) = \frac{1}{2\pi} \int_{-\infty}^{\infty} I_{\mathcal{A}\mathcal{B}}(\omega) \frac{e^{\frac{\omega}{\vartheta}} - 1}{\omega} d\omega.$$

Consequently, they obey the inequalities (28.43), or (28.40) and (28.41).

Sec. 29. Perturbation Theory for the Green's Functions*

We shall consider the chain of equations

$$E\langle\langle\mathcal{A}|\mathcal{B}\rangle\rangle = \frac{i}{2\pi}\langle[\mathcal{A}, \mathcal{B}]_\eta\rangle + \langle\langle[\mathcal{A}, \mathcal{H}]|\mathcal{B}\rangle\rangle. \tag{29.1}$$

*See Tyablikov and Bonch-Bruevich (1962).

We shall assume that the Hamiltonian $\mathcal{H}$ of a system can be represented in the form

$$\mathcal{H} = \mathcal{H}_0 + \varepsilon\mathcal{H}_1, \tag{29.2}$$

where the "zeroth-approximation Hamiltonian" $\mathcal{H}_0$ is selected so that

$$[\mathcal{A}, \mathcal{H}_0] = K\mathcal{A}; \tag{29.3}$$

K is a linear operator, and ε is some parameter. Then the equation of motion for the operator $\mathcal{A}$ is written as follows:

$$i\frac{d\mathcal{A}}{dt} = K\mathcal{A} + \varepsilon R_1\mathcal{A}_1, \tag{29.4}$$

where R_1 is also a linear operator, and $\mathcal{A}_1$ is a new operator made up from a product of a larger number of second quantization operators than $\mathcal{A}$. Consequently, the chain of equations (29.1) can be written in the form

$$\begin{aligned}(E-K)\langle\langle\mathcal{A}\,|\,\mathcal{B}\rangle\rangle &= \frac{i}{2\pi}\langle[\mathcal{A}, \mathcal{B}]_\eta\rangle + \varepsilon R_1\langle\langle\mathcal{A}_1\,|\,\mathcal{B}\rangle\rangle,\\ (E-K_1)\langle\langle\mathcal{A}_1\,|\,\mathcal{B}\rangle\rangle &= \frac{i}{2\pi}\langle[\mathcal{A}_1, \mathcal{B}]_\eta\rangle + \varepsilon R_2\langle\langle\mathcal{A}_2\,|\,\mathcal{B}\rangle\rangle, \dots\end{aligned} \tag{29.5}$$

or, using the obvious notation, in the form

$$\begin{aligned}L_1G_1 &= I_1 + \varepsilon R_1G_2,\\ L_2G_2 &= I_2 + \varepsilon R_2G_3,\end{aligned} \tag{29.6}$$

where $G_1, G_2, \dots$ are the first, second, etc., Green's functions.

As an example, we shall consider a system of particles with the pair interaction. The Hamiltonian of such a system has, in the second quantization representation, the form of Eq. (3.36). We shall take $\mathcal{H}_0$ to be the Hamiltonian of free particles and $\mathcal{H}_1$ to be the interaction Hamiltonian. Then, if $\mathcal{A}$ is the particle annihilation (or creation) operator, K of Eqs. (29.4) and (29.5) will represent the internal energy of a free particle, and R_1 of Eqs. (29.4) and (29.5) will be an integral operator whose kernel is a matrix element of the energy of interaction between two particles. The operators $K_1, K_2, \dots$ and $R_2, R_3, \dots$ have similar meaning.

In general, the quantities L_n and I_n are some functions of the parameter ε:

$$L_n = L_n(\varepsilon), \quad I_n = I_n(\varepsilon). \tag{29.7}$$

In the operator $L_n(\varepsilon)$, we can separate the operator L_n^0 of the internal energy of the free particles and the operator εL_n^1, which depends on the interaction energy, and which is due to the same terms in the Hamiltonian as the operators R_i. The quantities $I_n(\varepsilon)$ are the correlation functions. Sometimes, it is possible to represent $I_n(\varepsilon)$ in the form of a sum of the correlation function I_n^0 for the free particles and the so-called irreducible part, which is of the order of the interaction energy (ε). Therefore, usually

$$\begin{gathered} L_n(\varepsilon) = L_n^0 + \varepsilon L_n^1, \\ L_1 \equiv L_1^0, \quad L_n^0 = \Sigma L_1, \\ I_n(\varepsilon) = I_n^0 + O(\varepsilon). \end{gathered} \tag{29.8}$$

We shall use G_n^0 to denote the solutions of the following equations:

$$L_n^0 G_n^0 = I_n. \tag{29.9}$$

We shall now apply the perturbation theory method to the chain of equations (29.6). We shall formally consider ε to be a small parameter ($\varepsilon \ll 1$), and we shall call $\mathscr{H}_0$ the "free-particle" Hamiltonian and $\mathscr{H}_1$ the interaction operator. The poles of the Green's functions determine the energy of elementary excitations and they naturally depend on the interaction between particles. An attempt to obtain the solutions of the chain of equations (29.6) as formal expansions in powers of the small parameter ε is equivalent to the expansion of the functions in terms of the poles. We shall attempt to develop the theory of perturbations for the inverse function, constructing formal expansions in terms of ε for the inverse function and not for G. The Green's function will be obtained from the inverse function. Using this approach, we can hope to obtain at least a correct description of the behavior of the Green's function in the region of the poles.

We shall define the mass operator M_1 for the first Green's

---

*See also Kelin and Prange (1958).

function by the equation*

$$(L_1 - M_1)\, G_1 = I_1, \tag{29.10}$$

where L_1 is the free-particle energy operator, and M_1 includes the interaction. The comparison of Eq. (29.10) with the first equation in (29.6) shows that

$$M_1 = \varepsilon R_1 G_2 G_1^{-1}, \tag{29.11}$$

where G_1^{-1} is an operator which is the inverse of G_1. We shall use the notation

$$X_1 = R_1 G_2 I_1^{-1}. \tag{29.12}$$

Substituting G_1 of Eq. (29.10) into Eq. (29.11), and using Eq. (29.12), we obtain

$$M_1 = \varepsilon X_1 (L_1 - M_1). \tag{29.13}$$

Hence, we find M_1:

$$M_1 = (1 + \varepsilon X_1)^{-1}\, \varepsilon X_1 L_1 \tag{29.14}$$

or, expanding in a series of powers of the explicit parameter ε,

$$M_1 = \{\varepsilon X_1 - (\varepsilon X_1)^2 + (\varepsilon X_1)^3 - \ldots\}\, L_1, \tag{29.15}$$

where X_1 is defined by Eq. (29.12).

The poles of the mass operator M_1 are the zeros of the operator $1 + \varepsilon X_1$ [i.e., the zeros of $(G_1^0)^{-1} G_1$]. The poles of the operator M_1 are lost in the expansion (29.15). Therefore, the expansion (29.15) has meaning only far from the zeros of $1 + \varepsilon X_1$. Elsewhere, one must use the expression (29.14).

In the vicinity of the poles of M_1, we can write approximately

$$M_1 = \{1 - (1 + \varepsilon X_1)^{-1}\}\, L_1 \approx -(1 + \varepsilon X_1)^{-1} L_1. \tag{29.16}$$

Again, in the vicinity of the poles of X_1 ($\varepsilon\ X_1 \gg 1$), the formal expansions in terms of ε have no meaning. We must point out that, since $X_1 = R_1 G_2 I_1^{-1}$, the poles of X_1 correspond to the poles of G_2.

The formulas (29.14), (29.15) represent the mass operator M_1 in terms of the second Green's function. Since M_1 is a quantity which, in a certain sense, is the inverse of G_1, we can use the standard perturbation theory to calculate it.

From the system of equations (29.6), we find

$$G_2 = G_2^0 + \varepsilon L_2^{-1} R_2 G_3 = \\ = G_2^0 + \varepsilon L_2^{-1} R_2 G_3^0 + \varepsilon^2 L_2^{-1} R_2 L_3^{-1} R_3 G_4^0 + \ldots \quad (29.17)$$

Substitution of Eq. (29.17) into Eq. (29.12) gives X_1 in the form of a series in powers of ε:

$$X_1 = R_1 G_2^0 I_1^{-1} + \varepsilon R_1 L_2^{-1} R_2 G_3^0 I_1^{-1} + \\ + \varepsilon^2 R_1 L_2^{-1} R_2 L_3^{-1} R_3 G_4^0 I_1^{-1} + \ldots \quad (29.18)$$

Finally, using Eq. (29.15), we obtain the expansions of the mass operator in powers of the interaction energy:

$$M_1 = \varepsilon M_1' + \varepsilon^2 M_1'' + \ldots, \quad (29.19)$$

where

$$M_1' = R_1 G_2^0 I_1^{-1} L_1 = R_1 G_2^0 G_1^{0^{-1}}, \quad (29.20)$$

$$M_1'' = R_1 L_2^{-1} R_2 G_3^0 G_1^{0^{-1}} - R_1 G_2^0 G_1^{0^{-1}} L_1^{-1} R_1 G_2^0 G_1^{0^{-1}}, \ldots \quad (29.21)$$

The expansion (29.17) of the function G_2 represents the solution of the truncated form of the chain of equations (29.6)

$$L_2 G_2 = I_2 + \varepsilon R_2 G_3, \\ \cdots\cdots\cdots \\ L_n G_n = I_n. \quad (29.22)$$

In solving actual problems, we may find it more convenient to find an approximate solution for G_2, in the form of a series in powers of ε, directly from the system of equations (29.22).

For spatially uniform systems, the mass operator is diagonal in the momentum representation.

In fact, in the coordinate representation, the operators $\mathcal{A}$, $\mathcal{B}$, which are used to construct the Green's functions, are functions of points in the space: $\mathcal{A} = \mathcal{A}(x)$, $\mathcal{B} = \mathcal{B}(y)$, and first Green's functions is a function of a pair of points: $G_1 = G_1(x, y)$. Since G_1 is invariant under translations, it will be a function only of the difference of the coordinates: $G_1 = G_1(x - y)$. Therefore, in the coordinate representation Eq. (29.10) is written in the form

$$\{E - K(x)\} G_1(x - y) - \int M_1(x, x') G_1(x' - y)\, dx' = \frac{i}{2\pi} \sigma(x - y), \tag{29.23}$$

where K(x) is the free-particle internal-energy operator – for example, the kinetic energy operator – and $\sigma = 2\pi I_1/i$ is some function. This equation should be invariant (for uniform systems) under a shift of all the coordinates by an arbitrary vector. Consequently, the mass operator should depend only on the difference of coordinates: $M_1(x,x') = M_1(x - x')$. Therefore, Eq. (29.23) becomes

$$\{E - K(\xi)\} G_1(\xi) - \int M_1(\xi - \xi') G_1(\xi')\, d\xi' = \frac{i}{2\pi} \sigma(\xi), \tag{29.24}$$

where $\xi = x - y$, $\xi' = x' - y$. Going over in Eq. (29.24) to Fourier space transforms, we obtain

$$(E - K_\nu) G_\nu - M_\nu G_\nu = \frac{i}{2\pi} \sigma_\nu, \tag{29.25}$$

where

$$G_\nu = \int G_1(\xi) e^{i(\xi,\nu)}\, d\xi, \quad M_\nu = \int M_1(\xi) e^{i(\xi,\nu)}\, d\xi,$$
$$\sigma_\nu = \int \sigma(\xi) e^{i(\xi,\nu)}\, d\xi. \tag{29.26}$$

It follows that for spatially uniform systems the first Green's function and its mass operator are diagonal in momentum. In the case of crystal lattices, the integrals are replaced by the corresponding sums over the lattice sites.

By definition, the mass operator is a function of the variable E. We shall assume the explicit form of this dependence to be known, and M_ν to be a "small quantity," in the spirit of the perturbation theory. In order to stress the latter point, we shall put the formal small parameter ε in front of M_ν. From Eq. (29.25), we obtain the following expression for the function G_ν:

$$G_\nu(E) = \frac{i}{2\pi} \frac{\sigma_\nu}{E - K_\nu - \varepsilon M_\nu(E)}. \tag{29.27}$$

The energy and lifetime of elementary excitations are determined in accordance with Sec. 26b by the poles of the Green's function:

$$E = K_\nu + \varepsilon M_\nu (E). \tag{29.28}$$

In the zeroth approximation with respect to ε, the poles are real: $E = K_\nu$. Since εM_ν is considered to be a "small correlation," we shall seek the solution of Eq. (29.28) in the form $E = \omega + i\Gamma$, where Γ is a small real quantity. We shall assume that

$$M_\nu (\omega \pm i\Gamma) = M'_\nu (\omega, \Gamma) \mp i M''_\nu (\omega, \Gamma), \tag{29.29}$$

where M'_ν, M''_ν are real functions. To determine the energy ω of elementary excitations and their lifetime Γ, we obtain a system of equations:

$$\begin{aligned} \omega &= K_\nu + \varepsilon M'_\nu (\omega, \Gamma), \\ \Gamma &= -\varepsilon M''_\nu (\omega, \Gamma) \end{aligned} \tag{29.30}$$

or, approximately, assuming Γ to be a small quantity

$$\omega \approx K_\nu + \varepsilon M'_\nu (\omega, 0), \quad \Gamma \approx -\varepsilon M''_\nu (\omega, 0). \tag{29.31}$$

In accordance with Eq. (26.15), the spectral density is given by

$$\begin{aligned} \left(e^{\frac{\omega}{\theta}} - \eta\right) I_\nu (\omega) &= G_\nu (\omega + i\Gamma) - G_\nu (\omega - i\Gamma) = \\ &= \frac{\sigma_\nu}{\pi} \frac{M''_\nu}{\left(\omega - M'_\nu\right)^2 + \left(M''_\nu\right)^2}. \end{aligned} \tag{29.32}$$

Finally, we obtain the following expression for the Green's function:

$$G_\nu (E) \approx \frac{i}{2\pi} \frac{\sigma_\nu}{E - \left(K_\nu + M'_\nu\right) + i M''_\nu}. \tag{29.33}$$

We shall now consider some properties of the mass operator, which sometimes allow us to draw definite conclusions about the behavior of the energy of elementary excitations and their lifetime without a detailed analysis of the problem.

The formal expansions in terms of the parameter ε, which are used here, are convenient if the "interaction energy" $\mathscr{H}_1$ is sufficiently small, and if there are no singularities in the expressions given above. One must draw attention to two important cases to which these expansions are inapplicable: systems with the hard-sphere interaction and systems with the Coulomb interaction. In the former case, the interaction at short distances is not weak, but the radius of action of the forces is small; therefore, the expansions (29.19) can be rewritten as expansions in powers of the density. In the latter case, the collective (polarization) effects are important, and these were not allowed for sufficiently correctly in the successive expansions in powers of ε. However, even in this case, the mass operator may be retained in the form of Eq. (29.19), but we must understand M_1', M_1'', ... to be incompletely expanded expressions or, in other words, partial sums of series in the perturbation theory.

Sec. 30. Some Properties of the Mass Operator*

Spectral representations and some general relationships, similar to those obtained for the Green's functions, can be derived also for the mass operator. It is convenient to use the retarded and advanced Green's functions, which have simple analytic properties at any temperature.

We shall consider a spatially uniform system. We shall use G(E) to denote the Fourier transform of a one-particle retarded or advanced Green's function in the representation which is diagonal in momentum:

$$G(E) = \langle\langle \mathcal{A}_p \mid \mathcal{A}_p^+ \rangle\rangle_E, \tag{30.1}$$

where $\mathcal{A}_p$, $\mathcal{A}_p^+$ are one-particle creation and annihilation operators (Fermi, Bose, or Pauli).

We shall write Eq. (29.10) in the form

$$\left[E - E_p - M(E)\right] G(E) = \frac{i\sigma}{2\pi}, \quad \sigma = \langle [\mathcal{A}_p, \mathcal{A}_p^+]_\eta \rangle, \tag{30.2}$$

where E_p is the internal energy of a particle in the state p, $\sigma > 0$ (it is assumed that $\eta = -1$ for the Fermi operators and $\eta = 1$ for

* The case of zero temperature was considered by Luttinger (1961a) and Maleev (1961), and that of finite temperatures by Bonch-Bruevich (1962).

the Bose and Pauli operators). Hence, we obtain the following equation for the mass operator:

$$-M(E)=\frac{i\sigma}{2\pi}G^{-1}(E)-\frac{i\sigma}{2\pi}G_0^{-1}(E),$$
$$G_0(E)=(E-E_p)^{-1}\frac{i\sigma}{2\pi}. \qquad (30.3)$$

We shall assume that the spectrum of the system is limited $[I_{\mathcal{A}_p\mathcal{A}_p^+}[(\omega)=0]$ when $|\omega|\geq\omega_{max}]$. Then, according to Eqs. (26.13) and (26.8), we have the following expression for $|E|\to\infty$:

$$G(E)=\frac{i}{2\pi}\int_{-\infty}^{\infty}\left(e^{\frac{\omega}{\vartheta}}-\eta\right)I_{\mathcal{A}_p\mathcal{A}_p^+}(\omega)\frac{d\omega}{E-\omega}=$$
$$=\frac{i}{2\pi}\int_{-\infty}^{\infty}\left(e^{\frac{\omega}{\vartheta}}-\eta\right)I_{\mathcal{A}_p\mathcal{A}_p^+}(\omega)\frac{d\omega}{E}+O(E^{-2})=$$
$$=\frac{i}{2\pi E}\langle[\mathcal{A}_p\mathcal{A}_p^+]_{\eta}\rangle+O(E^{-2})=\frac{i\sigma}{2\pi E}+O(E^{-2}).$$

Hence, we obtain*

$$\lim_{|E|\to\infty}EG(E)=\frac{i\sigma}{2\pi}. \qquad (30.4)$$

Next, according to Eq. (30.3), we have

$$\lim_{|E|\to\infty}EG_0(E)=\frac{i\sigma}{2\pi}. \qquad (30.5)$$

Therefore, from the definition (30.2) and the formulas (30.4), (30.5), it follows that

$$\lim_{|E|\to\infty}E^{-1}M(E)=0. \qquad (30.6)$$

According to Eq. (27.2) for real values of E ($E=\omega+i\varepsilon$, $\varepsilon\to+0$), the expression for the spectral density can be written in the form

$$\pm\left(e^{\frac{\omega}{\vartheta}}-\eta\right)I_{\mathcal{A}_p\mathcal{A}_p^+}(\omega)=2\,\mathrm{Re}\,G(\omega). \qquad (30.7)$$

*Our conclusions about the behavior of G(E) at high values of E are qualitative. For more rigorous treatments, see Maleev (1961) and Bonch-Bruevich (1962).

We shall use the notation

$$M(\omega + i\varepsilon) = M'(\omega) - iM''(\omega)\,(\varepsilon \to +0). \tag{30.8}$$

where M', M" are real functions. Then, using Eq. (30.2), we can rewrite the expression (30.7) as follows:

$$\pm\left(e^{\frac{\omega}{\vartheta}} - \eta\right) I_{\mathcal{A}_p\mathcal{A}_p^+}(\omega) = 2\,\mathrm{Re}\,\frac{i\sigma}{2\pi}\,\frac{1}{(\omega - E_p - M') + iM''} =$$
$$= \frac{\sigma}{\pi}\,\frac{M''}{(\omega - E_p - M')^2 + (M'')^2}\,. \tag{30.9}$$

Using Eq. (30.6), we find that

$$\lim_{|\omega| \to \infty} \omega\left(e^{\frac{\omega}{\vartheta}} - \eta\right) I_{\mathcal{A}_p\mathcal{A}_p^+}(\omega) = 0, \tag{30.10}$$

and, therefore,

$$\lim_{|\omega| \to \infty} \frac{M''(\omega)}{M'(\omega)} = 0. \tag{30.11}$$

According to Eq. (30.3), the poles of the mass operator may be the zeros of $G_0(E)$, $G(E)$. Since $G_0^{-1}(E)$ of Eq. (30.3) has no poles, only the zeros of the total Green's function $G(E)$ can correspond to these poles.

Finally, we may show that the poles of the mass operator lie on the real axis and that the mass operator has spectral representations similar to the corresponding representations of the Green's functions.

Sec. 31. Reaction of a System to an External Stimulus*

One of the problems in statistical mechanics is the calculation of transport coefficients and of the complex susceptibility of a system under the action of an external force, as in the case of ferromagnetic resonance, electrical conduction, etc.

*See Callen and Welton (1951), Kubo and Tomita (1954), Kubo (1957), and Zubarev (1960).

Recently, certain workers have developed a general statistical-mechanics method for calculating transport coefficients as time correlation functions of dynamical variables, which describe a given system and the external stimulus applied to it. A more general approach to such problems follows from the development of the thermodynamics of irreversible processes [see, for example, Mori (1956), Kubo et al. (1957), and Zubarev (1961a)]. In this way, it is possible to obtain general expressions for the transport coefficients and to establish a number of exact relationships for these coefficients. We shall consider here only the simplest method of representing the transport coefficients in terms of the time correlation functions, and we shall link the expressions obtained with the retarded Green's functions. This will allow us later to obtain general formulas for the ferromagnetic resonance conditions (cf., Secs. 38 and 39).

Let a system be acted upon by a time-dependent perturbation. We shall represent the new Hamiltonian of the system in the form

$$\mathcal{H} = \mathcal{H}_0 + \mathcal{V}(t), \tag{31.1}$$

where $\mathcal{H}_0$ is the Hamiltonian of the unperturbed system, time-independent and $\mathcal{V}$ (t) is the perturbation. We shall assume that the system represented by the Hamiltonian $\mathcal{H}_0$ is in a state of statistical equilibrium. We shall also assume that the perturbation has the following form:

$$\mathcal{V}(t) = \sum_{\Omega} e^{\varepsilon t} e^{-i\Omega t} \mathcal{V}_{\Omega} \qquad (\varepsilon > 0), \tag{31.2}$$

where $\mathcal{V}_{\Omega}$ are some operators which are independent of time.

According to Eq. (8.2), the density matrix ρ of a system with the Hamiltonian given by Eq. (31.1) satisfies the equation

$$i \frac{d\rho}{dt} = [\mathcal{H}, \rho] = [\mathcal{H}_0, \rho] + [\mathcal{V}(t), \rho]. \tag{31.3}$$

Let the perturbation be applied at a time t_0, i.e.,

$$\mathcal{V}(t) = \begin{cases} 0, & t < t_0, \\ \mathcal{V}(t), & t > t_0. \end{cases} \tag{31.4}$$

We shall seek the solution of Eq. (31.3) in the form

$$\rho = \rho_0 + \Delta\rho, \tag{31.5}$$

$$\rho_0 = Q_0^{-1} \exp\left(-\frac{\mathcal{H}_0}{\vartheta}\right), \quad Q_0 = \mathrm{Sp}\exp\left(-\frac{\mathcal{H}_0}{\vartheta}\right) \tag{31.6}$$

with the initial condition

$$\rho\big|_{t=t_0} = \rho_0. \tag{31.7}$$

Clearly, ρ_0 is the equilibrium density matrix of the system represented by the Hamiltonian $\mathcal{H}_0$, and $\Delta\rho$ is an increment of ρ due to the action of the perturbation on the system.

Assuming the perturbation $\mathcal{V}(t)$ to be small, we can easily obtain the formal solution for $\Delta\rho$ in the form of a series in powers of $\mathcal{V}(t)$. Substituting Eq. (31.5) into Eq. (31.3), we obtain

$$\begin{gathered} i\frac{d}{dt}\Delta\rho = [\mathcal{H}_0, \Delta\rho] + [\mathcal{V}(t), \rho_0] + [\mathcal{V}(t), \Delta\rho], \\ \Delta\rho\big|_{t=t_0} = 0. \end{gathered} \tag{31.8}$$

We shall seek the solution of this equation in the form

$$\Delta\rho = e^{-i\mathcal{H}_0 t}\widetilde{\Delta\rho}\, e^{i\mathcal{H}_0 t}. \tag{31.9}$$

For $\widetilde{\Delta}\rho$, we shall obtain the following expression:

$$\begin{gathered} i\frac{d}{dt}\widetilde{\Delta\rho} = [\widetilde{\mathcal{V}}(t), \rho_0] + [\widetilde{\mathcal{V}}(t), \widetilde{\Delta\rho}], \\ \widetilde{\Delta\rho}\big|_{t=t_0} = 0, \end{gathered} \tag{31.10}$$

where

$$\widetilde{\mathcal{V}}(t) = e^{i\mathcal{H}_0 t}\mathcal{V}(t)\, e^{-i\mathcal{H}_0 t}. \tag{31.11}$$

Solving Eq. (31.10) by iteration (on the assumption that $\widetilde{\Delta}\rho$ is small), we finally obtain

$$\begin{gathered} \Delta\rho = \sum_{n=1}^{\infty}(-i)^n \int\limits_{t_0}^{t}\int\limits_{t_0}^{t_1}\cdots\int\limits_{t_0}^{t_{n-1}} dt_1\, dt_2 \ldots dt_n \times \\ \times e^{-i\mathcal{H}_0 t}[\widetilde{\mathcal{V}}(t_1), [\widetilde{\mathcal{V}}(t_2), \ldots [\widetilde{\mathcal{V}}(t_n), \rho]] \ldots] e^{i\mathcal{H}_0 t}. \end{gathered} \tag{31.12}$$

The average value of any dynamic variable $\mathcal{A}(t)$ is

$$\langle\mathcal{A}(t)\rangle = \mathrm{Sp}\{\mathcal{A}(t)\rho\} \qquad (\mathrm{Sp}\,\rho = 1) \tag{31.13}$$

or

$$\langle \mathcal{A}(t)\rangle = \mathrm{Sp}\,\{\mathcal{A}(t)\,\rho_0\} + \mathrm{Sp}\,\{\mathcal{A}(t)\,\Delta\rho\}, \tag{31.14}$$

where the second term on the right describes the increment in the average value of $\mathcal{A}$ (t) under the action of the perturbation $\mathcal{V}$ (t) applied to the system.

We shall use the fact that $\mathrm{Sp}\{\Delta\rho\} = 0$ by definition. Therefore, substituting Eq. (31.12) into Eq. (31.14), we obtain the following expression for the increment in the average value of the quantity $\mathcal{A}$ (t):

$$\delta\langle \mathcal{A}(t)\rangle = \mathrm{Sp}\,\{\mathcal{A}(t)\,\rho\} - \mathrm{Sp}\,\{\mathcal{A}(t)\,\rho_0\} = \sum_{n=1}^{\infty} \delta^{(n)}\langle \mathcal{A}(t)\rangle, \tag{31.15}$$

where

$$\delta^{(n)}\langle \mathcal{A}(t)\rangle = (-i)^n \int_{t_0}^{t} \cdots \int_{t_0}^{t_{n-1}} dt_1 \ldots dt_n \times$$
$$\times \langle [\ldots [\tilde{\mathcal{A}}(t), \tilde{\mathcal{V}}(t_1)], \ldots, \tilde{\mathcal{V}}(t_n)]\rangle, \tag{31.16}$$

and the value $\tilde{\mathcal{A}}$ (t) is defined in the same way as $\tilde{\mathcal{V}}$ (t) of Eq. (31.11).

Using Eq. (31.2), we shall rewrite Eq. (31.16) in the following form [see also Kubo (1957)]:

$$\delta^{(n)}\langle \mathcal{A}(t)\rangle = \sum_{\Omega_1,\ldots,\Omega_n} (-i)^n \int_{t_0}^{t} \cdots \int_{t_0}^{t_{n-1}} dt_1 \ldots dt_n \times$$
$$\times e^{\sum_{j=1}^{n} \varepsilon t_j - i \sum_{j=1}^{n} \Omega_j t_j} \langle [\ldots [\tilde{\mathcal{A}}(t), \tilde{\mathcal{V}}_{\Omega_1}(t_1)], \ldots, \tilde{\mathcal{V}}_{\Omega_n}(t_n)]\rangle, \tag{31.17}$$

where $\tilde{\mathcal{V}}_\Omega$ (t) is the Heisenberg representation of the operator $\mathcal{V}_\Omega$ of Eq. (31.2) for the unperturbed system:

$$\tilde{\mathcal{V}}_\Omega(t) = e^{i\mathcal{H}_0 t}\, \mathcal{V}_\Omega e^{-i\mathcal{H}_0 t}. \tag{31.18}$$

We note that the symbol < ... > in Eqs. (31.15)-(31.17) represents averaging over the density matrix ρ_0.

The calculation of the averages from the formulas (31.16)-(31.17) may be difficult. It is more convenient to calculate them in terms of the retarded Green's functions, to which they are linked by simple relationships.

For simplicity, we shall consider first the linear approximation for the perturbation $\mathcal{V}(t)$. The linear approximation formulas are widely used in the calculation of such quantities as the electrical conductivity, magnetic susceptibility, etc. In this approximation, the formula (31.17) for the increment in the average value when a perturbation is applied instantaneously has the following form:

$$\delta^{(1)}\langle\mathcal{A}(t)\rangle=\sum_{\Omega}(-i)\int_{t_0}^{t}dt_1e^{\varepsilon t_1-i\Omega t_1}\langle[\tilde{\mathcal{A}}(t),\ \tilde{\mathcal{V}}_\Omega(t_1)]\rangle=$$

$$=\sum_{\Omega}(-i)\int_{t_0}^{t}dt_1\,e^{\varepsilon t_1-i\Omega t_1}\theta(t-t_1)\langle[\tilde{\mathcal{A}}(t),\ \tilde{\mathcal{V}}_\Omega(t_1)]\rangle.$$

Using the definition of the retarded Green's function given by Eq. (25.7), we shall rewrite the above formula as follows:

$$\delta^{(1)}\langle\mathcal{A}(t)\rangle=\sum_{\Omega}(-i)\int_{t_0}^{t}dt_1\,e^{\varepsilon t_1-i\Omega t_1}\langle\langle\tilde{\mathcal{A}}(t)\,|\,\tilde{\mathcal{V}}_\Omega(t_1)\rangle\rangle^{(r)}. \tag{31.19}$$

Going over to the Fourier transforms of the Green's functions (25.17), and integrating with respect to t_1, we obtain

$$\delta^{(1)}\langle\mathcal{A}(t)\rangle=-\sum_{\Omega}e^{\varepsilon t-i\Omega t}\int_{-\infty}^{\infty}dE\,\langle\langle\tilde{\mathcal{A}}\,|\,\tilde{\mathcal{V}}_\Omega\rangle\rangle_E^{(r)}\times$$

$$\times\left\{\frac{1}{E-\Omega-i\varepsilon}-\frac{\exp[-i(E-\Omega)(t-t_0)-\varepsilon(t-t_0)]}{E-\Omega-i\varepsilon}\right\}.$$

We shall consider E to be a complex variable and apply the Cauchy theorem on residues. Since the retarded function is analytic in the upper half-plane of E [cf., Eq. (26.11)], it is convenient to close the contour in the same half-plane in order to calculate the first integral. Consequently, we obtain the following expression for the increment in the average value on the application of an interaction (perturbation) at a time t_0 ($t_0 > -\infty$):

$$\delta^{(1)}\langle\mathcal{A}(t)\rangle=-2\pi i\sum_{\Omega}e^{\varepsilon t-i\Omega t}\langle\langle\tilde{\mathcal{A}}\,|\,\tilde{\mathcal{V}}_\Omega\rangle\rangle_\Omega^{(r)}+$$

$$+\sum_{\Omega}e^{\varepsilon t_0-i\Omega t_0}\int_{-\infty}^{\infty}dE\,\langle\langle\tilde{\mathcal{A}}\,|\,\tilde{\mathcal{V}}_\Omega\rangle\rangle_E^{(r)}\frac{e^{-iE(t-t_0)}}{E-\Omega-i\varepsilon}. \tag{31.20}$$

The integration contour of the second integral should be closed because $(t - t_0) > 0$ in the lower half-plane (if this is at all possible).

If the perturbation is applied adiabatically, at a time $t_0 = -\infty$ the formula (31.20) assumes the form

$$\delta^{(1)}\langle\mathcal{A}(t)\rangle = -2\pi i \sum_{\Omega} e^{\varepsilon t - i\Omega t}\langle\langle\tilde{\mathcal{A}}\,|\,\tilde{\mathcal{V}}_{\Omega}\rangle\rangle_{\Omega}^{(r)}. \qquad (31.21)$$

If at a time t_0 the perturbation, which was applied earlier adiabatically, is instantaneously removed, the change in the average increment of $\mathcal{A}$ will be equal to the difference between the expressions (31.21) and (31.20):

$$\delta^{(1)}\langle\mathcal{A}(t)\rangle = -\sum_{\Omega} e^{\varepsilon t_0 - i\Omega t_0}\int_{-\infty}^{\infty} dE\,\langle\langle\tilde{\mathcal{A}}\,|\,\tilde{\mathcal{V}}_{\Omega}\rangle\rangle_E^{(r)}\frac{e^{-iE(t-t_0)}}{E-\Omega-i\varepsilon}. \qquad (31.22)$$

The formulas (31.20)-(31.22) represent the required expressions for the increment in the average value of the dynamic variable $\mathcal{A}$ under the action of a time-dependent external perturbation.

These expressions have simple physical meaning. If the perturbation is applied infinitely slowly and adiabatically, the average value of the dynamic variable $\mathcal{A}$ receives an increment $\delta\langle\mathcal{A}\rangle$ of Eq. (31.21). If the perturbation is applied instantaneously at a time $t_0 > -\infty$, natural oscillations are excited in the system, which lead to an additional change in the increment of the average value of $\mathcal{A}$, described by the second term of Eq. (31.20). Such natural oscillations are excited also when the perturbation is removed instantaneously, and they lead to a change in $\delta\langle\mathcal{A}\rangle$ in accordance with Eq. (31.22).

We shall now assume that the Green's function in the lower half-plane (on the nonphysical sheet) has a complex pole

$$E = E_\nu - i\Gamma_\nu \quad (\Gamma_\nu > 0).$$

Then, according to Eqs. (31.20) and (31.22), when a perturbation is applied instantaneously at a time t_0, we have

$$\delta^{(1)}\langle\mathcal{A}(t)\rangle = -2\pi i \sum_{\Omega} e^{\varepsilon t - i\Omega t}\langle\langle\tilde{\mathcal{A}}\,|\,\tilde{\mathcal{V}}_{\Omega}\rangle\rangle_{\Omega}^{(r)} +$$

$$+ 2\pi i \sum_{\Omega} e^{\varepsilon t_0 - i\Omega t_0} e^{-iE_\nu (t-t_0) - \Gamma_\nu (t-t_0)} \langle\langle \tilde{\mathcal{A}} \mid \tilde{\mathcal{V}}_\Omega \rangle\rangle_\Omega^{(r)} \qquad (31.23)$$

and when the perturbation is removed instantaneously at a time t_0, we find

$$\delta^{(1)} \langle \mathcal{A}(t) \rangle = -2\pi i \sum_{\Omega} e^{\varepsilon t_0 - i\Omega t_0} e^{-iE_\nu (t-t_0) - \Gamma_\nu (t-t_0)} \langle\langle \tilde{\mathcal{A}} \mid \tilde{\mathcal{V}}_\Omega \rangle\rangle_\Omega^{(r)}. \qquad (31.24)$$

Hence, we see that a change in $\langle \mathcal{A} \rangle$, which is due to the appearance of natural oscillations in the system, varies with the frequency E_ν, andits decay is represented by a decrement Γ_ν. This is in agreement with the interpretation of the real parts of the poles of the Green's functions as the energies of elementary excitations (or natural frequencies) of the system, and of the imaginary parts as the lifetimes of elementary excitations.

Let us assume that

$$\mathcal{V}_\Omega = \mathcal{B}_\Omega h,$$

where h is a number representing the intensity of the perturbation. For example, h may be the intensity of the alternating magnetic field in the ferromagnetic resonance condition or the electric field intensity when the electrical conductivity is being considered. We shall rewrite the expression (31.21) in the form

$$\delta^{(1)} \langle \mathcal{A}(t) \rangle = \sum_{\Omega} e^{-i\Omega t} \chi_{\mathcal{A}\mathcal{B}}(\Omega)\, h, \qquad (31.25)$$

where χ is the complex susceptibility of the system,

$$\chi_{\mathcal{A}\mathcal{B}}(\Omega) = -2\pi i \langle\langle \tilde{\mathcal{A}} \mid \tilde{\mathcal{B}}_\Omega \rangle\rangle_\Omega^{(r)} \qquad (31.26)$$

(the parameter ε, which ensures the adiabatic application of the perturbation, is assumed to be zero).

Using Eq. (25.17), the susceptibility can be given another form which is used quite frequently:

$$\chi_{\mathcal{A}\mathcal{B}}(\Omega) = -i \int_{-\infty}^{\infty} dt'\, e^{i\Omega (t-t')} \langle\langle \tilde{\mathcal{A}}(t) \mid \tilde{\mathcal{B}}_\Omega (t') \rangle\rangle^{(r)} =$$

$$= -i \int_{-\infty}^{\infty} dt' e^{i\Omega (t-t')}\, \theta(t-t') \langle [\tilde{\mathcal{A}}(t), \tilde{\mathcal{B}}_\Omega (t')] \rangle_\eta \qquad (31.27)$$

An expression of the Eq. (31.27) type, which relates the susceptibility of a system to the Fourier transform of the average value of the commutator $[\mathcal{A}(t), \mathcal{B}_{\mathfrak{Q}}(t')]$, is known as the fluctuation-dissipation theorem [see Callen and Welton (1951) and Kubo (1957)].

Since the complex susceptibility of a system is given in terms of the Green's function in accordance with Eq. (31.26), the susceptibility obeys the dispersion formulas and the general relationships established in Secs. 27 and 28.

Higher approximations can be obtained by including the terms containing higher powers of $\mathcal{V}$ in the expansion of Eq. (31.17).

For simplicity, we shall assume that the perturbation is applied adiabatically. Then Eq. (31.17) may be rewritten in the form [Tyablikov and P'u Fu-ch'o (1961)]

$$\delta^n \langle \mathcal{A}(t) \rangle = \sum_{\mathfrak{Q}_1, \ldots, \mathfrak{Q}_n} (-i)^n \int_{-\infty}^{\infty} \ldots \int_{-\infty}^{\infty} dt_1 \ldots dt_n e^{\varepsilon \sum_{j=1}^{n} t_j - i \sum_{j=1}^{n} \mathfrak{Q}_j t_j} \times$$
$$\times G^{(n)}_{\mathcal{A}; \mathcal{V}_{\mathfrak{Q}_1} \ldots \mathcal{V}_{\mathfrak{Q}_n}} (t - t_1, t_1 - t_2, \ldots, t_{n-1} - t_n), \qquad (31.28)$$

where $G^{(n)}_{\mathcal{A}; \mathcal{V}_{\mathfrak{Q}_1} \ldots \mathcal{V}_{\mathfrak{Q}_n}}$ is the n-time retarded Green's function of the type

$$G^{(n)}_{\mathcal{A}; \mathcal{V}_{\mathfrak{Q}_1} \ldots \mathcal{V}_{\mathfrak{Q}_n}} (t - t_1, t_1 - t_2, \ldots, t_{n-1} - t_n) =$$
$$= \theta(t - t_1)\, \theta(t_1 - t_2) \ldots \theta(t_{n-1} - t_n) \times$$
$$\times \langle [\ldots [\tilde{\mathcal{A}}(t), \tilde{\mathcal{V}}_{\mathfrak{Q}_1}(t_1)], \ldots, \tilde{\mathcal{V}}_{\mathfrak{Q}_n}(t_n)] \rangle. \qquad (31.29)$$

From the cyclic invariance of the spur (trace), it follows that the average value in Eq. (31.29) depends only on the differences $(t - t_1), \ldots, (t_{n-1} - t_n)$. The Fourier transform of this function is given by the equation

$$G^{(n)}_{\mathcal{A}; \mathcal{V}_{\mathfrak{Q}_1} \ldots \mathcal{V}_{\mathfrak{Q}_n}} (t - t_1, t_1 - t_2, \ldots, t_{n-1} - t_n) =$$
$$= \int_{-\infty}^{\infty} \ldots \int_{-\infty}^{\infty} dE_1 \ldots dE_n G^{(n)}_{\mathcal{A}; \mathcal{V}_{\mathfrak{Q}_1} \ldots \mathcal{V}_{\mathfrak{Q}_n}} (E_1, E_2, \ldots, E_n) \times$$
$$\times e^{-iE_1 (t - t_1) - \ldots - iE_n (t_{n-1} - t_n)}. \qquad (31.30)$$

From Eqs. (31.28) and (31.30), we obtain a formula which is a generalization of the formula (31.21):

$$\delta^{(n)}\langle \mathcal{A}(t)\rangle = \sum_{\Omega_1,\ldots,\Omega_n}(-2\pi i)^n e^{-i\sum_{j=1}^{n}\Omega_j t} \times$$
$$\times G^{(n)}_{\mathcal{A};\,\mathcal{V}_{\Omega_1}\ldots\mathcal{V}_{\Omega_n}}(\Omega_1+\Omega_2+\ldots+\Omega_n+i\varepsilon n,\ \Omega_2+\ldots$$
$$\ldots+\Omega_n+i\varepsilon(n-1),\ \ldots,\ \Omega_{n-1}+\Omega_n+2i\varepsilon,\ \Omega_n+i\varepsilon). \qquad (31.31)$$

To determine many-time functions we can, as in the case of two-time functions, set up chains of equations by differentiating with respect to one of the time arguments, and then seek the solutions of such a chain. We can derive the analogs of the spectral representations for many-time functions. Such representations were considered by Bonch-Bruevich (1959) for noncommutator many-time retarded and advanced Green's functions.

Chapter VIII

Applications of the Green's Function Method

In the present chapter, we shall discuss the applications of the Green's function method to a number of typical problems in the theory of magnetism: the calculation of the magnetization of a ferromagnet over a wide range of temperatures, low-temperature expansions, the theory of ferromagnetic resonance, and the scattering of neutrons by spin systems. In concluding the chapter, we indicate some further applications of the method.

Sec. 32. Spin Waves at Finite Temperatures*

Here, and in the next two sections, we shall consider the problem of the approximate determination of the magnetization of an isotropic ferromagnet over a wide range of temperatures. For simplicity, we shall consider first only the case of spin S = ½.

The Hamiltonian of an isotropic ferromagnet, expressed in terms of the Pauli operators [cf., Eqs. (14.14)-(14.16)], has the form

$$\widetilde{\mathscr{H}} = \mathscr{E}_0 + \mathscr{H}, \tag{32.1}$$

where

$$\mathscr{E}_0 = -\frac{1}{2} N\mu H - \frac{1}{8} NJ(0), \tag{32.2}$$

$$\mathscr{H} = \left[\mu H + \frac{1}{2} J(0)\right] \sum_f n_f - \frac{1}{2} \sum_{f_1, f_2} I(f_1 - f_2)\, b_{f_1}^{+} b_{f_2} - \\ - \frac{1}{2} \sum_{f_1, f_2} I(f_1 - f_2)\, n_{f_1} n_{f_2} \tag{32.3}$$

---

*See Bogolyubov and Tyablikov (1959) and Tyablikov (1959b).

and

$$J(\nu) = \sum_f I(f)\, e^{i(f,\nu)}, \quad I(0) = 0. \tag{32.4}$$

The equations of motion for the operators b have, according to Eq. (5.13), the following form:

$$i\frac{db_f}{dt} = \left[\mu H + \frac{1}{2}J(0)\right]b_f - \frac{1}{2}\sum_{f'} I(f-f')\, b_{f'} + \\ + \sum_{f'} I(f-f')\, n_f b_{f'} - \sum_{f'} I(f-f')\, b_f n_{f'}. \tag{32.5}$$

We shall introduce the Green's functions $\langle\langle b_f | b_g^+\rangle\rangle$, $\langle\langle n_{f_1} b_{f_2} | b_g^+\rangle\rangle, \ldots$ For these functions, we obtain, in accordance with Eq. (25.18), the chain of equations

$$E\langle\langle b_f | b_g^+\rangle\rangle = \frac{i\sigma_{1/2}}{2\pi}\Delta(f-g) + \left[\mu H + \frac{1}{2}J(0)\right]\langle\langle b_f | b_g^+\rangle\rangle - \\ - \frac{1}{2}\sum_{f'} I(f-f')\langle\langle b_{f'} | b_g^+\rangle\rangle + \sum_{f'} I(f-f')\langle\langle n_f b_{f'} | b_g^+\rangle\rangle - \\ - \sum_{f'} I(f-f')\langle\langle n_{f'} b_f | b_g^+\rangle\rangle, \ldots$$

$$\sigma_{1/2} = \langle b_f b_f^+ - b_f^+ b_f\rangle = \langle 1 - 2n_f\rangle. \tag{32.6}$$

We are not giving equations for the higher Green's functions, because we shall consider only the lowest approximation. The chain of equations (32.6) will be decoupled in an approximate way, expressing the second Green's functions in terms of the first functions:

$$\langle\langle n_{f_1} b_{f_2} | b_g^+\rangle\rangle \to \langle n_{f_1}\rangle \langle\langle b_{f_2} | b_g^+\rangle\rangle. \tag{32.7}$$

We note that, in view of the translational invariance, the quantities $\langle n_f\rangle$ are independent of the site index and are related to the average magnetization per site by

$$\bar{m} = \mu\langle S_f^z\rangle = \frac{\mu}{2}\sigma_{1/2};\ \sigma_{1/2} = \langle 1 - 2n_f\rangle = 1 - 2\bar{n}. \tag{32.8}$$

Consequently, we obtain the following equation for the first function:

$$\left\{E-\left[\mu H+\frac{1}{2}\sigma_{1/2}J(0)\right]\right\}\langle\langle b_f \mid b_g^+\rangle\rangle+ \\ +\sum \frac{1}{2}\sigma_{1/2} I(f-f')\langle\langle b_{f'} \mid b_g^+\rangle\rangle=\frac{i\sigma_{1/2}}{2\pi}\Delta(f-g). \qquad (32.9)$$

When applied to the theory of magnetism, the proposed decoupling method (32.7) represents an improved variant of the approximate second quantization.

In fact, in the approximate second quantization method, the Pauli operators b_f, b_g^+ are approximately regarded as the Bose operators, and the contribution of the third term in the Hamiltonian (32.3) is ignored. If equations for the Green's functions are written with all these restrictions, they will correspond to Eq. (32.9) for $\sigma_{1/2} = 1$ or $\langle n_f \rangle = 0$. The chain will be decoupled at the first of these equations.

Next, we can attempt to improve the approximate second quantization method by rearranging the Hamiltonian of Eq. (32.3) as follows:

$$\mathcal{H}=\Delta\mathcal{E}_0+\mathcal{H}_0+\mathcal{H}_1,$$

where

$$\Delta\mathcal{E}_0=\frac{1}{2}NJ(0)\,\bar{n}^2,$$

$$\mathcal{H}_0=\left[\mu H+\left(\frac{1}{2}-\bar{n}\right)J(0)\right]\sum_f n_f-\sum_{f_1,f_2}\left(\frac{1}{2}-\bar{n}\right)I(f_1-f_2)\,b_{f_1}^+b_{f_2},$$

$$\mathcal{H}_1=-\frac{1}{2}\sum_{f_1,f_2}I(f_1-f_2)(n_{f_1}-\bar{n})(n_{f_2}-\bar{n})- \\ -\sum_{f_1,f_2}\bar{n}I(f_1-f_2)\,b_{f_1}^+b_{f_2}.$$

At sufficiently low temperatures, the small parameters are the average values of the spin deviations $\bar{n}$ and the differences $n_f - \bar{n}$. Therefore, the operator $\mathcal{H}_1$ may be regarded as a small perturbation compared with $\mathcal{H}_0$. We can easily see that, in the spin-wave approximation (neglecting the operator $\mathcal{H}_1$), the chain of equations is decoupled at the first equation and becomes identical with Eq. (32.9).

We note also that the decoupling achieved by means of Eq. (32.7) is equivalent to neglecting fluctuations of the spin deviations.

Since the problem is translationally invariant, the function

$$G_{f-g}(E) = \langle\langle b_f \mid b_g^+ \rangle\rangle \tag{32.10}$$

depends only on the difference between the lattice site coordinates. Therefore, Eq. (32.9) is solved easily by means of the Fourier transformation:

$$G_h(E) = \frac{1}{N} \sum_{\nu} e^{i(h,\nu)} G_{\nu}(E). \tag{32.11}$$

For $G_\nu(E)$, we obtain the expression

$$G_{\nu}(E) = \frac{i\sigma_{1/2}}{2\pi} \frac{1}{E - E_{\nu}}, \tag{32.12}$$

where E_ν is the energy of elementary excitations or spin waves,

$$E_{\nu} = \mu H + \frac{1}{2}\sigma_{1/2} J(0)(1-\gamma_{\nu}) = \mu H + \frac{1}{2}\sigma_{1/2} J(0)\mathcal{E}_{\nu},$$
$$\gamma_{\nu} = J(\nu)/J(0), \tag{32.13}$$

and the quantities J(0), $J(\nu)$, and $\sigma_{1/2}$ are defined by Eqs. (32.4) and (32.8).

We shall confine ourselves to the zeroth approximation with respect to the interaction energy of elementary excitations. Therefore, the one-particle Green's function of Eq. (32.12) has only simple poles on the real axis. In this approximation, the lifetimes of elementary excitations are infinite.

Comparing the expression (32.13) with the expression (15.6), we see that the difference is due to the appearance in Eq. (32.13) of a multiplier $\sigma_{1/2}$, which represents the relative magnetization. Since this magnetization is temperature-dependent, it follows that the energy E_ν is also. Thus, we find a common characteristic of the Green's function method, which is a temperature-dependent spectrum of elementary excitations.

The first attempts to introduce the relative magnetization corrections into the spin-wave spectrum were made by Heber

(1954) and Schafroth (1954). These two used the Holstein–Primakoff transformation and attempted to improve the results by allowing thermodynamically for the terms which are usually neglected. More successful were the attempts of Brout and Haken (1960), Englert (1960), and Ginzburg and Fain (1960), who used other methods to obtain an expression of the (32.13) type for the energy of spin waves.

The expression for elementary excitations (spin waves) given by Eq. (32.13) includes the as-yet-unknown value of the relative magnetization $\sigma_{1/2}$. We shall obtain an equation for this magnetization.

We shall determine the spectral density for the function (32.12). Substituting Eq. (32.12) into Eq. (26.15), and using (26.16), we obtain

$$I_\nu(\omega) = \sigma_{1/2}\overline{N}_\nu \delta(\omega - E_\nu), \tag{32.14}$$

$$\overline{N}_\nu = \left(e^{E_\nu/\vartheta} - 1\right)^{-1}. \tag{32.15}$$

The spectral density for the function (32.10) has the form

$$I_{fg}(\omega) = \frac{1}{N}\sum_\nu e^{i(f-g,\,\nu)} I_\nu(\omega) = \frac{\sigma_{1/2}}{N}\sum_\nu e^{i(f-g,\,\nu)}\overline{N}_\nu \delta(\omega - E_\nu). \tag{32.16}$$

We shall now use the formulas (26.5), which give the average value of a product of two operators in terms of the spectral density. Substituting Eq. (32.16) into Eq. (26.5), we find

$$\langle b_g^+ b_f\rangle = \frac{\sigma_{1/2}}{N}\sum_\nu e^{i(f-g,\,\nu)}\overline{N}_\nu. \tag{32.17}$$

Hence, for $g = f$, we obtain

$$\overline{n} = \frac{\sigma_{1/2}}{N}\sum_\nu \overline{N}_\nu \quad \left(\overline{n} = \langle b_f^+ b_f\rangle\right). \tag{32.18}$$

Noting that $\sigma_{1/2} = 1 - 2\overline{n}$, we can simplify the above equation:

$$\frac{1}{\sigma_{1/2}} = \frac{1}{N}\sum_\nu \coth\frac{E_\nu}{2\vartheta} \tag{32.19}$$

or

$$\left.\begin{aligned} \sigma_{1/2} &= \frac{1}{1+2P_{1/2}}, \\ P_{1/2} &= \frac{1}{N}\sum_{\nu} \overline{N}_{\nu}. \end{aligned}\right\} \tag{32.20}$$

Thus, in the approximation considered, the spectrum of elementary excitations (32.13) depends on temperature through the relative magnetization. To determine this magnetization, we have the transcendental equations (32.19) or (32.20). We shall now consider the solutions of these equations.

Sec. 33. Magnetization at Finite Temperatures*

The determination of the magnetization $\sigma_{1/2}$ as a function of temperature and field reduces, in the approximation used, to the solution of the transcendental equations (32.19) or (32.20). We shall show that these equations give a sufficiently satisfactory interpolation for the magnetization over the whole range of temperatures, namely: in the limit, as $\vartheta \to 0$, the first terms of the expansion of $\sigma_{1/2}$ in powers of ϑ are identical with the corresponding terms of expansion in the spin-wave approximation (cf., Sec. 15); when $\vartheta \leq \vartheta_C$ (ϑ_C is the Curie temperature), the results obtained here are identical with those found by the molecular field approximation (cf., Sec. 21); finally, when $\vartheta > \vartheta_C$, the expansion for $\sigma_{1/2}$ in powers of ϑ^{-1} is found to be identical with the first terms of the perturbation theory series for high temperatures (cf., Sec. 24).

Using the formula (A2.14), we shall transform the sums in Eqs. (32.19) and (32.20) into integrals:

$$\frac{1}{\sigma_{1/2}} = \frac{v}{(2\pi)^3}\int \coth \frac{\mu H + \frac{1}{2}\sigma_{1/2} J(0)\,\mathcal{E}_\nu}{2\vartheta}\, d\nu, \tag{33.1}$$

$$\sigma_{1/2} = \frac{1}{1+2P_{1/2}}, \quad P_{1/2} = \frac{v}{(2\pi)^3}\int \overline{N}_\nu\, d\nu, \tag{33.2}$$

where v = V/N is volume per lattice site. The integrals in Eqs. (33.1) and (33.2) are taken over the first reduced zone.

*Tyablikov (1959b).

We shall determine the temperature and field dependences of the magnetization at various temperatures.

a) Low Temperatures ($\vartheta \to 0$). In this case, it is convenient to use the equation for $\sigma_{1/2}$ in the form (33.2). The quantity $P_{1/2}$ describes the difference of the magnetization from its saturation value and, as $\vartheta \to 0$, it may be regarded as a small correction, so that the equation for $\sigma_{1/2}$ can be rewritten in the form of a series:

$$\sigma_{1/2} = 1 - 2P_{1/2} + (2P_{1/2})^2 - \dots \tag{33.3}$$

The only difference between the values of $P_{1/2}$ in Eqs. (33.2) and (15.10) is the inclusion of the dependence of the spin-wave energy on the magnetization $\sigma_{1/2}$ in the case of Eq. (33.2). Therefore, using an expansion of the (15.12) type, we can write

$$P_{1/2} = \sum_{n=1}^{\infty} e^{-n\frac{\mu H}{\vartheta}} \frac{v}{(2\pi)^3} \int e^{-n\frac{\sigma_{1/2}}{2\vartheta} J_0 \varepsilon_\nu} d\nu. \tag{33.4}$$

As $\vartheta \to 0$, the main contribution to Eq. (33.4) is made by the spin waves with low values of ν. Therefore, repeating the treatment of Sec. 15 which follows Eq. (15.12), we obtain, in the lowest powers of temperature,

$$P_{1/2} = \frac{v}{(\overline{\delta^2})^{3/2}} \left(\frac{\vartheta}{\frac{\pi}{3} J(0) \sigma_{1/2}} \right)^{3/2} Z_{3/2}\left(\frac{\mu H}{\vartheta}\right) + O(\vartheta^{5/2}). \tag{33.5}$$

where

$$\overline{\delta^2} = \frac{\sum f^2 I(f)}{\sum I(f)}.$$

Substituting Eq. (33.5) into Eq. (33.3), and using terms up to $\vartheta^{3/2}$, inclusive, we obtain an approximate expression for $\sigma_{1/2}$:

$$\sigma_{1/2} \approx 1 - 2P_{1/2} \approx 1 - \frac{2v}{(\overline{\delta^2})^{3/2}} \left(\frac{\vartheta}{\frac{\pi}{3} J(0)} \right)^{3/2} Z_{3/2}\left(\frac{\mu H}{\vartheta}\right). \tag{33.6}$$

We can easily see that Eq. (33.6) is identical with Bloch's formula (1930, 1932) given by Eq. (15.15) and obtained by the approximate second quantization method.

We shall calculate the terms in $\sigma_{1/2}$ which contain higher powers of temperature. We shall confine ourselves to the simplest cubic lattices (simple cubic, bcc, and fcc) and the nearest-neighbor approximation. In this case,

$$\overline{\delta^2} = \delta^2, \qquad J(0) = Iz,$$

where δ is the distance between the nearest neighbors, I is the value of the exchange integral for the nearest neighbors, and z is the number of such neighbors. Next, by analogy with Eq. (15.19), we obtain

$$\begin{aligned}\sigma_{1/2} = 1 - 2\Big[\rho Z_{3/2}\left(\frac{h}{\tau}\right)\left(\frac{\tau}{\sigma_{1/2}}\right)^{3/2} + \frac{3\pi}{4}\rho Z_{5/2}\left(\frac{h}{\tau}\right)\left(\frac{\tau}{\sigma_{1/2}}\right)^{5/2} + \\ + \pi^2\omega\rho Z_{7/2}\left(\frac{h}{\tau}\right)\left(\frac{\tau}{\sigma_{1/2}}\right)^{7/2} + \ldots\Big] + \\ + 4\left[\rho Z_{3/2}\left(\frac{h}{\tau}\right)\left(\frac{\tau}{\sigma_{1/2}}\right)^{3/2} + \ldots\right]^2 - \ldots,\end{aligned} \tag{33.7}$$

where

$$\tau = \frac{3\vartheta}{\pi I z}, \qquad h = \frac{3\mu H}{\pi I z}, \qquad \rho = \frac{v}{\delta^3}, \tag{33.8}$$

and ω is a numerical factor which depends on the lattice geometry [Eq. (15.21)].

We shall solve the above equation by the iteration method. We shall use $\sigma_{1/2} = 1$ as the zeroth approximation and then find the higher approximations of $\sigma_{1/2}$ in terms of powers of temperature:

$$\begin{aligned}\sigma_{1/2} = 1 - 2\rho Z_{3/2}\left(\frac{h}{\tau}\right)\tau^{3/2} - \frac{3\pi}{2}\rho Z_{5/2}\left(\frac{h}{\tau}\right)\tau^{5/2} - \\ - 4\rho^2 Z^2_{3/2}\left(\frac{h}{\tau}\right)\tau^{6/2} - 2\pi^2\omega\rho Z_{7/2}\left(\frac{h}{\tau}\right)\tau^{7/2} - \ldots\end{aligned} \tag{33.9}$$

The reasons for the appearance of terms of the order of $\tau^{3/2}$ and higher are the following. First there is the dependence of the spin-wave energy on the wave vector (32.13), a result of the condition which leads to trivial expansions of the (15.19) type. Secondly, these terms arise due to the characteristic features of the kinematic conditions for the spin operators, which lead to a dependence of $\sigma_{1/2}$ on $P_{1/2}$ of the (33.2) or (33.3) type. Thirdly, such terms result from the dependence of the spin-wave energy on temperature through $\sigma_{1/2}$ alone. At low temperatures, the last result is not accurate; its accuracy is governed by the approximate nature of the decoupling of Eq. (32.7). Therefore, the expansion (33.9) is valid only to terms of the order of $\tau^{5/2}$, inclusive.

b) High Temperatures ($\vartheta \leq \vartheta_C$, H = 0). Equation (33.1) for $\sigma_{1/2}$ does not have the solution $\sigma_{1/2} \neq 0$ when H = 0 and $\vartheta \to \infty$. In fact, if $\sigma_{1/2}$ is finite at some sufficiently high value of ϑ, then

$$\coth \frac{\sigma_{1/2} J(0) \mathscr{E}_\nu}{4\vartheta} \approx \frac{4\vartheta}{\sigma_{1/2} J(0) \mathscr{E}_\nu}.$$

Substituting this expression into Eq. (33.1), we see that the right-hand side of the expression can be made as large as we please by a suitable selection of ϑ, while the left-hand side remains finite. It follows that at sufficiently high values of ϑ and H = 0, the magnetization cannot remain finite.

We shall now assume that (for H = 0) $\sigma_{1/2} \neq 0$ if $\vartheta < \vartheta_C$, and $\sigma_{1/2} \to 0$ if $\vartheta \to \vartheta_C$, where ϑ_C is some constant having the meaning of the Curie temperature.

Let $\sigma_{1/2} \ll 1$; then $[\sigma_{1/2} J(0) \mathscr{E}_\nu / 4\vartheta] \ll 1$ and the hyperbolic cotangent can be replaced by the expansion [see Ryzhik and Gradshtein (1951)]

$$\coth \xi = \frac{1}{\xi} + \sum_{n=1}^{\infty} \frac{2^{2n} B_{2n}}{(2n)!} \xi^{2n-1}, \tag{33.10}$$

where B_{2n} are Bernoulli's numbers ($B_2 = 1/6$, $B_4 = -1/30$, $B_6 = 1/42$, ...).

Substituting Eq. (33.10) into (33.1), we obtain

$$\frac{1}{\sigma_{1/2}} = \frac{\tau}{\sigma_{1/2}} C + \sum_{n=1}^{\infty} \frac{2^{2n} B_{2n}}{(2n)!} C_{2n-1} \left(\frac{\sigma_{1/2}}{\tau}\right)^{2n-1} \tag{33.11}$$

where τ is dimensionless temperature, defined by

$$\tau = \frac{4\vartheta}{J(0)} \tag{33.12}$$

and

$$C=\frac{v}{(2\pi)^3}\int\frac{d\nu}{\mathcal{E}_\nu},\quad C_m=\frac{v}{(2\pi)^3}\int\mathcal{E}_\nu^m\,d\nu. \tag{33.13}$$

In the nearest-neighbor approximation, we can easily show that the first three coefficients C_m have the form

$$C_1=1,\quad C_2=1+\frac{1}{z},\quad C_3=1+\frac{3}{z}, \tag{33.14}$$

where z is the number of nearest neighbors.

We shall rewrite Eq. (33.11) in the form

$$\frac{\sigma_{1/2}}{\tau}=\sqrt{\frac{3}{\tau}(1-C\tau)}\left\{1+3\sum_{n=2}^{\infty}\frac{2^{2n}B_{2n}}{(2n)!}C_{2n-1}\left(\frac{\sigma_{1/2}}{\tau}\right)^{2n-2}\right\}^{-1/2} \tag{33.15}$$

and solve it by the iteration method. If $\sigma_{1/2}=0$ is regarded as the zeroth approximation, then, in the first approximation,

$$\frac{\sigma_{1/2}}{\tau}=\sqrt{\frac{3}{\tau}(1-C\tau)}\,. \tag{33.16}$$

This expression can be rewritten as follows:

$$\sigma_{1/2}=\sqrt{3\tau\left(1-\frac{\tau}{\tau_C}\right)}\approx\sqrt{3\tau_C\left(1-\frac{\tau}{\tau_C}\right)}, \tag{33.17}$$

where

$$\tau_C^{-1}=C=\frac{v}{(2\pi)^3}\int\frac{d\nu}{\mathcal{E}_\nu} \tag{33.18}$$

is the reciprocal of the value of temperature at which the magnetization disappears; τ_C may be identified with the Curie temperature. In the conventional units, this temperature is

$$\vartheta_C=\frac{1}{4}J(0)\tau_C=\frac{J(0)}{4C}. \tag{33.19}$$

On comparing Eqs. (33.17) and (21.19), we find that near the Curie point Eq. (33.1) gives the same temperature dependence as the molecular field approximation. The expression for the Curie temperature in the form given by Eq. (33.18) was obtained first by Lax (1955), who considered a modified Ising model – the "spherical model."

The constant C depends on the crystal lattice geometry. For the simple cubic, bcc, and fcc lattices, in the nearest-neighbor approximation, this constant is [see, for example, Ryzhik and Gradshtein (1951)], respectively,

$$C = 1.5164; \quad 1.393; \quad 1.345. \tag{33.20}$$

The reduced Curie temperatures are, respectively,

$$\tau_C = 0.659; \quad 0.718; \quad 0.743. \tag{33.21}$$

The following corrections to Eq. (33.16) can also be easily found. Thus, in the second approximation,

$$\sigma_{1/2} = \tau\zeta\left(1 + \frac{1}{30}\,\frac{z+3}{z}\,\zeta^2\right), \quad \zeta = \sqrt{\frac{3}{\tau}\left(1 - \frac{\tau}{\tau_C}\right)} \tag{33.22}$$

etc. (z is the number of nearest neighbors).

c) High Temperatures ($\vartheta > \vartheta_C$, $H \neq 0$). In this case, we use expansions of a different type, namely,

$$\coth \frac{\mu H + \frac{1}{2}\sigma_{1/2} J(0)\,\mathcal{E}_\nu}{2\vartheta} = \coth\left(\frac{h}{\tau} + \frac{\sigma_{1/2}\mathcal{E}_\nu}{\tau}\right) = \\ = \frac{1}{t_0}\left\{1 + (1 - t_0^2)\sum_{n=1}^{\infty}(-1)^n\left(\frac{t_1}{t_0}\right)^n\right\}, \tag{33.23}$$

where

$$t_0 = \tanh\frac{h}{\tau}, \quad t_1 = \tanh\frac{\sigma_{1/2}\mathcal{E}_\nu}{\tau}, \tag{33.24}$$

and τ, h are dimensionless quantities:

$$\tau = \frac{4\vartheta}{J(0)}, \quad h = \frac{2\mu H}{J(0)}. \tag{33.25}$$

Then Eq. (33.1) assumes the form

$$\frac{1}{\sigma_{1/2}} = \frac{1}{t_0}\left\{1 + (1 - t_0^2)\sum_{n=1}^{\infty}\frac{(-1)^n}{t_0^n}\,\frac{v}{(2\pi)^3}\int t_1^n\,d\nu\right\}. \tag{33.26}$$

Next, expanding t_1 as a series in powers of τ^{-1}, we obtain

$$t_1 = \sum_{k=1}^{\infty} \frac{2^{2k}(2^{2k}-1)}{(2k)!} B_{2k} \left(\frac{\sigma_{1/2}\mathscr{E}_v}{\tau}\right)^{2k-1}. \tag{33.27}$$

We shall substitute Eq. (33.27) into Eq. (33.26) and retain terms up to τ^{-2}, inclusive. Then the approximate equation for $\sigma_{1/2}$ becomes

$$\frac{1}{\sigma_{1/2}} = \frac{1}{t_0}\left\{1 - C_1 \frac{1-t_0^2}{t_0}\frac{\sigma_{1/2}}{\tau} + C_2 \frac{1-t_0^2}{t_0^2}\left(\frac{\sigma_{1/2}}{\tau}\right)^2 - \ldots\right\}, \tag{33.28}$$

where the coefficients C_n are given by the formulas (33.13) and (33.14). As the zeroth approximation, we shall use

$$\sigma_{1/2} = t_0 = \tanh\frac{h}{\tau}. \tag{33.29}$$

Next, the iteration method yields easily an expression for $\sigma_{1/2}$ with any required degree of accuracy with respect to τ^{-1}. In the second approximation, we have

$$\sigma_{1/2} = t_0 + t_0(1-t_0^2)\tau^{-1} + t_0(1-t_0^2)\left(\frac{z-1}{z} - 2t_0^2\right)\tau^{-2} + \ldots \tag{33.30}$$

The expansion (33.30) converges well above the Curie temperature. In this region, we can use the dimensionless ratio τ_C/τ as a small parameter, which makes it possible to apply the standard perturbation theory (cf., Sec. 24). We shall now compare the expression for $\sigma_{1/2}$ of Eq. (33.30), found from the approximate equation (33.1), with the expression (24.24), obtained by the perturbation theory for high temperatures. Since $t_0 = \tanh(h/\tau) \ll 1$, then, approximately,

$$\Delta\sigma_{1/2} = \sigma_{(33.30)} - \sigma_{(24.24)} \approx \frac{t_0}{z\tau^2}.$$

The relative error is small,

$$\frac{\Delta\sigma_{1/2}}{\sigma_{1/2}} \approx \frac{1}{z\tau^2}, \tag{33.31}$$

and we see that in this case the approximate equation (33.1) gives a sufficiently satisfactory description of the temperature and field dependences of the magnetization.

In the vicinity of the Curie point, the (33.30)-type expansions converge slowly, and here we should look for more convenient expansions. However, the most important problem in this range of temperatures is to obtain more exact solutions of the system of equations for the Green's functions than the solutions given by Eqs. (33.1) or (33.2).

Going back to our previous results, we see that the approximate equation for the relative magnetization $\sigma_{1/2}$ given by Eqs. (33.1) or (33.2), containing the principal terms of the expansions for the limiting cases $\vartheta \to 0$, $\vartheta \leq \vartheta_C$ (H = 0), and $\vartheta \gg \vartheta_C$, leads to the results of the spin-wave theory in the molecular field approximation and of the perturbation theory at high temperatures. We can, therefore, assume that it also gives a satisfactory description at intermediate temperatures.

Sec. 34. Average Energy of Ferromagnets*

In this section, we shall show how to calculate the average energy using the first Green's function.

We shall use the Hamiltonian of a system in the form of Eqs. (14.14)-(14.16):

$$\tilde{\mathscr{H}} = \mathscr{E}_0 + \mathscr{H} = \mathscr{E}_0 + \mathscr{H}_2 + \mathscr{H}_4, \tag{34.1}$$

where

$$\mathscr{E}_0 = -\frac{1}{2} N\mu H - \frac{1}{8} NJ(0), \tag{34.2}$$

$$\mathscr{H}_2 = \mu H \sum_f n_f + \frac{1}{2} \sum_{f_1, f_2} I(f_1 - f_2) n_{f_1} - \\ - \frac{1}{2} \sum_{f_1, f_2} I(f_1 - f_2) b_{f_1}^+ b_{f_2}, \tag{34.3}$$

$$\mathscr{H}_4 = -\frac{1}{2} \sum_{f_1, f_2} I(f_1 - f_2) n_{f_1} n_{f_2}. \tag{34.4}$$

We shall multiply the equation of motion for the operator b_f of Eq. (32.5) by b_f^+ on the left-hand side and sum over all values of f. By averaging the equation obtained, we find that

---

*Zubarev (1961b) and Baym and Sessler (1963).

$$\sum_f \left\langle b_f^+ i \frac{db_f}{dt} \right\rangle = \langle \mathscr{H}_2 \rangle + 2 \langle \mathscr{H}_4 \rangle. \tag{34.5}$$

On the other hand,

$$\langle \tilde{\mathscr{H}} \rangle = \mathscr{E}_0 + \langle \mathscr{H}_2 \rangle + \langle \mathscr{H}_4 \rangle. \tag{34.6}$$

Eliminating from Eqs. (34.5) and (34.6) the average value of the fourth term, we obtain the following expression for the average energy of the system:

$$\langle \tilde{\mathscr{H}} \rangle = \mathscr{E}_0 + \frac{1}{2} \langle \mathscr{H}_2 \rangle + \frac{1}{2} \sum_f \left\langle b_f^+ i \frac{db_f}{dt} \right\rangle. \tag{34.7}$$

We shall use the fact that, according to Eq. (26.2),

$$\langle b_g^+ (t') b_f (t) \rangle = \int_{-\infty}^{\infty} I_{fg} (\omega) e^{-i\omega (t-t')} d\omega,$$

where I_{fg} (ω) is the spectral density. Differentiating this expression with respect to t, and going to the limit $t' \to t$, we obtain

$$\left\langle b_g^+ i \frac{db_f}{dt} \right\rangle = \int_{-\infty}^{\infty} I_{fg} (\omega) \, \omega \, d\omega, \tag{34.8}$$

and also

$$\langle b_g^+ b_f \rangle = \int_{-\infty}^{\infty} I_{fg} (\omega) \, d\omega. \tag{34.9}$$

We shall substitute the expression for $\mathscr{H}_2$ of Eq. (34.3) into Eq. (34.7), and use the formulas (34.8) and (34.9). This gives, for the average energy,

$$2 \langle \tilde{\mathscr{H}} \rangle = 2\mathscr{E}_0 + \left(\mu H + \frac{1}{2} J (0) \right) \sum_f \int_{-\infty}^{\infty} I_{ff} (\omega) \, d\omega - \\ - \frac{1}{2} \sum_{f_1, f_2} I (f_1 - f_2) \int_{-\infty}^{\infty} I_{f_1 f_2} (\omega) \, d\omega + \sum_f \int_{-\infty}^{\infty} I_{ff} (\omega) \, \omega \, d\omega. \tag{34.10}$$

We shall introduce the Fourier transformation for the spectral density:

$$I_{fg}(\omega) = \frac{1}{N} \sum_{\nu} I_{\nu}(\omega)\, e^{i(f-g,\,\nu)}. \tag{34.11}$$

Then Eq. (34.10) is transformed into

$$\langle \tilde{\mathscr{H}} \rangle = \mathscr{E}_0 + N \frac{1}{2N} \sum_{\nu} \int_{-\infty}^{\infty} (E_{\nu}^{B} + \omega) I_{\nu}(\omega)\, d\omega, \tag{34.12}$$

where E_{ν}^{B} is the Bloch energy of a spin wave:

$$E_{\nu}^{B} = \mu H + \frac{1}{2}[J(0) - J(\nu)] = \mu H + \frac{1}{2} J(0)\, \mathscr{E}_{\nu}. \tag{34.13}$$

In particular, if the approximate expression (32.14) is used for $I_{\nu}(\omega)$, the expression for the average energy can be written in the following form [Wortis (1963)]:

$$\langle \tilde{\mathscr{H}} \rangle = \mathscr{E}_0 + N \frac{\sigma_{1/2}}{2N} \sum_{\nu} (E_{\nu}^{B} + E_{\nu}) \overline{N}_{\nu}, \tag{34.14}$$

where $\overline{N}_{\nu}$ are given by Eqs. (34.2) and (32.15), while E_{ν}^{B}, E_{ν} are given by Eqs. (34.13) and (32.13); $\sigma_{1/2}$ is the relative magnetization.

Substituting Eq. (34.14) into Eq. (8.24a), we obtain, in the same approximation, an expression for the free energy:

$$F = \mathscr{E}_0 - \vartheta \int_0^{\vartheta} \frac{d\vartheta_1}{\vartheta_1^2} \frac{\sigma_{1/2}}{2} \sum_{\nu} (\mathscr{E}_{\nu}^{B} + \mathscr{E}_{\nu}) \overline{N}_{\nu}.$$

In using this formula, we must bear in mind that $\sigma_{1/2}$ and E_{ν} depend also on the integration variable (E_{ν} depends through the relative magnetization $\sigma_{1/2}$). If we neglect the temperature dependence of $\sigma_{1/2}$, and consequently the dependence of E_{ν}, then Eq. (34.15) can easily be reduced to the formula for the free energy of an ideal gas (8.32):

$$F = \mathscr{E}_0 - \vartheta \int_0^{\vartheta} \frac{d\vartheta_1}{\vartheta_1^2} \sum_{\nu} \mathscr{E}_{\nu}^{B} \overline{N}_{\nu} = \mathscr{E}_0 + \vartheta \sum_{\nu} \ln\left[1 - \exp(-E_{\nu}^{B}/\vartheta)\right].$$

This result could have been predicted earlier. In our case, the divergence of a spin-wave gas from an ideal gas is allowed for only by the presence of the factors $\sigma_{1/2}$ in the spin-wave energy E_ν. Since we have neglected this divergence, we naturally obtained the same results as for an ideal gas of particles of energy E_ν.

Sec. 35. Isotropic Ferromagnets, $S \geq 1/2$ Case*

We have seen that the solution of the approximate equation for the magnetization in the case $S = 1/2$ (cf., Secs. 32 and 33) gives a satisfactory interpolation for the temperature dependence over the whole range of temperatures. We shall extend these results to the case of arbitrary spin.

We shall introduce the spin variables

$$S_f^{\pm} = S_f^x \pm iS_f^y; \quad S_f^z. \tag{35.1}$$

In the case of spin $S = 1/2$, the operators S_f^+ and S_f^- are identical with the Pauli operators b_f and b_f^+. The operators of Eq. (35.1) satisfy the commutation relationships (5.5) and (5.6), as well as the relationships (5.2), (5.3), and (5.7).

We shall write the Hamiltonian of a system in terms of the variables given by Eq. (35.1):

$$\mathscr{H} = -\mu H \sum_f S_f^z - \frac{1}{2} \sum_{f_1, f_2} I(f_1 - f_2)(2S_{f_1}^+ S_{f_2}^- + S_{f_1}^z S_{f_2}^z). \tag{35.2}$$

The equations of motion for the operators of Eq. (35.1) have the form

$$\pm i \frac{dS_f^{\pm}}{dt} = \mu H S_f^{\pm} + \sum_{f'} I(f - f')(S_{f'}^z S_f^{\pm} - S_f^z S_{f'}^{\pm}),$$

$$i \frac{dS}{dt} = -\frac{1}{2} \sum_{f'} I(f - f')(S_f^+ S_{f'}^- - S_f^- S_{f'}^+). \tag{35.3}$$

In the case of spin $S \geq 1$, it is not sufficient to use only the first Green's function $\langle\langle S_f^+ | S_f^- \rangle\rangle$ for the determination of the magnetization $\langle S_f^z \rangle = S\sigma_S$. We can use the Green's function to find the correlation function $\langle S_f^- S_f^+ \rangle$, which, in accordance with

*See Tahir-Kheli and ter Haar (1962a), Hewson and ter Haar (1963a), Tyablikov (1963d), and Callen (1963b).

Eq. (5.7), relates the average values of the z-component of the spin to the square of this component. Therefore, in the present case, we must introduce a system of the Green's functions from which we will be able to determine 2S linearly independent quantities $\langle S_f^z \rangle, \ldots, \langle (S_f^z)^{2S} \rangle$. The higher degrees, according to Eq. (5.3), are not linearly independent [Tahir-Kheli and ter Haar (1962a)].

We shall derive a system of equations for the average values of the powers of S_f^z by multiplying the first of the equations in (5.7) by $(S_f^z)^n$ on the left:

$$\langle (S_f^z)^n S_f^- S_f^+ \rangle = S(S+1)\langle (S_f^z)^n \rangle - \langle (S_f^z)^{n+1} \rangle - \langle (S_f^z)^{n+2} \rangle,\quad n = 0, 1, \ldots, 2S-1. \tag{35.4}$$

For $n = 2S - 1$, the last term on the right-hand side can be represented, by means of Eq. (5.3), in the form

$$\langle (S_f^z)^{2S+1} \rangle = \sum_{r=1}^{2S} a_r^{(S)} \langle (S_f^z)^r \rangle \tag{35.5}$$

[$a_r^{(S)}$ are some numbers] and, therefore, the chain of equations for the averages is decoupled automatically.

The quantities on the left-hand side of Eq. (35.4) will be found in terms of the Green's functions of the type

$$\langle\langle S_f^+(t) | [S_g^z(t')]^n S_g^-(t') \rangle\rangle, \quad n = 0, 1, \ldots, 2S-1. \tag{35.6}$$

Differentiating with respect to the argument t, and going over to the Fourier time transforms, we obtain a chain of equations:

$$\begin{aligned} E\langle\langle S_f^+ | (S_g^z)^n S_g^- \rangle\rangle = \\ = \frac{i}{2\pi}\Delta(f-g)\langle [S_g^+, (S_g^z)^n S_g^-] \rangle + \mu H \langle\langle S_f^+ | (S_g^z)^n S_g^- \rangle\rangle + \\ + \sum_{f'} I(f-f')\{\langle\langle S_{f'}^z S_f^+ | (S_g^z)^n S_g^- \rangle\rangle - \langle\langle S_f^z S_{f'}^+ | (S_g^z)^n S_g^- \rangle\rangle\}, \ldots \\ n = 0, 1, \ldots, 2S-1. \end{aligned} \tag{35.7}$$

As in the case of $S = \frac{1}{2}$, we shall decouple the chain of equations (35.7) by assuming that

$$\langle\langle S_{f_1}^z S_{f_2}^+ | (S_g^z)^n S_g^- \rangle\rangle \rightarrow S\sigma_S \langle\langle S_{f_2}^+ | (S_g^z)^n S_g^- \rangle\rangle, \tag{35.8}$$

where σ_S is the relative magnetization:

$$\langle S_f^z \rangle = S\sigma_S. \tag{35.9}$$

We thus obtain a system of 2S independent equations for the functions (35.6), which is solved by the Fourier transformation

$$S_f^+ = \frac{1}{\sqrt{N}} \sum_\nu S_\nu^+ e^{-i(f,\nu)}.$$

These operations give the following expression for the functions (35.6):

$$\langle\langle S_f^+ \mid (S_g^z)^n S_g^- \rangle\rangle = \frac{i}{2\pi} \frac{1}{N} \sum_\nu \frac{e^{i(g-f,\nu)}}{E - E_S(\nu)} \langle [S_g^+, (S_g^z) S_g^-] \rangle,$$
$$n = 0, 1, \ldots, 2S-1, \tag{35.10}$$

where $E_S(\nu)$ is the energy of the elementary excitations:

$$E_S(\nu) = \mu H + S\sigma_S J(0)(1-\gamma_\nu) = \mu H + S\sigma_S J(0)\mathscr{E}_\nu,$$
$$\gamma_\nu = \frac{J(\nu)}{J(0)}; \tag{35.11}$$

$J(\nu)$ is the Fourier transform of the exchange integral. Using Eq. (26.15), we find the spectral density for the function (35.10):

$$I_{fg}^{(n)}(\omega) = \frac{1}{N} \sum e^{i(g-f,\nu)} \overline{N}_\nu^{(S)} \delta(\omega - E_S(\nu)) \langle [S_g^+, (S_g^z)^n S_g^-] \rangle \tag{35.12}$$

and, using Eq. (26.5), we find the average values

$$\langle (S_g^z)^n S_g^- S_g^+ \rangle = \langle [S_g^+, (S_g^z)^n S_g^-] \rangle P_S, \quad n = 0, 1, \ldots, 2S-1. \tag{35.13}$$

Here, the following notation is used:

$$P_S = \frac{1}{N} \sum_\nu \overline{N}_\nu^{(S)}, \tag{35.14}$$

$$\overline{N}_\nu^{(S)} = \left(e^{\frac{1}{\vartheta} E_S(\nu)} - 1 \right)^{-1}. \tag{35.15}$$

Noting that $S_g^+ (S_g^z)^n = (S_g^z - 1)^n S_g^+$, we obtain the following expression for the commutator on the right-hand side of Eq. (35.13):

$$[S_g^+, (S_g^z)^n S_g^-] = 2S_g^z (S_g^z - 1)^n +$$
$$+ [S(S+1) - S_g^z - (S_g^z)^2] [(S_g^z - 1)^n - (S_g^z)^n]. \tag{35.16}$$

We shall now substitute Eqs. (35.13) and (35.16) into Eq. (35.4). Consequently, we obtain a system of equations for the averages $\langle (S_g^z)^n \rangle$ in the form

$$S(S+1)\langle (S_g^z)^n\rangle - \langle (S_g^z)^{n+1}\rangle - \langle (S_g^z)^{n+2}\rangle =$$
$$= P_S\{\langle [S(S+1) - S_g^z - (S_g^z)^2][(S_g^z - 1)^n - (S_g^z)^n]\rangle +$$
$$+ 2\langle S_g^z(S_g^z - 1)^n\rangle\}, \quad n = 0, 1, \ldots, 2S-1. \tag{35.17}$$

The system of equations (35.17), (35.5) is sufficient for the determination of all the necessary averages. We shall now consider the cases of S = ½, 1, ..., 3.

Case of S = $^1/_2$. Here, n = 0 and the system (35.17), (35.5) reduces to only two equations:

$$\frac{3}{4} - \frac{1}{2}\sigma_{1/2} - \langle (S_g^z)^2\rangle = \sigma_{1/2}P_{1/2},$$
$$\langle (S_g^z)^2\rangle = \frac{1}{4}. \tag{35.18}$$

Hence, we obtain an equation for $\sigma_{1/2}$,

$$\frac{1}{2}\sigma_{1/2} = \frac{1/2}{1+2P_{1/2}}, \tag{35.19}$$

which, as can be easily seen, is identical with Eq. (32.20).

Case of S = 1. The system of equations (35.17), (35.5) now has the form

$$2 - \sigma_1 - \langle (S_g^z)^2\rangle = 2\sigma_1 P_1,$$
$$2\sigma_1 - \langle (S_g^z)^2\rangle - \langle (S_g^z)^3\rangle = P_1\{\langle (S_g^z)^2\rangle - \sigma_1 - 2\},$$
$$\langle (S_g^z)^3\rangle = \sigma_1. \tag{35.20}$$

Hence, we obtain an equation for σ_1 and an expression for the average square of the z-component of spin:

$$\sigma_1 = \frac{1+2P_1}{1+3P_1+3P_1^2}, \tag{35.21}$$

$$\langle (S_g^z)^2\rangle = \frac{2P_1}{1+3P_1} + \frac{1+P_1}{1+3P_1}\sigma_1. \tag{35.22}$$

Case of S > 1. Similarly, we can easily obtain equations for σ_S with S = $^3/_2$, 2, ... Omitting the elementary steps, we shall quote an equation for σ_S in the general form [Praveczki (1963)]:

$$S\sigma_S = \frac{(S-P_S)(1+P_S)^{2S+1} + (1+S+P_S)P_S^{2S+1}}{(1+P_S)^{2S+1} - P_S^{2S+1}}$$

or

$$S\sigma_S = \frac{S + (S+1)\sum_{n=1}^{2S} \frac{2S-n}{n} C_{2S+1}^n P_S^n}{1 + \sum_{n=1}^{2S} C_{2S+1}^n P_S^n}, \tag{35.23}$$

where C_l^k is the number of combinations of l elements, taken k at a time; P_S is the function (35.14). For $S = 1, \frac{3}{2}, \ldots, 3$, this equation was obtained by Tahir-Kheli and ter Haar (1962a).

At low temperatures, the solution of Eq. (35.23) is in the form of a series of powers of temperature. Expanding the right-hand side of Eq. (35.23) in powers of P_S:

$$S\sigma_S = \begin{cases} \frac{1}{2} - P_{1/2} + 2P_{1/2}^2 + O(P_{1/2}^3), & S = \frac{1}{2}, \quad (35.24) \\ S - P_S + O(P_S^3), & S \geq 1. \quad (35.25) \end{cases}$$

Next, we shall allow for the fact that, at low temperatures, P_S, and consequently σ_S, are expansions of the type given by Eq. (15.19), but with the fractional powers of τ replaced by the same powers of τ/σ_S. We shall consider the nearest-neighbor approximation for the simplest cubic lattices: simple cubic, bcc, and fcc. Solving Eqs. (35.24), (35.25) in the same way as in Sec. 33, we obtain the following expansions:

$$\frac{1}{2}\sigma_{1/2} = \frac{1}{2} - \rho Z_{3/2}\left(\frac{h}{\tau}\right)\tau^{3/2} - \frac{3\pi}{4}\rho Z_{5/2}\left(\frac{h}{\tau}\right)\tau^{5/2} - 2\rho^2 Z_{3/2}^2\left(\frac{h}{\tau}\right)\tau^{6/2} - \pi^2\omega\rho Z_{7/2}\left(\frac{h}{\tau}\right)\tau^{7/2} - \ldots, \tag{35.26}$$

$$S\sigma_S = S - \rho Z_{3/2}\left(\frac{h}{\tau}\right)\tau^{3/2} - \frac{3\pi}{4}\rho Z_{5/2}\left(\frac{h}{\tau}\right)\tau^{5/2} - \frac{3}{2S}\rho^2 Z_{3/2}^2\left(\frac{h}{\tau}\right)\tau^{6/2} - \pi^2\omega\rho Z_{7/2}\left(\frac{h}{\tau}\right)\tau^{7/2} - \ldots, \tag{35.27}$$

where the following notation is used:

$$\tau = \frac{\vartheta}{\frac{2\pi}{3}SIz}, \quad h = \frac{\mu H}{\frac{2\pi}{3}SIz}. \tag{35.28}$$

We note that, in the case of both spin $S = \frac{1}{2}$ and $S \geq 1$, the first corrections in the nearest-neighbor approximation begin with the term $\tau^{6/2}$. For spin S = $^1/_2$, it appears, first, because of the term $2P_{1/2}^2$ in the expansion (35.34) and, secondly, due to the dependence of the spin-wave energy on temperature through $\sigma_{1/2}$ (see also Sec. 33). For spins S ≥ 1, the corrections of the order of $\tau^{6/2}$ appear due only to the dependence of the spin-wave energy on σ_S. Corrections of this type are due to the error introduced by the decoupling given by Eq. (35.8). At high temperatures ($\vartheta \leq \vartheta_C$, H = 0), the expansions in powers of τ converge slowly, and it is more convenient to obtain a solution in the form of a series in powers of some other quantity which can be regarded as small. In this range of temperatures, $\sigma_S \ll 1$ and the quantity P_S^{-1} can be regarded as small. In fact, we shall write Eq. (35.14) in the form

$$1+2P_S=\frac{1}{N}\sum_\nu \coth\frac{\sigma_S\mathscr{E}_\nu}{\tau}\quad\left(\mathscr{E}_\nu=1-\frac{J(\nu)}{J(0)}\right),$$

where we use the notation

$$\tau=\frac{2\vartheta}{SJ(0)}. \tag{35.29}$$

Using Eq. (33.10), we find

$$P_S^{-1}=\frac{2\sigma_S}{C\tau}\left(1+\frac{1}{C}\frac{\sigma_S}{\tau}+\frac{3-C}{3C^2}\frac{\sigma_S^2}{\tau^2}+\ldots\right), \tag{35.30}$$

where C is defined by Eq. (33.13). We shall expand the right-hand side of Eq. (35.23) as a series in powers of P_S^{-1}:

$$S\sigma_S=\frac{S(S+1)}{3}\times$$
$$\times P_S^{-1}\frac{1+\frac{2\cdot 3!}{4!}(2S-1)P_S^{-1}+\frac{3\cdot 3!}{5!}(2S-1)(2S-2)P_S^{-2}+\ldots}{1+\frac{2S}{2!}P_S^{-1}+\frac{2S(2S-1)}{3!}P_S^{-2}+\ldots}=$$
$$=\frac{S(S+1)}{3}P_S^{-1}-\frac{S(S+1)}{6}P_S^{-2}+\frac{S(S+1)}{3}\frac{9-2S-2S^2}{30}P_S^{-3}-\ldots \tag{35.31}$$

We shall substitute Eq. (35.30) into Eq. (35.31) and use only the terms up to σ_S^3, inclusive. In this approximation, we easily obtain the following expression for σ_S:

$$\sigma_S=\sqrt{\Gamma_S\tau_C\left(1-\frac{\tau}{\tau_C}\right)}, \tag{35.32}$$

where τ_C is the Curie temperature,

$$\tau_C=\frac{2(S+1)}{3C}; \tag{35.33}$$

Γ_S is a constant which depends on the value of the spin and on the lattice geometry:

$$\Gamma_S = \frac{10C(S+1)}{4S^2+4S+5C-3}, \tag{35.34}$$

where

$$C = \frac{1}{N}\sum_{\nu} \mathscr{E}_{\nu}^{-1}. \tag{35.34a}$$

We can easily check that for spin S = ½ the formulas (35.32) and (35.33) reduce to the formulas (33.17) and (33.18).

In the usual units, the Curie temperature has the form

$$\vartheta_C = \frac{S\tau_C}{2} J(0) = \frac{S(S+1)}{3C} J(0), \tag{35.35}$$

where C is defined in accordance with (35.34a). An expression of the Eq. (35.35) type was obtained earlier for the Curie temperature by Brown and Luttinger (1955), using the perturbation theory at high temperatures.

At temperatures above the Curie point ($\vartheta > \vartheta_C$, $H \neq 0$), we can obtain an expansion for σ_S in powers of the reciprocal of temperature, similar to the expansion (33.30). Then, calculating the susceptibility for H = 0, we obtain [Tahir-Kheli and ter Haar (1962a)]

$$\chi_S = \left[\frac{\partial}{\partial H}(S\mu\sigma_S)\right]_{H=0} =$$
$$= \mu^2 \frac{S(S+1)}{3\vartheta}\left\{1 + \frac{\vartheta_{Cp}}{\vartheta} + \frac{z-1}{z}\left(\frac{\vartheta_{Cp}}{\vartheta}\right)^2 + O(\vartheta^{-3})\right\}, \tag{35.36}$$

where ϑ_{Cp} is the paramagnetic Curie temperature:

$$\vartheta_{Cp} = \frac{S(S+1)}{3} J(0). \tag{35.37}$$

With an accuracy to values of the order of z^{-1}, the formula (35.36) may be given the form of the Curie–Weiss law (1.1):

$$\chi = \frac{\mu^2 S(S+1)}{3(\vartheta - \vartheta_{Cp})}. \tag{35.38}$$

The expression (35.36) for the susceptibility is very similar to the result obtained by Brown and Luttinger (1955) from the perturbation theory. In the approximation considered here, the

Table 10. Values of the Curie Temperature τ_C and of the Quantity ϑ_C/Iz for Cubic Lattices [Values of τ_C were calculated using Eq. (35.33), and the ratio ϑ_C/Iz was found using Eq. (35.35); I is the exchange integral for the nearest neighbors, z is the number of these neighbors, and C is a constant, given by Eq. (35.34a), which depends on the type of lattice]

Lattice type		Spin					
		1/2	1	3/2	2	5/2	3
Simple cubic	τ_C	0.659	0.879	1.099	1.319	1.539	1.759
	ϑ_C/Iz	0.165	0.440	0.824	1.319	1.923	2.638
BCC	τ_C	0.718	0.972	1.196	1.436	1.675	1.914
	ϑ_C/Iz	0.179	0.479	0.897	1.436	2.094	2.871
FCC	τ_C	0.743	0.991	1.239	1.487	1.735	1.983
	ϑ_C/Iz	0.186	0.496	0.929	1.487	2.168	2.974

Table 11. Values of the Constant Γ_S [Values of Γ_S calculated using Eq. (35.34); dependence on the type of lattice is through the constant C of Eq. (35.34a)]

Lattice type	Spin					
	1/2	1	3/2	2	5/2	3
Simple cubic	3	2.410	1.936	1.592	1.341	1.154
BCC	3	2.328	1.836	1.494	1.251	1.072
FCC	3	2.294	1.796	1.455	1.216	1.040

Curie temperature (35.35) and the paramagnetic Curie temperature (35.37) are related by

$$C\vartheta_C = \vartheta_{Cp}. \tag{35.39}$$

A comparison of the data in Table 9 (p. 201) and Table 10 shows that the values of the Curie temperature calculated using Eq. (35.33) are close to the values calculated using the perturbation theory at high temperatures.

It should be mentioned that the decoupling of the (35.8) type is not the only possible one. Callen (1963b) proposed a decoupling for the second functions, which are, obviously, more accurate at higher values of the spin. They are also of definite interest, because they can be used to obtain more accurate approximate solutions.

According to Eqs. (5.7) and (5.5), we have

$$S_f^z = S(S+1) - (S_f^z)^2 - S_f^- S_f^+. \tag{35.40}$$

$$S_f^z = \frac{1}{2}(S_f^+ S_f^- - S_f^- S_f^+). \tag{35.41}$$

Multiplying the first of these relationships by an arbitrary parameter α, andthe second by $(1-\alpha)$, and adding we obtain

$$S_f^z = \alpha S(S+1) - \alpha (S_f^z)^2 - \frac{1+\alpha}{2} S_f^- S_f^+ + \frac{1-\alpha}{2} S_f^+ S_f^-. \tag{35.42}$$

We shall consider the Green's function

$$\langle\langle S_f^z S_{f'}^+ \mid \ldots \rangle\rangle \quad (f' \neq f). \tag{35.43}$$

We shall substitute into (35.43) the expression (35.42) and we shall neglect the fluctuation $(S_f^z)^2$. The functions $\langle\langle S_f^- S_f^+ S_{f'}^+ | \ldots \rangle\rangle$ will be decoupled as follows:

$$\langle\langle S_f^- S_f^+ S_{f'}^+ \mid \ldots \rangle\rangle \to \langle S_f^- S_f^+ \rangle \langle\langle S_{f'}^+ \mid \ldots \rangle\rangle + \\ + \langle S_f^- S_{f'}^+ \rangle \langle\langle S_f^+ \mid \ldots \rangle\rangle. \tag{35.44}$$

Consequently, we shall obtain the following decoupling for the function (35.43):

$$\langle\langle S_f^z S_{f'}^+ \mid \ldots \rangle\rangle \to \\ \to \langle S_f^z \rangle \langle\langle S_{f'}^+ \mid \ldots \rangle\rangle - \alpha \langle S_f^- S_{f'}^+ \rangle \langle\langle S_f^+ \mid \ldots \rangle\rangle \quad (f' \neq f). \tag{35.45}$$

According to Callen, α should be taken in the form

$$\alpha = (2S)^{-1} \sigma_S. \tag{35.46}$$

In fact, at low temperatures, the decoupling (35.45) should correspond to the spin-wave approximation. Using the approximate formulas (12.19) for S_f^+, S_f^z and then applying the Wick–Bloch–de Dominicis theorem (cf., Sec. 10), we see that the required correspondence is indeed obtained if $\alpha \to \frac{1}{2}S$ when $\vartheta \to 0$. At high temperatures $\langle S_f^z \rangle \approx 0$ and, therefore, the right-hand side of Eq. (35.45) should vanish together with $\langle S_f^z \rangle$. The quantity α in the form given by Eq. (35.46) satisfies both these conditions; we note that the selected α is not single-valued.

Sec. 36. Numerical Integration of the Magnetization Equation

The equation for the magnetization (33.1) for $h = 0$ has the form

$$\frac{1}{\sigma_{1/2}} = \frac{1}{\pi^3} \int \int_0^\pi \int \coth \frac{\sigma_{1/2} \mathscr{E}_\nu}{\tau} \, d\nu, \quad \tau = \frac{4\vartheta}{Iz}. \tag{36.1}$$

It was solved numerically by S. P. Lomnev for the case of a simple cubic lattice in the nearest-neighbor approximation.* The results of these calculations are listed in Table 12.

The third column of Table 12 contains, for comparison, the values of the magnetization calculated using Eq. (33.9), which represents the spin-wave approximation:

$$\sigma_{1/2} = 1 - 2\zeta\left(\frac{3}{2}\right)\left(\frac{3\tau}{4\pi}\right)^{3/2} - \frac{3\pi}{2}\zeta\left(\frac{5}{2}\right)\left(\frac{3\tau}{4\pi}\right)^{5/2}, \quad \tau = \frac{4\vartheta}{Iz}, \tag{36.2}$$

where ζ (p) is given by Eq. (15.17).

Figure 11 shows the $\sigma_{1/2}(\tau)$ curves calculated using Eq. (36.1), the approximate formula (36.2), and the molecular field equation (21.17), as well as the experimental data for iron.

For small wave numbers, the spin-wave energy is approximately

$$E_S(\nu) = \alpha_S \nu^2, \tag{36.3}$$

where $\alpha_S = S\sigma_S \cdot$ const. In experiments using inelastic neutron scattering by spin waves [see, for example, Izyumov's review (1963a)], one measures the parameter α_S in the dispersion law (36.3). From measurements made at different temperatures, one can establish whether α_S depends on temperature and, if it does, the nature of this dependence. Figure 12 shows the values of the

*The present author is grateful to S. P. Lomnev for carrying out these calculations on an electronic computer.

Table 12. Temperature Dependence of the Relative Magnetization $\sigma_{1/2}$ [Values of $\sigma_{1/2}$ calculated using Eq. (36.1) (second column) and Eq. (36.2) (third column)]

τ/τ_C	$\sigma_{1/2}$ (36.1)	$\sigma_{1/2}$ (36.2)
0	1	1
0,15164	0.981	0,9802
0.30328	0,941	0.9424
0.45496	0,881	0,8912
0.60656	0.790	0,8280
0,75820	0,654	0,7534
0,90984	0.430	0,6677
0.98560	0.20—0.21	0.6207
1		0.6110

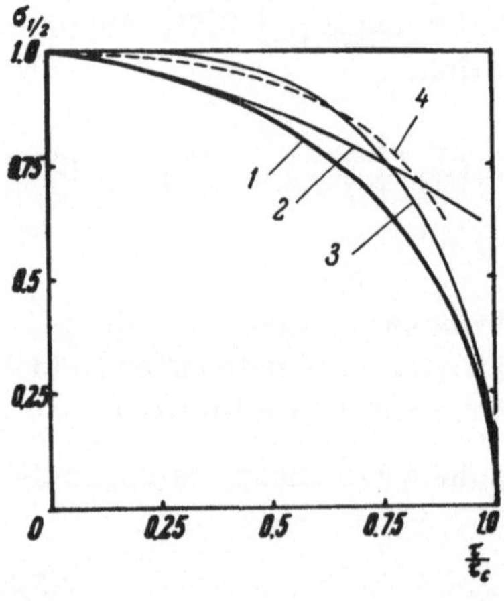

Fig. 11. Dependence of the relative magnetization $\sigma_{1/2}$ on the relative temperature τ/τ_C: 1) according to Eq. (36.1); 2) according to Eq. (36.2); 3) according to the molecular field equation (21.17); 4) experimental data for iron [Potter (1934)].

relative magnetization of magnetite according to Pauthenet's data (1952), and the parameter α_S according to Brockhouse and Watanabe (1962).

It is evident that the measured values of α_S lie on a single curve. Thus, the formula (36.3) generally represents, more or

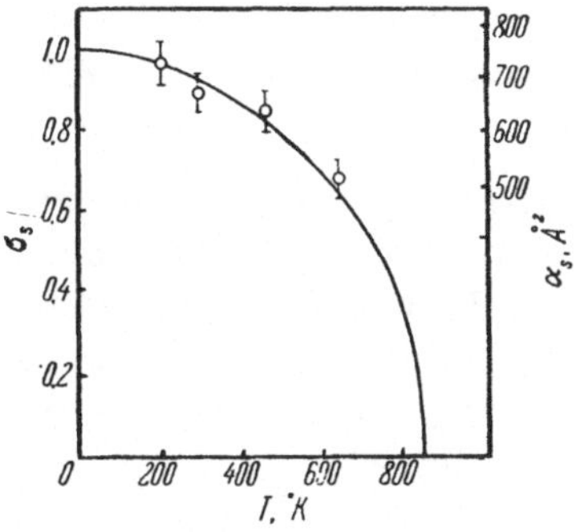

Fig. 12. Temperature dependence of the spin-wave energy of magnetite: continuous curve represents experimental values of the magnetization σ_S, while circles give experimental values of the parameter $\alpha_S = S\sigma_S \cdot \text{const}$ in the approximate formula (36.3) for the spin-wave energy (the vertical segments indicate the experimental errors; T_C is the Curie temperature).

less satisfactorily, the form of the temperature dependence of the spin-wave energy. According to Hatherly et al. (1963), the spin-wave energy at low temperatures varies as $\vartheta^{5/2}$ and not as $\vartheta^{3/2}$. This result is in agreement with the theoretical considerations, since, at $\vartheta \to 0$, the $\vartheta^{3/2}$-type dependence of $E_S(\nu)$ is due to the approximate nature of the solution and, in fact, we have a dependence of the $\vartheta^{5/2}$type (cf., Secs. 33, 35, and 37).

Sec. 37. Low-Temperature Expansions

We have already considered the approximate solutions of the equation for the magnetization, which are valid over a wide range of temperatures. Naturally, the question arises how accurate are the solutions obtained and how to find the higher approximations. Dyson (1956a,b) obtained successive expansions, in powers of temperature, for the free energy, specific heat, and magnetization at low temperatures; these can be regarded as the standard equations.

We shall confine ourselves to low temperatures and consider the problem of obtaining higher approximations in the Green's function method.

In the case of arbitrary spin, the magnetization is a function of the sums P_S (35.14) of the occupation numbers of the spin wave. At low temperatures, P_S may be regarded as a small quantity $[P_S = O(\vartheta^{3/2})]$ and we can expand the expression for the magnetization σ_S as a series in powers of P_S. With an accuracy to the third-order terms, we have

$$S\sigma_S = \begin{cases} \frac{1}{2} - P_{1/2} + 2P_{1/2}^2 + O(P_{1/2}^3), & S = 1/2, \\ S - P_S + O(P_S^3), & S \geq 1. \end{cases} \tag{37.1}$$

In a rough approximation, using the decoupling formulas (32.7) or (35.8), the quantities P_S themselves depend on σ_S through the spin-wave energy $E_S(\nu)$, in accordance with Eq. (32.13) or Eq. (35.11).

In the spin-wave approximation, the energy of elementary excitations has the form

$$E_S(\nu) = \mu H + S[J(0) - J(\nu)]. \tag{37.2}$$

We shall rewrite Eq. (35.11) as follows:

$$E_S(\nu) = \mu H + S[J(0) - J(\nu)] - S(1 - \sigma_S)[J(0) - J(\nu)]. \tag{37.3}$$

The last term in Eq. (37.3) may be interpreted as the mass operator in the approximation corresponding to the decoupling formulas (32.7) or (35.8). Then the mass operator depends on the temperature as a parameter; at low temperatures, this dependence is of the $\vartheta^{3/2}$ type.

We shall consider the interaction of spin waves as a small perturbation and apply the perturbation theory of Sec. 29 to solve the system of equations for the Green's functions. In the lowest approximation, the expression (37.2) for the energy is replaced with

$$E_S(\nu) = \mu H + S[J(0) - J(\nu)] + M_S(\nu), \tag{37.4}$$

where $M_S(\nu)$ is the mass operator for the first functions in the case of arbitrary spin S ($\geq 1/2$). As before, the magnetization σ_S

is given by the formulas of the (37.1) type. It is found that when $S = 1/2$, the mass operator depends on temperature as $\vartheta^{5/2}$. As pointed out by Dyson (1956a), the kinematic interaction does not affect the dynamic properties of a system. Therefore, we can expect that the temperature dependence will remain roughly the same in this approximation even for $S \geq 1$. Consequently, the first corrections to the magnetization due to the spin-wave interaction will be of the order of $\vartheta^{6/2}$ for spin $S = 1/2$ (due to the term $2P_{1/2}^2$) and of the order of $\vartheta^{8/2}$ for spin $S \geq 1$.

The direct solution, using the perturbation theory of Sec. 29, of the chains of equations for the spin operators in the case of $S = 1/2$ [Tyablikov (1963a, b, c; 1964)] and $S \geq 1$ [Bross (1964a, b)] gives temperature dependences of the type referred to earlier. A similar result was obtained by Praveczki (1963). However, the application of the same perturbation theory method to the Hamiltonian of Dyson's ideal spin waves (1956a, b) at any value of S ($\geq 1/2$) gives the dependence $\vartheta^{5/2}$ for the mass operator and, consequently, it yields an expansion for the magnetization which does not contain a term of the order of $\vartheta^{6/2}$ [Tyablikov and Yakovlev (1962, 1963)]. The latter result, first established by Dyson (1956a, b), was obtained also by other methods using the Hamiltonian for ideal spin waves [see, for example, Opechowski (1960), Szaniecki (1962), Tahir-Kheli and ter Haar (1962b), and Wortis (1963)]. In the opinion of Hewson and ter Haar (1963c), the difference is due to an error committed by replacing the exact second Green's function in spin operators with an approximate function, although the error for the corresponding correlation function may be of the order of ϑ^3. The problem of the causes of the discrepancies of the results for the case of $S = 1/2$ has not yet been investigated in detail.

Sec. 38. Ferromagnetic Resonance*

If a spin system is subjected to a constant magnetic field H and, at right angles to the latter, an alternating radio-frequency field h(t), the transfer of energy from the field h(t) to the spin

*See Kubo and Tomita (1954), Kubo (1957), and Tyablikov (1960a, b).

system increases sharply at an alternating-field frequency close to the frequency of free precession of the spins about the direction of the vector H. We shall consider this phenomenon in more detail.

Let a spin system consist of N identical spins, located at lattice sites f. The Hamiltonian of the spin system, which includes the operator representing the energy of interaction between the system and the constant field H, will be denoted by $\mathcal{H}_0$. Let the alternating field h(t) have the form

$$h(t) = \sum_{\Omega} e^{-i\Omega t} h_{\Omega}(f). \tag{38.1}$$

Then the operator for the energy of interaction of the spin system with the alternating field is written as follows:

$$\mathcal{V}(t) = -\mu \sum_{f,\Omega} e^{-i\Omega t}(h_{\Omega}(f), S_f). \tag{38.2}$$

As a rule, the radio-frequency wavelength in the experiments on ferromagnetic resonance is usually much greater than the dimensions of the sample and, therefore, the field may be regarded as being uniform. In this case,

$$\mathcal{V}(t) = -\mu \sum_{\Omega} e^{-i\Omega t}(h_{\Omega}, S), \tag{38.3}$$

where $S^{\alpha} = \sum_{f} S^{\alpha}_{f}$ is the α-component of the total spin of the system.

We shall confine ourselves to an investigation of the spatially uniform case. Moreover, we shall assume that the operator $\mathcal{V}(t)$ is a small perturbation compared with the operator $\mathcal{H}_0$

According to Eq. (31.21) (linear approximation), the increment in the average value of the magnetization vector under the action of the perturbation (38.3), applied adiabatically, is

$$\delta\langle\mathcal{M}^{\alpha}(t)\rangle = \mu\delta\langle S^{\alpha}(t)\rangle = 2\pi i\mu^2 \sum_{\Omega,\beta} e^{-i\Omega t}\langle\langle S^{\alpha}|S^{\beta}\rangle\rangle^{(r)}_{\Omega} h^{\beta}_{\Omega} \tag{38.4}$$

or, using Eq. (31.25),

$$\delta\langle\mathcal{M}^{\alpha}(t)\rangle = \sum_{\Omega,\beta} e^{-i\Omega t}\chi_{\alpha\beta}(\Omega)\, h^{\beta}_{\Omega}. \tag{38.5}$$

and the complex susceptibility has the form

$$\chi_{\alpha\beta}(\Omega) = 2\pi i \mu^2 \langle\langle S^\alpha | S^\beta \rangle\rangle^{(r)} \tag{38.6}$$

The following form is also used to write down $\mathscr{V}(t)$:

$$\mathscr{V}(t) = -\mu \sum_{\Omega>0} \{e^{-i\Omega t}(h_\Omega, S) + e^{i\Omega t}(h^*_\Omega, S)\}, \tag{38.7}$$

where, by definition, $h^\alpha_{-\Omega} = h^{*\alpha}_\Omega$. Then, instead of the formula (38.5), we have

$$\delta\langle\mathscr{M}^\alpha(t)\rangle = \sum_{(\beta,\, \Omega>0)} \{\chi_{\alpha\beta}(\Omega) h^\beta_\Omega e^{-i\Omega t} + \chi_{\alpha\beta}(-\Omega) h^{*\beta}_\Omega e^{i\Omega t}\}, \tag{38.8}$$

where the susceptibility χ is, as before, given by the expression (38.6).

Since χ can be expressed in a linear form in terms of the Green's functions, the poles Ω_R of these functions should determine the behavior of χ in the region of resonance. The characteristic features of the behavior of χ are determined by the explicit form of the Green's functions. We shall consider here only some general properties of the susceptibility tensor, which are not associated with any special problem.

From the relationships (28.9) and (28.12) for the Green's functions, it follows that in the complex plane

$$\langle\langle S^\alpha | S^\beta \rangle\rangle_E = \langle\langle S^\beta | S^\alpha \rangle\rangle_{-E}$$

and on the real axis

$$\langle\langle S^\alpha | S^\beta \rangle\rangle^{(r)}_\Omega = -\{\langle\langle S^\alpha | S^\beta \rangle\rangle^{(r)}_{-\Omega}\}^*.$$

Hence, for the susceptibility tensor χ of Eq. (38.6) we obtain

$$\chi_{\alpha\beta}(E) = \chi_{\beta\alpha}(-E) \quad (E \text{ complex}), \tag{38.9}$$

$$\chi_{\alpha\beta}(\Omega) = \chi^*_{\alpha\beta}(-\Omega) \quad (\Omega \text{ real}). \tag{38.10}$$

To simplify these formulas, we shall assume that the radio-frequency field lies in the (x, y) plane. We shall introduce the operators

$$S^\pm = S^x \pm iS^y \tag{38.11}$$

and the Green's functions

$$G_{11}(\Omega)=\langle\langle S^+|S^+\rangle\rangle_\Omega^{(r)},\quad G_{12}(\Omega)=\langle\langle S^+|S^-\rangle\rangle_\Omega^{(r)},$$
$$G_{21}(\Omega)=\langle\langle S^-|S^+\rangle\rangle_\Omega^{(r)},\quad G_{22}(\Omega)=\langle\langle S^-|S^-\rangle\rangle_\Omega^{(r)}. \tag{38.12}$$

In terms of the variables $S^\pm$, the tensor χ of Eq. (38.6) is written as follows:

$$\chi(\Omega)=\begin{pmatrix} X(\Omega)+X_a(\Omega) & i[\mathcal{G}(\Omega)+\mathcal{G}_a(\Omega)] & 0 \\ -i[\mathcal{G}(\Omega)-\mathcal{G}_a(\Omega)] & X(\Omega)-X_a(\Omega) & 0 \\ 0 & 0 & \chi_{st}\end{pmatrix}, \tag{38.13}$$

where

$$X=\nu(G_{12}+G_{21}),\quad X_a=\nu(G_{11}+G_{22}),$$
$$\mathcal{G}=\nu(G_{12}-G_{21}),\quad \mathcal{G}_a=\nu(-G_{11}+G_{22}), \tag{38.14}$$

$\nu = (i/2)\pi\mu^2$, and χ_{st} is the static susceptibility.

The Green's functions in the formulas (38.12)-(38.14) are considered only along the real axis. In this case, the relationship (28.12) applies and, therefore,

$$G_{12}(\Omega)=-G_{21}^*(-\Omega),\quad G_{11}(\Omega)=-G_{22}^*(-\Omega).$$

Hence, it follows that

$$X(\Omega)=X^*(-\Omega),\qquad X_a(\Omega)=X_a^*(-\Omega),$$
$$\mathcal{G}(\Omega)=-\mathcal{G}^*(-\Omega),\quad \mathcal{G}_a(\Omega)=-\mathcal{G}_a^*(-\Omega). \tag{38.15}$$

For the real and imaginary parts of the susceptibility, we obtain, on the basis of Eqs. (38.10) and (38.15), the relationships

$$\chi'_{\alpha\beta}(\Omega)=\chi'_{\alpha\beta}(-\Omega),\quad \chi''_{\alpha\beta}(\Omega)=-\chi''_{\alpha\beta}(-\Omega) \tag{38.16}$$

and, similarly, the relationships

$$X'(\Omega)=X'(-\Omega),\qquad X''(\Omega)=-X''(-\Omega),$$
$$\mathcal{G}'(\Omega)=-\mathcal{G}'(-\Omega),\quad \mathcal{G}''(\Omega)=\mathcal{G}''(-\Omega), \tag{38.17}$$

where X_a, $\mathcal{G}_a$ obey the same relationships as X and $\mathcal{G}$.

If the spin system is such that the unperturbed equations of motion for the operators $S^{\pm}$ allow time inversion, then, according to Eq. (28.30),

$$G_{11}(\Omega) = G_{22}(\Omega)$$

and, consequently,

$$X_a(\Omega) = 2\nu G_{11}(\Omega), \quad \mathcal{G}_a(\Omega) = 0. \tag{38.18}$$

The behavior of the susceptibility χ at resonance is determined by singularities of the Green's functions. We shall assume that a Green's function has a pole at a point $\Omega = \Omega_R$.

We shall introduce a matrix $G(\Omega)$ whose elements are the Green's functions (38.12), and an inverse of this matrix $F(\Omega)$ (cf., Sec. 28). Using the formulas (28.33)-(28.38) and the definition (38.14), we easily obtain

$$\begin{aligned} X(\Omega) &= \nu \frac{-F_{12}(\Omega) + F_{12}^{*}(-\Omega)}{2\Omega_R \Lambda(\Omega)} R(\Omega), \\ X_a(\Omega) &= \nu \frac{F_{11}(\Omega) - F_{11}^{*}(-\Omega)}{2\Omega_R \Lambda(\Omega)} R(\Omega), \end{aligned} \tag{38.19}$$

$$\begin{aligned} \mathcal{G}(\Omega) &= \nu \frac{-F_{12}(\Omega) - F_{12}^{*}(-\Omega)}{2\Omega_R \Lambda(\Omega)} R(\Omega), \\ \mathcal{G}_a(\Omega) &= \nu \frac{F_{11}(\Omega) + F_{11}^{*}(-\Omega)}{2\Omega_R \Lambda(\Omega)} R(\Omega), \end{aligned} \tag{38.20}$$

where

$$R(\Omega) = \frac{1}{\Omega - \Omega_R + i\varepsilon} - \frac{1}{\Omega + \Omega_R + i\varepsilon}. \tag{38.20a}$$

From the formulas (38.19) and (38.20), it follows that

$$X^2(\Omega) - X_a^2(\Omega) - \mathcal{G}^2(\Omega) + \mathcal{G}_a^2(\Omega) = \frac{\pi^2\mu^4}{2\Omega_R \Lambda(\Omega)} R(\Omega) \tag{38.21}$$

or

$$\chi_{xx}(\Omega)\chi_{yy}(\Omega) - \chi_{xy}(\Omega)\chi_{yx}(\Omega) = \frac{\pi^2\mu^4}{2\Omega_R \Lambda(\Omega)} R(\Omega). \tag{38.22}$$

Approximate expressions, valid in the region of a pole, are obtained by replacing Ω with Ω_R in the multipliers of $R(\Omega)$ in the formulas (38.19) and (38.20):

$$X(\Omega) \approx C(1+K_1K_2)R(\Omega), \quad X_a(\Omega) \approx -C(K_1+K_2)R(\Omega),$$

$$\mathcal{G}(\Omega) \approx C(1-K_1K_2)R(\Omega), \quad \mathcal{G}_a(\Omega \approx -C(K_1-K_2)R(\Omega), \tag{38.23}$$

where C, K_1, and K_2 are constants:

$$C=\nu\frac{F^*_{12}(-\Omega_R)}{2\Omega_R\Lambda(\Omega_R)}, \quad K_1=-\frac{F_{11}(\Omega_R)}{F^*_{12}(-\Omega_R)}, \quad K_2=\frac{F^*_{11}(-\Omega_R)}{F^*_{12}(-\Omega_R)}. \tag{38.24}$$

If the equations of motion for the operators $S^{\pm}$ (38.11) are invariant under time inversion, then, in addition to (28.12), we also have the condition (28.30), and $F_{11}(\Omega) = F^*_{11}(-\Omega)$. Hence, it follows that

$$K_1=K_2, \quad \mathcal{G}_a(\Omega)=0. \tag{38.24a}$$

The energy absorbed by the spin system from the radio-frequency field in unit time is numerically equal to the work done by the field in unit time. We shall define the latter as follows:

$$\frac{d}{dt}A(t)=\sum_{\alpha} h^{\alpha}(t)\frac{d}{dt}\delta\langle\mathcal{M}^{\alpha}(t)\rangle =$$

$$=-i\sum_{\substack{\alpha,\beta\\ \Omega,\Omega'>0}} \Omega'\{h^{\alpha}_{\Omega}e^{-i\Omega t}+h^{*\alpha}_{\Omega}e^{i\Omega t}\}\times$$

$$\times\{\chi_{\alpha\beta}(\Omega')h^{\beta}_{\Omega'}e^{-i\Omega' t}-\chi_{\alpha\beta}(-\Omega')h^{*\beta}_{\Omega'}e^{i\Omega' t}\}.$$

The average energy absorbed by the system in unit time (i.e., the absorbed power) is

$$W=-\lim_{T\to\infty}\frac{1}{T}\int_0^T\frac{d}{dt}A(t)\,dt.$$

Using Eq. (38.10), we now obtain the following expression:

$$W=-i\sum_{\substack{\alpha,\beta\\ \Omega>0}}\Omega\{\chi^*_{\alpha\beta}(\Omega)h^{\alpha}_{\Omega}h^{*\beta}_{\Omega}-\chi_{\alpha\beta}(\Omega)h^{*\alpha}_{\Omega}h^{\beta}_{\Omega}\}. \tag{38.25}$$

In the case of a linearly polarized field $h^{\alpha}_{\Omega} = h^{*\alpha}_{\Omega}$ and the formula (38.25) becomes

$$W=-\sum_{\substack{\alpha,\beta\\ \Omega>0}}2\Omega\chi''_{\alpha\beta}(\Omega)h^{\alpha}_{\Omega}h^{\beta}_{\Omega} \quad (\chi''_{\alpha\beta}=\mathrm{Im}\,\chi_{\alpha\beta}) \tag{38.26}$$

It follows that the absorption of energy by the spin system from a linearly polarized field is determined by the imaginary part of the susceptibility tensor. In the region of the poles of the Green's functions (38.6) or (38.12), the susceptibility rises rapidly and the absorption has a "resonance" nature.

We shall now attempt to include in our discussion the damping in the spin system. We shall assume that, in some approximation, the appropriate Green's functions have poles in the lower half-plane $E = \pm\ \Omega_R - i\Gamma$ $(\Gamma > 0)$. Then the singular part of $R(\Omega)$ of Eq. (38.20a) can be formally written in the form

$$R(\Omega) = \frac{1}{\Omega - \Omega_R + i\Gamma} - \frac{1}{\Omega + \Omega_R + i\Gamma} = 2\Omega_R \frac{\left(\Omega^2 - \Omega_R^2 - \Gamma^2\right) - 2i\Omega\Gamma}{\left(\Omega^2 - \Omega_R^2 - \Gamma^2\right)^2 + (2\Omega\Gamma)^2}. \tag{38.27}$$

The quantity Γ represents the damping of those spin waves which are excited at ferromagnetic resonance.

Substituting Eq. (38.27) into Eqs. (38.19) and (38.20), we obtain the results of the classical theory of Landau and Lifshits (1935b) or Blembergen (1950) [see also "Ferromagnetic Resonance" (1961)]. The damping is determined by the imaginary part of the Green's function pole. The result of the substitution can be represented in the form

$$\chi_{\alpha\beta}(\Omega) = \chi^0_{\alpha\beta} R(\Omega), \tag{38.28}$$

where $\chi^0_{\alpha\beta}$ are some constants, and $R(\Omega)$ is the expression (38.27). If we introduce Eqs. (38.28) and (38.27) into Eq. (38.26), the power absorbed by the system will be given by the approximate formula

$$W \approx \sum_{\substack{\alpha,\beta \\ \Omega > 0}} \chi^0_{\alpha\beta} h^\alpha_\Omega h^\beta_\Omega \frac{8\Omega^2\Omega_R\Gamma}{\left(\Omega^2 - \Omega_R^2 - \Gamma^2\right)^2 + (2\Omega\Gamma)^2}. \tag{38.29}$$

When the frequency of the external field approaches the natural frequency of the system, Ω_R, the absorption increases and reaches, at $\Omega \approx \Omega_R$, its maximum value, which is of the order of

$$W_{\max} \sim 2\Omega_R \Gamma^{-1} \sum_{\substack{\alpha,\beta \\ \Omega > 0}} \chi^0_{\alpha\beta} h^\alpha_\Omega h^\beta_\Omega. \tag{38.30}$$

The phenomenological introduction of the damping in the form of the imaginary parts of the poles of the Green's functions is equivalent to the allowance for the redistribution of energy among various degrees of freedom in the system. Consequently, there is some dissipation of the energy absorbed at resonance. In the classical theory of ferromagnetic resonance, this process is due to the presence of friction terms in the equations of motion for the average magnetization components. Basically, the two approaches are equivalent and, as we have seen, they lead to the same results.

Sec. 39. Resonance in Isotropic Ferromagnets*

In the preceding section we obtained a general expression for the magnetic susceptibility tensor χ in terms of the Green's functions and the spin operators. The explicit form of χ is governed by the actual form of the spin Hamiltonian. In the present section, we shall consider the simplest case of resonance in an isotropic ferromagnet.

A constant magnetic field H will be assumed to be directed along the z-axis, and a radio-frequency field h(t) will be taken to lie in the (x, y) plane and will be assumed to be linearly polarized.

For an infinite medium the spin Hamiltonian $\mathscr{H}_0$ is in the form of Eq. (6.8) or Eq. (35.2). In the case of a finite sample, we must also allow for the demagnetization energy (7.16). For an ellipsoidal sample, the tensor of the demagnetization factors is diagonal.

The operator for the energy of the spin system itself has the following form in the variables $S_f^{\pm}$, S_f^z:

$$\mathscr{H}_0 = -\mu H \sum_f S_f^z - \frac{1}{2} \sum_{f_1 f_2} I(f_1 - f_2) S_{f_1}^+ S_{f_2}^- - \\ - \frac{1}{2} \sum_{f_1 f_2} I(f_1 - f_2) S_{f_1}^z S_{f_2}^z. \quad (39.1)$$

The demagnetization-energy operator for an ellipsoidal sample has the form

*See Tyablikov (1960a).

$$\mathscr{H}_{\text{demag}} = \frac{\mu^2}{8N}(N_x - N_y)\sum_{f_1, f_2}(S_{f_1}^+ S_{f_2}^+ + S_{f_1}^- S_{f_2}^-) + \\ + \frac{\mu^2}{4N}(N_x + N_y)\sum_{f_1 f_2} S_{f_1}^+ S_{f_2}^- + \frac{\mu^2}{2N} N_z \sum_{f_1, f_2} S_{f_1}^z S_{f_2}^z \tag{39.2}$$

where N_α are the demagnetization factors.

The Green's functions (38.12), in terms of which the susceptibility is defined, are obtained from the functions $\langle\langle S_f^+ | S_g^- \rangle\rangle$ by summation over the indices f and g. We shall now give the equations of motion for the operators $S_f^\pm$:

$$i\frac{dS_f^+}{dt} = \mu H S_f^z + \sum_{f'}\left[I(f - f') - \frac{\mu^2}{N} N_z\right] S_{f'}^z S_f^+ - \\ - \sum_{f'}\left[I(f - f') - \frac{\mu^2}{2N}(N_x + N_y)\right] S_f^z S_{f'}^+ + \\ + \frac{\mu^2}{2N}(N_x - N_y)\sum_{f'} S_f^z S_{f'}^-. \tag{39.3}$$

Next, in accordance with the results of Sec. 35, we must consider the system of equations of the (35.7) type:

$$E\langle\langle S_f^+ | n\rangle\rangle = I_0(n)\Delta(f - g) + \mu H\langle\langle S_f^+ | n\rangle\rangle + \\ + \sum_{f'}\left[I(f - f') - \frac{\mu^2}{N} N_z\right]\langle\langle S_{f'}^z S_f^+ | n\rangle\rangle - \\ - \sum_{f'}\left[I(f - f') - \frac{\mu^2}{2N}(N_x + N_y)\right]\langle\langle S_f^z S_{f'}^+ | n\rangle\rangle + \\ + \frac{\mu^2}{2N}(N_x - N_y)\sum_{f'}\langle\langle S_f^z S_{f'}^- | n\rangle\rangle,$$

$$E\langle\langle S_f^- | n\rangle\rangle = -\mu H\langle\langle S_f^- | n\rangle\rangle - \\ - \sum_{f'}\left[I(f - f') - \frac{\mu^2}{N} N_z\right]\langle\langle S_{f'}^z S_f^- | n\rangle\rangle + \\ + \sum_{f'}\left[I(f - f') - \frac{\mu^2}{2N}(N_x + N_y)\right]\langle\langle S_f^z S_{f'}^- | n\rangle\rangle - \\ - \frac{\mu^2}{2N}(N_x - N_y)\sum_{f'}\langle\langle S_f^z S_{f'}^+ | n\rangle\rangle, \ \ldots; \ n = 0, 1, \ldots, 2S - 1, \tag{39.4}$$

where for brevity we have introduced the notation

$$\langle\langle A | n\rangle\rangle = \langle\langle A \,|\, (S_g^z)^n S_g^- \rangle\rangle, \quad I_0(n) = \frac{i}{2\pi}\langle [A, (S_g^z)^n S_g^-]\rangle. \tag{39.5}$$

We shall now terminate the chain of equations by means of an interpolation decoupling of the second functions through the first,

$$\langle\langle S^z_{f'} S^{\pm}_f \mid n\rangle\rangle \to S\sigma_S \langle\langle S^{\pm}_f \mid n\rangle\rangle, \quad \langle S^z_f\rangle = S\sigma_S \tag{39.6}$$

and we shall apply the Fourier transformation with respect to the index f:

$$S^{\pm}_f = \frac{1}{\sqrt{N}} \sum_{\nu} S^{\pm}_{\nu} e^{\mp i(f,\nu)}.$$

As a result of these operations, we obtain

$$\begin{aligned} E\langle\langle S^+_\nu \mid n\rangle\rangle &= \frac{e^{i(g,\nu)}}{\sqrt{N}} I_0(n) + \\ &+ \left\{\mu H + S\sigma_S [J(0) - \mu^2 N_z - S\sigma_S\left[J(\nu) - \frac{\mu^2}{2}(N_x + N_y)\Delta(\nu)\right]\right\} \times \\ &\times \langle\langle S^+_\nu \mid n\rangle\rangle + S\sigma_S \frac{\mu^2}{2}(N_x - N_y)\Delta(\nu)\langle\langle S^-_\nu \mid n\rangle\rangle, \\ E\langle\langle S^-_\nu \mid n\rangle\rangle &= -\left\{\mu H + S\sigma_S [J(0) - \mu^2 N_z] - \right. \\ &- \left. S\sigma_S\left[J(\nu) - \frac{\mu^2}{2}(N_x + N_y)\Delta(\nu)\right]\right\}\langle\langle S^-_\nu \mid n\rangle\rangle - \\ &- S\sigma_S \frac{\mu^2}{2}(N_x - N_y)\Delta(\nu)\langle\langle S^+_\nu \mid n\rangle\rangle, \\ &\qquad n = 0, 1, \ldots, 2S - 1. \end{aligned} \tag{39.7}$$

Equations (39.7) are solved separately for $\nu \neq 0$ and $\nu = 0$. For $\nu \neq 0$ we have

$$(E - E_S(\nu))\langle\langle S^+_\nu \mid n\rangle\rangle = \frac{e^{i(g,\nu)}}{\sqrt{N}} I_0(n), \quad \langle\langle S^-_\nu \mid n\rangle\rangle = 0, \tag{39.8}$$

$$E_S(\nu) = \mu H + S\sigma_S [J(0) - J(\nu)] - S\sigma_S \mu^2 N_z. \tag{39.9}$$

Hence, we see that if $\nu \neq 0$ the problem reduces to that solved in Sec. 35. Repeating step by step the considerations presented in Sec. 35, we obtain again the same formulas for the determination of the magnetization. On going over from summation with respect to ν to integration, we can extend integration to $\nu = 0$, since the corresponding terms will have statistical weight equal to zero. The only difference is that now the spin-wave energy is given by Eq. (39.9) and not by Eq. (35.11). The solution will apply only in sufficiently strong fields:

$$H - S\sigma_S \mu N_z \geq 0. \qquad (39.10)$$

Otherwise, at sufficiently low values of ν, the energy of elementary excitations would have been negative, which is impossible.

For $\nu = 0$, we have

$$\begin{gathered}\left\{E - \left[\mu H + S\sigma_S \frac{\mu^2}{2}(N_x + N_y - 2N_z)\right]\right\} \langle\langle S_0^+ | n \rangle\rangle - \\ - S\sigma_S \frac{\mu^2}{2}(N_x - N_y) \langle\langle S_0^- | n \rangle\rangle = \frac{I_0(n)}{\sqrt{N}}, \\ - S\sigma_S \frac{\mu^2}{2}(N_x - N_y) \langle\langle S_0^+ | n \rangle\rangle - \\ - \left\{E + \left[\mu H + S\sigma_S \frac{\mu^2}{2}(N_x + N_y - 2N_z)\right]\right\} \langle\langle S_0^- | n \rangle\rangle = 0.\end{gathered} \qquad (39.11)$$

Hence, we easily find that

$$\langle\langle S_0^+ | n \rangle\rangle = \frac{I_0(n)}{\sqrt{N}} \frac{E + E_0}{E^2 - E_R^2}, \quad \langle\langle S_0^- | n \rangle\rangle = - \frac{I_0(n)}{\sqrt{N}} \frac{E_1}{E^2 - E_R^2}, \qquad (39.12)$$
$$n = 0, 1, \ldots, 2S - 1,$$

where the following notation is used:

$$E_0 = \mu H + S\sigma_S \frac{\mu^2}{2}(N_x + N_y - 2N_z), \quad E_1 = S\sigma_S \frac{\mu^2}{2}(N_x - N_y), \qquad (39.13)$$

$$E_R^2 = E_0^2 - E_1^2 = \mu^2 \{H + S\sigma_S \mu (N_x - N_z)\} \times \\ \times \{H + S\sigma_S \mu (N_y - N_z)\}. \qquad (39.14)$$

At resonance, the susceptibility is given in terms of the functions (38.12). We shall define the latter in terms of the functions (39.12), assuming that n = 0 and bearing in mind that

$$S^{\pm} = \sum_f S_f^{\pm}, \quad S_0^{\pm} = \frac{S^{\pm}}{\sqrt{N}}, \quad I_0(0) = 2S\sigma_S.$$

In this way, we finally obtain

$$\begin{gathered}G_{12}(\Omega) = \langle\langle S^+ | S^- \rangle\rangle_\Omega^{(r)} = iN \frac{S\sigma_S}{\pi} \frac{\Omega + i\varepsilon + E_0}{(\Omega + i\varepsilon)^2 - E_R^2}, \\ G_{22}(\Omega) = \langle\langle S^- | S^- \rangle\rangle_\Omega^{(r)} = - iN \frac{S\sigma_S}{\pi} \frac{E_1}{(\Omega + i\varepsilon)^2 - E_R^2}, \ldots\end{gathered} \qquad (39.15)$$

Since the equations of motion (39.3) for the operators $S_f^{\pm}$ are invariant under time inversion (28.24), it follows from Eqs. (28.30), (28.31) that

$$G_{11}(\Omega) = G_{22}(\Omega). \tag{39.16}$$

We shall now consider some consequences of these results.

Resonance in an Infinite Medium. In an infinite medium, all the demagnetization factors vanish, $N_\alpha = 0$ ($\alpha = x, y, z$), and, consequently,

$$\begin{gathered} G_{12}(\Omega) = -G_{21}^{*}(-\Omega) = iN\frac{S\sigma_S}{\pi}\frac{1}{\Omega - E_R + i\varepsilon}, \\ G_{11}(\Omega) = G_{22}(\Omega) = 0, \quad E_R = \mu H. \end{gathered} \tag{39.17}$$

Substituting Eq. (39.17) into Eqs. (38.13) and (38.14), we obtain the following formulas for the real and imaginary parts of the susceptibility tensor components:

$$\begin{aligned} \chi'_{xx}(\Omega) &= P\frac{\chi_0}{1-\left(\frac{\Omega}{E_R}\right)^2}, \\ \chi''_{xx}(\Omega) &= \frac{\pi\chi_0}{2}E_R[\delta(\Omega - E_R) - \delta(\Omega + E_R)], \\ \chi'_{yx}(\Omega) &= \frac{\pi\chi_0}{2}E_R[\delta(\Omega - E_R) + \delta(\Omega + E_R)], \\ \chi''_{yx}(\Omega) &= -\frac{\Omega}{E_R}P\frac{\chi_0}{1-\left(\frac{\Omega}{E_R}\right)^2}, \\ \chi_{yy}(\Omega) &= \chi_{xx}(\Omega), \quad \chi_{xy}(\Omega) = -\chi_{yx}(\Omega), \end{aligned} \tag{39.18}$$

where

$$\chi_0 = N\frac{S\sigma_S\mu^2}{E_R} = N\frac{S\sigma_S\mu}{H} = \frac{M_\vartheta}{H}, \quad E_R = \mu H, \tag{39.19}$$

P is the principal value, and M_ϑ is the magnetization of a sample at a temperature ϑ.

In the case under consideration, the resonance frequency E_R is equal to the frequency of the Larmor precession of the spins about the direction of the constant field H. Since E_R is the spin-wave energy for the wave vector $\nu = 0$ [cf. Eq. (39.9)] this means also that at resonance (in the linear approximation) only the spin waves with this wave vector are excited.

The profile of the resonance line is governed by the imaginary parts of the components X and $\mathcal{G}$ of the tensor χ (38.13). In the present case,

$$X''(\Omega)=\chi''_{xx}(\Omega), \quad \mathcal{G}''(\Omega)=-\chi'_{xy}(\Omega) \qquad (39.20)$$
$$[X_a(\Omega)=\mathcal{G}_a(\Omega)=0].$$

and the line has a δ-shaped profile.

If the spin system is damped, this may be allowed for, as mentioned earlier, by displacing the poles of the Green's functions (38.12) to the complex plane [see also remarks made in connection with Eq. (38.27)]. In this case,

$$G_{12}(\Omega)=iN\frac{S\sigma_S}{\pi}\frac{1}{\Omega-E_R+i\Gamma},$$
$$G_{21}(\Omega)=-iN\frac{S\sigma_S}{\pi}\frac{1}{\Omega+E_R+i\Gamma}, \qquad (39.21)$$

where Γ is the magnitude of the damping for spin waves with $\nu = 0$. Instead of the formulas (39.18), we now have

$$\chi'_{xx}(\Omega)\approx-\chi_0\frac{E_R^2(\Omega^2-E_R^2)}{(\Omega^2-E_R^2)^2+(2\Omega\Gamma)^2},$$
$$\chi''_{xx}(\Omega)\approx\chi_0\frac{E_R^2\cdot 2\Omega\Gamma}{(\Omega^2-E_R^2)^2+(2\Omega\Gamma)^2},$$
$$\chi'_{yx}(\Omega)\approx\chi_0\frac{2E_R\Omega^2\Gamma}{(\Omega^2-E_R^2)^2+(2\Omega\Gamma)^2},$$
$$\chi''_{yx}(\Omega)\approx\chi_0\frac{E_R\Omega(\Omega^2-E_R^2)}{(\Omega^2-E_R^2)^2+(2\Omega\Gamma)^2}, \qquad (39.22)$$

where the damping is regarded as small ($\Gamma \ll E_R$) and quantities of higher orders in Γ are omitted. We can easily see that in this case the line profile is of the Lorentz type.

If we assume that the demagnetization factors are equal to zero, we neglect the influence of boundaries. However, it is well known that the resonance frequencies of ferromagnets depend strongly on the shape of the sample. The boundary effects can be neglected if the total magnetization is equal to zero (for example, in antiferromagnets) or if a sample is spherical. In the latter case, $N_x = N_y = N_z$, and we can easily see from the formulas (39.11) that the same results are obtained as for an infinite medium.

The Damping of Free Precession. We shall consider the behavior of a spin system when an external radio-frequency field h(t), regarded as a perturbation, is switched off at a time t = 0. We shall assume that a constant magnetic field H is directed along the z-axis and that the radio-frequency field is linearly polarized and directed along the x-axis. For simplicity, we shall assume that the sample is spherical.

The change in the components of the magnetic moment along the x- and y-axes is, according to Eq. (31.22), given by the formulas

$$\delta^{(1)}\langle\mathcal{M}^{\alpha}(t)\rangle = \mu^2 \sum_{\beta} \int_{-\infty}^{\infty} dE \frac{e^{-iEt}}{E-\Omega-i\varepsilon} \langle\langle S^{\alpha} | S^{\beta}\rangle\rangle_E^{(r)} h^{\beta}. \tag{39.23}$$

We shall assume that $h^x = h$, $h^y = h^z = 0$, and that

$$\begin{aligned} \langle\langle S^x | S^x\rangle\rangle &= \frac{1}{4}\{\langle\langle S^- | S^+\rangle\rangle + \langle\langle S^+ | S^-\rangle\rangle\}, \\ \langle\langle S^y | S^x\rangle\rangle &= \frac{i}{4}\{\langle\langle S^- | S^+\rangle\rangle - \langle\langle S^+ | S^-\rangle\rangle\}. \end{aligned} \tag{39.24}$$

For the functions which are on the right-hand side of these formulas, we shall use the expressions (39.21). Consequently, the formulas (39.23) will become

$$\delta^{(1)}\langle\mathcal{M}^x(t)\rangle =$$
$$= iN\frac{S\sigma_S\mu^2 h}{4\pi} \int_{-\infty}^{\infty} dE \frac{e^{-iEt}}{E-\Omega-i\varepsilon}\left\{\frac{1}{E+E_R+i\Gamma} - \frac{1}{E-E_R+i\Gamma}\right\},$$
$$\delta^{(1)}\langle\mathcal{M}^y(t)\rangle =$$
$$= iN\frac{S\sigma_S\mu^2 h}{4\pi} \int_{-\infty}^{\infty} dE \frac{e^{-iEt}}{E-\Omega-i\varepsilon}\left\{\frac{i}{E+E_R+i\Gamma} - \frac{i}{E-E_R+i\Gamma}\right\}.$$

We are interested in the behavior of the system after the perturbation stops, i.e., when t > 0. Therefore, the contour of integration in the complex E-plane should be closed in the lower half-plane, where the integrands have simple poles at the points $E = \pm E_R - i\Gamma$. Using the theorem on residues, we obtain:

$$\begin{aligned}
&\delta^{(1)}\langle \mathcal{M}^x(t)\rangle = \\
&= \frac{1}{2} NS\sigma_S\mu^2 h e^{-\Gamma t}\left\{\frac{e^{iE_R t}}{\Omega + E_R + i\Gamma} - \frac{e^{-iE_R t}}{\Omega - E_R + i\Gamma}\right\}, \\
&\delta^{(1)}\langle \mathcal{M}^y(t)\rangle = \\
&= \frac{i}{2} NS\sigma_S\mu^2 h e^{-\Gamma t}\left\{\frac{e^{iE_R t}}{\Omega + E_R + i\Gamma} + \frac{e^{-iE_R t}}{\Omega - E_R + i\Gamma}\right\}
\end{aligned} \tag{39.25}$$

(the formal parameter ε is omitted, since it has the same sign as the damping Γ, which is small but finite).

From the formulas (39.25), we see that the average values of the components of the magnetic moment along the x- and y-axes oscillate at the frequency $E_R = \mu H$, which is equal to the frequency of the Larmor precession of the spin moments about the direction of the constant field H. The amplitude of the oscillations decreases with time as $\exp(-\Gamma t)$, where Γ is the magnitude of the damping or the reciprocal of the lifetime of the spin-wave states with $\nu = 0$.

Ellipsoidal Sample. In this case, the susceptibility is given by the general expression (38.13). In the approximation considered, the functions G_{11}, G_{12}, ... are given by the formulas (39.15) and the resonance frequency E_R is given by the formula (39.14). We shall substitute Eq. (39.15) into Eqs. (38.13) and (38.14), and we shall separate the real and imaginary parts of χ, using Eq. (26.16). As a result, we obtain

$$\begin{aligned}
&\chi'_{xx}(\Omega) = \sqrt{\frac{A_y}{A_x}}\, P\frac{\chi_0}{1-\left(\frac{\Omega}{E_R}\right)^2}, \\
&\chi''_{xx}(\Omega) = \frac{\pi\chi_0}{2} E_R \sqrt{\frac{A_y}{A_x}}\{\delta(\Omega - E_R) - \delta(\Omega + E_R)\}, \\
&\chi'_{yx}(\Omega) = \frac{\pi\chi_0}{2} E_R\{\delta(\Omega - E_R) + \delta(\Omega + E_R)\}, \\
&\chi''_{yx}(\Omega) = -\frac{\Omega}{E_R} P\frac{\chi_0}{1-\left(\frac{\Omega}{E_R}\right)^2},
\end{aligned} \tag{39.26}$$

where the following notation is used:

$$\chi_0 = N\frac{S\sigma_S\mu^2}{E_R}, \quad A_\alpha = \mu H + S\sigma_S\mu^2(N_\alpha - N_z) \quad (\alpha = x,\ y), \tag{39.27}$$

$$E_R = \sqrt{A_x A_y} = \\ = \mu\sqrt{[H + S\sigma_S\mu(N_x - N_z)]\,[H + S\sigma_S\mu(N_y - N_z)]}. \tag{39.28}$$

The formula (39.28) is the well-known Kittel formula (1947, 1948) for arbitrary temperatures. In this case, it is necessary to use, in general, the values of σ_S obtained by solving the system of equations (39.8).

Sec. 40. Elastic Scattering of Neutrons*

The diffraction pattern produced by the scattering of slow neutrons in crystals allows us to investigate the crystallographic and magnetic structure of crystals and to obtain data on the phonon and spin-wave spectra. The behavior of substances near their phase transition points can also be examined in this way.

We shall present the elements of the theory of neutron scattering in crystals of strongly magnetic substances using the time formalism of Van Hove (1954a, b), which is closely related to the Green's function method.

Let us assume that a monochromatic beam of neutrons, which does not vary with time, is incident on a crystal. We shall write the total Hamiltonian of the system in the form

$$\mathscr{H} = \mathscr{H}_0 + \mathscr{V} = \mathscr{H}_{SS} + \mathscr{H}_1 + \mathscr{V}, \tag{40.1}$$

where $\mathscr{H}_{SS}$ is the operator for the energy of the spin system, which includes the lattice vibrations, $\mathscr{H}_1 = -(2m_n)^{-1}\Delta$ is the operator for the energy of a free neutron, m_n is the neutron mass, and $\mathscr{V}$ is the operator for the energy of interaction between a neutron and the spin system. We shall use $\mathscr{E}_\nu$, C_ν to denote the eigenvalues and the eigenfunctions of $\mathscr{H}_{SS}$, E_p, ψ_p to denote the eigenvalues and eigenfunctions of $\mathscr{H}_1$, and p to denote the neutron momentum.

The probability of a transition from a state (ν, p) to a state

*See, for example, Bacon (1957, 1963) and Izyumov (1963a).

(ν', p') in the Born approximation is (40.2)

$$W_{\nu p,\,\nu' p'} = 2\pi\,|(\nu' p'\,|\,\mathcal{V}\,|\,\nu p)|^2\,\delta(\mathcal{E}_\nu + E_p - \mathcal{E}_{\nu'} - E_{p'}),$$

where

$$(\nu' p'\,|\,\mathcal{V}\,|\,\nu p) = (C^*_{\nu'}\psi^*_{p'},\ \mathcal{V} C_\nu \psi_p). \tag{40.3}$$

The probability of a neutron transition from a state p to a state p' is found by averaging the expression (40.2) over the initial states of the scatterer (crystal) having the density matrix

$$\rho_S = Q_S^{-1}\exp\left(-\frac{1}{\vartheta}\mathcal{H}_{SS}\right),\quad Q_S = \operatorname{Sp}\exp\left(-\frac{1}{\vartheta}\mathcal{H}_{SS}\right) \tag{40.4}$$

and by summation over all finite values of ν':

$$W_{pp'} = \frac{2\pi}{Q_S}\sum_{\nu\nu'} e^{-\frac{\mathcal{E}_\nu}{\vartheta}}(\nu p\,|\,\mathcal{V}\,|\,\nu' p')(\nu' p'\,|\,\mathcal{V}\,|\,\nu p)\times$$
$$\times\,\delta(\mathcal{E}_\nu + E_p - \mathcal{E}_{\nu'} - E_{p'}). \tag{40.5}$$

We shall use the notation

$$(\nu p\,|\,\mathcal{V}\,|\,\nu' p') = (\nu\,|\,\mathcal{V}_{pp'}\,|\,\nu'),$$
$$\mathcal{V}_{pp'} = (\psi^*_p,\ \mathcal{V}\psi_{p'}). \tag{40.6}$$

Substituting (40.6) into (40.5) and using the δ-function representation in the form of an integral, we obtain

$$W_{pp'} = \frac{1}{Q_S}\sum_{\nu\nu'} e^{-\frac{\mathcal{E}_\nu}{\vartheta}}(\nu\,|\,\mathcal{V}_{pp'}\,|\,\nu')\times$$
$$\times(\nu'\,|\,\mathcal{V}_{p'p}\,|\,\nu)\int_{-\infty}^{\infty} dt e^{-i(\mathcal{E}_\nu + E_p - \mathcal{E}_{\nu'} - E_{p'})} =$$
$$= \int_{-\infty}^{\infty} dt e^{i(E_{p'} - E_p)t}\frac{1}{Q_S}\sum_{\nu\nu'} e^{-\frac{\mathcal{E}_\nu}{\vartheta}}\times$$
$$\times(\nu\,|\,\mathcal{V}_{pp'}\,|\,\nu')\,e^{i\mathcal{E}_{\nu'}t}(\nu'\,|\,\mathcal{V}_{p'p}\,|\,\nu)\,e^{-i\mathcal{E}_\nu t}.$$

The sum under the integral sign is equal to the statistical average, having the density matrix (40.4), of the operator product $\mathcal{V}_{pp'}$. We shall bear in mind that

$$e^{i\mathcal{E}_{\nu'}t}(\nu'\,|\,\mathcal{V}_{p'p}\,|\,\nu)\,e^{-i\mathcal{E}_\nu t} =$$
$$= (C^*_{\nu'},\ e^{i\mathcal{E}_{\nu'}t}\mathcal{V}_{p'p}e^{-i\mathcal{E}_\nu t}C_\nu) = (C_{\nu'},\ e^{i\mathcal{H}_{SS}t}\mathcal{V}_{p'p}e^{-i\mathcal{H}_{SS}t}C_\nu) =$$
$$= \left(\nu'\,\middle|\,e^{i\mathcal{H}_{SS}t}\mathcal{V}_{p'p}e^{-i\mathcal{H}_{SS}t}\,\middle|\,\nu\right).$$

Consequently,

$$\frac{1}{Q_S}\sum_{\nu\nu'} e^{-\frac{\mathcal{E}_\nu}{\vartheta}} (\nu|\mathcal{V}_{pp'}|\nu')(\nu'|e^{i\mathcal{H}_{SS}t}\mathcal{V}_{p'p}e^{-i\mathcal{H}_{SS}t}|\nu) =$$

$$= \frac{1}{Q_S}\sum_{\nu} e^{-\frac{\mathcal{E}_\nu}{\vartheta}} (\nu|\mathcal{V}_{pp'}e^{i\mathcal{H}_{SS}t}\mathcal{V}_{p'p}e^{-i\mathcal{H}_{SS}t}|\nu) =$$

$$= \mathrm{Sp}\,\{\mathcal{V}_{pp'}e^{i\mathcal{H}_{SS}t}\mathcal{V}_{p'p}e^{-i\mathcal{H}_{SS}t}\rho_S\}.$$

Using the definition (25.3), the probability of neutron scattering can be written as a Fourier integral of the correlation function with respect to time:

$$W_{pp'} = \int_{-\infty}^{\infty} dt e^{i(E_{p'}-E_p)t} \langle \mathcal{V}^+_{p'p}(0)\mathcal{V}_{p'p}(t)\rangle, \tag{40.7}$$

where $\mathcal{V}_{p'p}(t)$ is the Heisenberg representation of the operator $\mathcal{V}_{p'p}$ using the Hamiltonian $\mathcal{H}_{SS}$:

$$\mathcal{V}_{p'p}(t) = e^{i\mathcal{H}_{SS}t}\mathcal{V}_{p'p}e^{-i\mathcal{H}_{SS}t}, \tag{40.8}$$

and the angular brackets in Eq. (40.7) represent the averaging using the density matrix of (40.4). The formulas (40.7) and (40.8) give the required representation for the probability of neutron scattering.

Using the representation (26.2) for the correlation functions, the expression (40.9) can be written in another form:

$$W_{pp'} = 2\pi I_{\mathcal{V}_{p'p}\mathcal{V}^*_{p'p}}(\omega), \quad \omega = E_{p'} - E_p, \tag{40.9}$$

where $I_{\mathcal{V}_{p'p}\mathcal{V}^*_{p'p}}$ is the spectral density.

The probability of neutron scattering in an element of phase space $(2\pi)^{-3}dp'^x dp'^y dp'^z$, which is referred to the flux density, is

$$\delta^2\Sigma_{pp'} = \frac{m_n}{p} W_{pp'}(2\pi)^{-3} dp'^x\, dp'^y\, dp'^z = \frac{m_n^2}{(2\pi)^3}\frac{p'}{p} W_{pp'}\,\delta E_{p'}\,\delta O,$$

where δO is an element of the solid angle. Hence, we obtain the following expression for the differential effective scattering cross section per unit solid angle and per unit energy of the scattered neutron:

$$\frac{\delta^2\Sigma_{pp'}}{\delta O\,\delta E_{p'}}=\frac{m_n^2}{(2\pi)^3}\frac{p'}{p}W_{pp'}. \qquad (40.10)$$

Using the formulas (40.7) or (40.9), we shall rewrite Eq. (40.10) in the following form:

$$\frac{\delta^2\Sigma_{pp'}}{\delta O\,\delta E_{p'}}=\frac{m_n^2}{(2\pi)^2}\frac{p'}{p}\frac{1}{2\pi}\int_{-\infty}^{\infty}dte^{i\omega t}\langle\mathcal{V}^+_{p'p}(0)\mathcal{V}_{p'p}(t)\rangle= \qquad (40.11)$$

$$=\frac{m_n^2}{(2\pi)^2}\frac{p'}{p}I_{\mathcal{V}_{p'p}\mathcal{V}^+_{p'p}}(\omega)\quad(\omega=E_{p'}-E_p). \qquad (40.12)$$

Integrating these expressions with respect to the scattered neutron energy, we obtain the differential scattering cross section per unit solid angle:

$$\frac{\delta\Sigma_{pp'}}{\delta O}=\frac{m_n^2}{(2\pi)^2}\int\frac{p'}{p}W_{pp'}\,dE_{p'} \qquad (40.13)$$

or, according to Eqs. (40.11) and (40.12),

$$\frac{\delta\Sigma_{pp'}}{\delta O}=\frac{m_n^2}{(2\pi)^2}\frac{1}{2\pi}\int dE_{p'}\frac{p'}{p}\int_{-\infty}^{\infty}dte^{i\omega t}\langle\mathcal{V}^+_{p'p}(0)\mathcal{V}_{p'p}(t)\rangle= \qquad (40.14)$$

$$=\frac{m_n^2}{(2\pi)^2}\int dE_{p'}\frac{p'}{p}I_{\mathcal{V}_{p'p}\mathcal{V}^+_{p'p}}(\omega)\quad(\omega=E_{p'}-E_p). \qquad (40.15)$$

The formulas for the probability of a transition and for the differential elastic-scattering cross sections will be obtained by separating the term independent of time from the correlation function. We shall represent this time-independent term by

$$\langle\mathcal{V}^+_{p'p}(0)\mathcal{V}_{p'p}(t)\rangle_c. \qquad (40.16)$$

Substituting Eq. (40.16) into Eqs. (40.7), (40.11), and (40.14) we shall find

$$W_{pp'}=2\pi\langle\mathcal{V}^+_{p'p}(0)\mathcal{V}_{p'p}(t)\rangle_c\,\delta(E_{p'}-E_p), \qquad (40.17)$$

$$\frac{\delta^2\Sigma_{pp'}}{\delta O\,\delta E_{p'}}=\frac{m_n^2}{(2\pi)^2}\langle\mathcal{V}^+_{p'p}(0)\mathcal{V}_{p'p}(t)\rangle_c\,\delta(E_{p'}-E_p), \qquad (40.18)$$

$$\frac{\delta\Sigma_{pp'}}{\delta O}=\frac{m_n^2}{(2\pi)^2}\langle\mathcal{V}^+_{p'p}(0)\mathcal{V}_{p'p}(t)\rangle_c. \qquad (40.19)$$

If the incident neutron beam is not polarized, the expressions for the probability of a transition and for the differential cross sections should be averaged out over all the states of neutron spins in the incident beam. This averaging operation will be denoted by wavy overlining, for example, $\widetilde{W}_{pp'}$.

The energy of the interaction of a neutron with the spin system, $\mathcal{V}$ of Eq. (40.1), consists of the energy of the interaction of the neutron with the nuclei of atoms in the crystal lattice, $\mathcal{V}^{(l)}$, and of the energy of the magnetic interaction of the neutron with the electrons in the partly filled shells, $\mathcal{V}^{(m)}$. Let the scatterer consist of N unit cells, each of which contains n atoms. We shall denote the number of a cell by f and the number of an atom in a cell by j. The radius vector of an atom at a site (fj) will be written in the form

$$r_{fj} = r^0_{fj} + u_{fj} = f + j + u_{fj}, \tag{40.20}$$

where r^0_{fj} is the equilibrium value of the radius vector of the atom, u_{fj} is the displacement of this atom from its equilibrium position (the radius vectors of the unit cells and of the atoms in a cell will be denoted, for brevity, by the same letters as the numbers of the unit cells and of the atoms in the cells: $r_f \to f, r_j \to j$). The matrix elements of the operators $\mathcal{V}^{(l)}$, $\mathcal{V}^{(m)}$ expressed in terms of the wave functions of free neutrons, have the form*

$$\mathcal{V}^{(l)}_{p'p} = \sum_{fj} \alpha_j (S_n, I_{fj}) e^{i(q, r_{fj})}, \tag{40.21}$$

$$\mathcal{V}^{(m)}_{p'p} = -4\pi \frac{\gamma_n r_0}{m_n} \sum_{fj} F_j(q) e^{i(q, r_{fj})} (S_{fj}, S_n - e(e, S_n)), \tag{40.22}$$

where

$$q^\alpha = p^\alpha - p'^\alpha, \quad e^\alpha = \frac{q^\alpha}{q} \quad (\alpha = 1, 2, 3), \tag{40.23}$$

and where S_n is the neutron spin operator, γ_n the magnetic moment of a neutron expressed in nuclear magnetons ($\gamma_n = -1.93$),

---

*See, for example, the monograph by Akhiezer and Pomeranchuk (1950), Halpern and Johnson's paper (1939), or the review by Izyumov (1963a).

S_{fj} the operator for the electron spin in an unfilled atomic shell fj, I_{fj} the operator of the nuclear spin of the atom fj, α_j (S_n, I_{fj}) some function of S_n and I_{fj}, $r_0 = e^2/mc^2$ the classical radius of the electron, $F_j(q)$ the so-called magnetic form-factor of an atom, which represent the distribution of the spin density in the atom. Usually, α_j is written in the form

$$\alpha_j(S_n, I_{fj}) = a_j + b_j(S_n, I_{fj}), \tag{40.24}$$

where a_j, b_j are constants.

If the neutrons are unpolarized, there is no interference between the lattice (40.21) and the magnetic (40.22) scattering. Therefore, the total differential cross section is equal to the sum of the differential cross sections for each of these mechanisms:

$$\frac{\delta\Sigma_{pp'}}{\delta O} = \frac{\delta\Sigma_{pp'}^{(l)}}{\delta O} + \frac{\delta\Sigma_{pp'}^{(m)}}{\delta O}. \tag{40.25}$$

We must stress that the unit cells of the crystallographic and magnetic lattices are not necessarily identical. Therefore, in the formulas (40.16)-(40.20) one should distinguish between the numbers of sites fj in the crystallographic and magnetic lattices. We shall not do this, since we are dealing with the scattering of only unpolarized neutrons, when the contributions of the scattering by the crystal lattice (the lattice scattering) and the magnetic lattice (the magnetic scattering) are independent. In the final results, we shall use primes to denote the quantities referring to the magnetic lattice. In neutron experiments, both the lattice and magnetic scattering processes are observed simultaneously. We shall consider both scattering mechanisms in order to clarify the differences between their diffraction patterns.

Substituting Eq. (40.21) into Eq. (40.19), and averaging over the spin states of neutrons in the incident beam, we obtain for the lattice scattering

$$\frac{\delta\Sigma_{pp'}^{(l)}}{\delta O} = \frac{m_n^2}{(2\pi)^2} \sum_{fj,\, f'j'} \widetilde{\alpha_j\alpha_{j'}} \langle e^{-i(q,\, r_{fj}(0))} e^{i(q,\, r_{f'j'}(t))} \rangle_c. \tag{40.26}$$

For the first factor under the summation sign, we obtain, using

Eq. (40.24), the expression

$$\widetilde{\alpha_j \alpha_{j'}} = a_j a_{j'} + b_j b_{j'} \widetilde{(S_n, I_{fj})(S_n, I_{f'j'})}.$$

We shall take into account the fact that S_n^α is the Pauli matrix and $\widetilde{S_n^\alpha S_n^\beta} = S_p(S_n^\alpha S_n^\beta) = {}^1\!/_4\,\delta_{\alpha\beta}$. Consequently,

$$\widetilde{\alpha_j \alpha_{j'}} = a_j a_{j'} + \frac{1}{4} b_j b_{j'} (I_{fj}, I_{f'j'}).$$

The effective scattering cross section (40.26) depends on the distribution of the orientations of the nuclear spins in a crystal. We shall assume that the nuclei are unpolarized, i.e., that their spin moments are randomly oriented. Then Eq. (40.26) will be identical with the average over all the nuclear spin orientations. Consequently, we shall replace $\widetilde{\alpha_j \alpha_{j'}}$ with this average value:

$$\widetilde{\alpha_j \alpha_{j'}} \to a_j a_{j'} + \frac{1}{4} b_j b_{j'} \, \mathrm{Sp}(I_{fj}, I_{f'j'}) =$$
$$= a_j a_{j'} + \frac{1}{4} b_j^2 I_j (I_j + 1) \delta_{ff'} \delta_{jj'}, \qquad (40.27)$$

where we have allowed for the fact that $(I_{fj}, I_{fj}) = I_j (I_j + 1)$.

Substituting Eq. (40.27) into Eq. (40.26), we obtain

$$\frac{\delta \Sigma_{pp'}^{(l)}}{\delta O} = \frac{m_n^2}{(2\pi)^2} \sum_{fj,\, f'j'} \Big\{ a_j a_{j'} e^{i(q,\, f'-f)} e^{i(q,\, j'-j)} +$$
$$+ \frac{1}{4} b_j^2 I_j (I_j + 1) \delta_{ff'} \delta_{jj'} \Big\} \langle e^{-i(q,\, u_{fj}(0))} e^{i(q,\, u_{f'j'}(t))} \rangle_c. \qquad (40.28)$$

The time-independent term $< \ldots >_c$ in the correlation function, due to the atomic displacements, will be selected as follows:

$$\langle e^{-i(q,\, u_{fj}(0))} e^{i(q,\, u_{f'j'}(t))} \rangle_c \approx \langle e^{-i(q,\, u_{fj}(0))} \rangle \langle e^{i(q,\, u_{f'j'}(t))} \rangle. \qquad (40.29)$$

Since the average value of the function of the atomic displacements should be independent of the numbers of cells, we can write

$$e^{-W_j(q)} = \langle e^{-i(q,\, u_{fj}(t))} \rangle = \langle e^{-i(q,\, u_{fj}(0))} \rangle. \qquad (40.30)$$

The quantity $W_j(q)$ is called the Debye–Waller thermal factor.

This separation of the time-independent term corresponds to the principle of correlation weakening proposed by Bogolyubov (1961).

Let us consider the sum

$$\sum_{ff'} e^{-i(q,\,f)} e^{i(q,\,f')} \langle e^{-i(q,\,u_{fj}(0))} e^{i(q,\,u_{f'j'}(t))} \rangle.$$

The main contribution to this sum is made by the terms representing points f and f', which are sufficiently far apart. However, at large distances, the correlation between particle displacements may be regarded as small and we can represent the sum approximately by

$$\sum_{ff'} e^{-i(q,\,f)} \langle e^{-i(q,\,u_{fj}(0))} \rangle e^{i(q,\,f')} \langle e^{i(q,\,u_{f'j'}(t))} \rangle.$$

Hence, it becomes possible to use the decoupling given by Eq. (40.29).

Substituting Eqs. (40.29) and (40.30) into Eq. (40.28), we shall transform the sum over the lattice sites:

$$\begin{aligned}\frac{\delta\Sigma^{(l)}_{pp'}}{\delta O} &= \frac{m_n^2}{(2\pi)^2} \sum_{fj,\,f'j'} a_j a_{j'} e^{i(q,\,f'-f)+i(q,\,j'-j)} e^{-W_j(q)-W_{j'}(q)} + \\ &\quad + \frac{m_n^2}{(2\pi)^2} \sum_{fj} \frac{b_j^2}{4} I_j(I_j+1) e^{-2W_j(q)} = \\ &= \frac{m_n^2}{(2\pi)^2} \left| \sum_f e^{-i(q,\,f)} \right|^2 \left| \sum_j a_j e^{-i(q,\,j)} e^{-W_j(q)} \right|^2 + \\ &\quad + N \frac{m_n^2}{(2\pi)^2} \sum_j \frac{b_j^2}{4} I_j(I_j+1) e^{-2W_j(q)}.\end{aligned}$$

We shall introduce the notation

$$\begin{gathered} e^{-2W(q)} = \sum_j e^{-2W_j(a)}, \\ \Phi_0^{(l)} = 2\pi \frac{m_n^2}{v} \left| \sum_j a_j e^{-i(q,\,j)} e^{-W_j(q)} \right|^2 e^{2W(q)}, \\ \Phi_1^{(l)} = \frac{m_n^2}{(2\pi)^2} \sum_j \frac{b_j^2}{4} I_j(I_j+1) e^{-2W_j(q)} e^{2W(q)}. \end{gathered} \tag{40.31}$$

In particular, if a unit cell contains only one atom, then

$$\Phi_0^{(l)} = 2\pi \frac{m_n^2 a^2}{v}, \quad \Phi_1^{(l)} = \frac{m_n^2}{(2\pi)^2} \frac{b^2}{4} I(I+1). \tag{40.32}$$

The sum over all unit cells is [cf., (A2.19)]

$$\left|\sum_j e^{-i(f,\,q)}\right|^2 = N\,\frac{(2\pi)^3}{v}\sum_k \delta(q-2\pi g_k), \qquad (40.33)$$

where g_k are the reciprocal lattice vectors, N is the number of unit cells, and v is the volume of a unit cell.

Using the formulas (40.31) and (40.33), we shall write the final form of the expression for the differential effective cross section of the lattice scattering:

$$\frac{\delta\Sigma^{(l)}_{pp'}}{\delta O} = N\left\{\Phi_0^l\sum_k \delta(q-2\pi g_k)+\Phi_1^{(l)}\right\}e^{-2W(q)}, \qquad (40.34)$$

where N is the number of unit cells in the crystal lattice, and g_k are the reciprocal lattice vectors.

The first term in Eq. (40.34) describes the coherent scattering of neutrons. It gives sharp maxima at the scattering angles defined by the conditions

$$\begin{gathered} q^\alpha = p^\alpha - p'^\alpha = 2\pi g_k^\alpha \quad (\alpha = 1,\ 2,\ 3), \\ |p| = |p'|. \end{gathered} \qquad (40.35)$$

We can easily see that this is nothing else but the Bragg scattering condition for the diffraction maxima (cf., Appendix A3). The second term in the braces of Eq. (40.34) describes the incoherent scattering of neutrons.

The overall diffraction pattern is distorted somewhat by the weak angular dependence of the Debye–Waller factors.

We shall now consider the magnetic scattering of slow neutrons.

In order to obtain an expression for the differential effective cross section for the magnetic scattering, we shall substitute Eq. (40.22) into Eq. (40.19) and average the result over all the orientations of the neutron spins in the incident beam:

$$\begin{gathered} \frac{\delta\Sigma^{(m)}_{pp'}}{\delta O} = 4r_0^2\gamma_n^2 \sum_{fj,\,f'j'} F_j(q)\,F_{j'}(q)\,\langle e^{-i(q,\,r_{fj}(0))}e^{i(q,\,r_{f'j'}(t))}\,\times \\ \times\,\widetilde{(S_{fj}(0),\ S_n - e(e,\ S_n))\,(S_{f'j'}(t),\ S_n - e(e,\ S_n))}\rangle_c. \end{gathered} \qquad (40.36)$$

Since $\overline{S_n^{\alpha} S_n^{\beta}} = \frac{1}{4}\delta_{\alpha\beta}$, the average over the orientations of spins in the incident beam, which is part of Eq. (40.36), is equal to

$$\frac{1}{4}\sum_{\alpha\beta}(\delta_{\alpha\beta} - e_\alpha e_\beta)\, S_{fj}^{\alpha}(0)\, S_{f'j'}^{\beta}(t). \tag{40.37}$$

Substituting Eq. (40.37) into Eq. (40.36), we obtain

$$\frac{\delta\Sigma_{pp'}^{(m)}}{\delta O} = r_0^2\gamma_n^2 \sum_{\alpha\beta}\sum_{fj,\, f'j'}(\delta_{\alpha\beta} - e_\alpha e_\beta)\, F_j(q)\, F_{j'}(q) \times$$
$$\times \langle e^{-i(q,\, r_{fj}(0))} e^{i(q,\, r_{f'j'}(t))} S_{fj}^{\alpha}(0)\, S_{f'j'}^{\beta}(t)\rangle_c. \tag{40.38}$$

We shall neglect the spin–phonon scattering in the crystal. Then the spin and phonon variables in the Hamiltonian $\mathscr{H}_{SS}$ of Eq. (40.1) can be separated and the statistical average in Eq. (40.38) splits up into a product of the averages

$$\langle e^{-i(q,\, r_{fj}(0))} e^{i(q,\, r_{f'j'}(t))}\rangle_c\, \langle S_{fj}^{\alpha}(0)\, S_{f'j'}^{\beta}(t)\rangle_c.$$

Consequently, the formula (40.38) transforms into

$$\frac{\delta\Sigma_{pp'}^{(m)}}{\delta O} = r_0^2\gamma_n^2 \sum_{fj,\, f'j'} F_j(q)\, F_{j'}(q)\, \langle e^{-i(q,\, r_{fj}(0))} e^{i(q,\, r_{f'j'}(t))}\rangle_c \times$$
$$\times \sum_{\alpha\beta}(\delta_{\alpha\beta} - e_\alpha e_\beta)\, \langle S_{fj}^{\alpha}(0)\, S_{f'j'}^{\beta}(t)\rangle_c. \tag{40.39}$$

The time-independent term in the correlation function of the coordinates of the atoms is determined as in the case of the lattice scattering [cf., Eq. (40.29)]:

$$\langle e^{-i(q,\, r_{fj}(0))} e^{i(q,\, r_{f'j'}(t))}\rangle_c \approx e^{i(q,\, f'-f)} e^{i(q,\, j'-j)} e^{-W_j(q) - W_{j'}(q)}, \tag{40.40}$$

where f, j, and $W_j(q)$ are introduced in accordance with Eqs. (40.20) and (40.30). The corresponding term in the correlation function of the spin operators will be taken in the form

$$\langle S_{fj}^{\alpha}(0)\, S_{f'j'}^{\beta}(t)\rangle_c \approx \langle S_{fj}^{\alpha}(0)\rangle \langle S_{f'j'}^{\beta}(0)\rangle. \tag{40.41}$$

As before [see the remark made after Eq. (40.30)], we can easily show that the selection of the time-independent part of the form of Eq. (40.41) can be justified by the principle of correlation weakening. In fact, if we use the expression (40.40), the main contribution to Eq. (40.39) will be made by the terms for which $|f - f'| \to \infty$. Since the correlation between spins should vanish as the spins move far apart, we can use the expression (40.41) under the summation sign.

We shall bear in mind that the value of the spin at a site (fj) depends only on the index j. We shall transform the spin operators (5.8) by selecting the new quantization axis γ to be directed along the average magnetization vector. Then

$$\langle S^{\alpha}_{fj}\rangle = \gamma^{\alpha}_{j}\langle S'^{z}_{fj}\rangle = \gamma^{\alpha}_{j} S_j \sigma_{S_j}, \quad \langle S'^{x}_{fj}\rangle = \langle S'^{y}_{fj}\rangle = 0, \tag{40.42}$$

where σ_{S_j} is the relative magnetization of atoms of type j, and S'^{α}_{fj} (α = x, y, z) are the new spin variables.

Substituting Eqs. (40.40)-(40.42) into Eq. (40.39) and using Eq. (40.33), we finally obtain

$$\begin{aligned} \frac{\delta \Sigma^{(m)}_{pp'}}{\delta O} = r_0^2 \gamma_n^2 N' \frac{(2\pi)^3}{v} \sum_k \delta(q - 2\pi g'_k) \times \\ \times \sum_{jj'} [(\gamma_j, \gamma_{j'}) - (e, \gamma_j)(e, \gamma_{j'})] \times \\ \times S_j S_{j'} \sigma_{S_j} \sigma_{S_{j'}} F_j(q) F_{j'}(q) e^{i(q, j'-j)} e^{-W_j(q) - W_{j'}(q)}, \end{aligned} \tag{40.43}$$

where N' is the number of unit cells in the magnetic lattice, and g'_k are the radius vectors of the sites in the reciprocal magnetic lattice.

We shall introduce the notation

$$\begin{aligned} \Phi^{(m)} = (2\pi)^3 \frac{r_0^2 \gamma_n^2}{v} \sum_{jj'} [(\gamma_j, \gamma_{j'}) - (e, \gamma_j)(e, \gamma_{j'})] \times \\ \times S_j S_{j'} \sigma_{S_j} \sigma_{S_{j'}} F_j(q) F_{j'}(q) e^{i(q, j'-j)} e^{-W_j(q) - W_{j'}(q)} e^{2W(q)}, \end{aligned} \tag{40.44}$$

where W(q) is defined by the formula (40.30).

In special cases, the formula (40.44) may be simplified. In the case of ferromagnetic ordering, $\gamma^{\alpha}_{j} = \gamma^{\alpha}$ ($\alpha = 1, 2, 3$) and therefore, Eq. (40.44) assumes the form

$$\Phi^{(m)} = (2\pi)^3 \frac{r_0^2 \gamma_n^2}{v} [1 - (e, \gamma)^2] \times$$
$$\times \left| \sum_j S_j \sigma_{s_j} F_j(q) e^{-i(q, j)} e^{-W_j(q)} \right|^2 e^{2W(q)} = \varphi_f^{(m)} [1 - (e, \gamma)^2]. \quad (40.45)$$

In the case of the antiferromagnetic ordering of a system consisting of two equivalent sublattices, and in the case of the antiparallel orientation of spins (cf., Fig. 2a), we have j = 1,2 and $\gamma_1^\alpha = -\gamma_2^\alpha = \gamma^\alpha$ (α = 1,2,3), and the formula (40.44) is written as follows:

$$\Phi^{(m)} = 2 \frac{(2\pi)^3}{v} r_0^2 \gamma_n^2 [1 - (e, \gamma)^2] [1 - \cos(q, j_1 - j_2)] \times$$
$$\times [S\sigma_s F(q)]^2 = \varphi_a^{(m)} [1 - (e, \gamma)^2] [1 - \cos(q, j_1 - j_2)], \quad (40.46)$$

where j_1, j_2 are the radius vectors of atoms in a unit magnetic cell.

Using the notation of Eq. (40.44), the differential effective cross section for the elastic magnetic scattering is written in the form

$$\frac{\delta \Sigma_{pp'}^{(m)}}{\delta O} = N' \Phi^{(m)} \sum_k \delta(q - 2\pi g_k') e^{-2W(q)}. \quad (40.47)$$

In the special case of the collinear ordering of spins, for example, ferromagnetic ordering, it is convenient to separate out the factor with (e, γ) to make the result clearer:

$$\frac{\delta \Sigma_{pp'}^{(m)}}{\delta O} = N' \varphi_f^{(m)} [1 - (e, \gamma)^2] \sum_k \delta(q - 2\pi g_k'). \quad (40.48)$$

Here, N' is the number of unit cells in the magnetic lattice, and g_k' the reciprocal magnetic lattice vector.

The diffraction maxima are, in this case, observed at angles given by the conditions

$$p^\alpha - p'^\alpha = 2\pi g_k'^\alpha, \quad |p| = |p'| \quad (\alpha = 1, 2, 3). \quad (40.49)$$

The vectors g_k and g_k' in the formulas (40.35) and 40.48) are, in general, different. Consequently, the magnetic scattering leads to the appearance of additional diffraction maxima. The intensity of these maxima varies with temperature approximately as the square of the average magnetization [cf., Eq. (40.44). At temperatures $\vartheta > \vartheta_C$, the magnetization vanishes and these maxima disappear. The intensity of the diffraction maxima depends on the angle through the form-factors, thermal factors, and, to a considerable degree, through a factor depending on the orientation of the scattering vector q with respect to the direction cosines γ_j of the magnetization vectors of the magnetic sublattices.

In conclusion, we would mention that the thermal Debye–Waller factor of Eq. (40.30) can also be calculated using the Green's functions. The calculation methods can be found in the references cited above.

Sec. 41. Inelastic Scattering of Neutrons*

In this section, we shall consider the inelastic magnetic scattering of neutrons. Using the general formula (40.11) for the differential effective cross section and a somewhat different method of calculating the correlation function in the expression for the effective cross section, we shall simultaneously obtain expressions for the inelastic and elastic scattering of neutrons. Partial repetition of the results of the preceding section seems desirable for methodological reasons.

We shall assume that the incident neutron beam is not polarized. Then, according to Eq. (40.11), we shall have

$$\frac{\delta^2 \Sigma_{pp'}}{\delta O\, \delta E_{p'}} = \frac{m_n^2}{(2\pi)^2} \frac{p'}{p} \frac{1}{2\pi} \int_{-\infty}^{\infty} dt\, e^{i\omega t} \langle \widetilde{\mathcal{V}^{+}_{p'p}(0)\, \mathcal{V}_{p'p}(t)} \rangle \qquad (41.1)$$

$$(\omega = E_{p'} - E_p)$$

(the notation is the same as in the preceding section).

We shall substitute into Eq. (41.1) the expression (40.22) for the matrix elements of the magnetic interaction, and average out over all orientations of the incident neutron spins [cf., Eq. (40.37)]. Consequently, we obtain

$$\frac{\delta^2 \Sigma^{(m)}_{pp'}}{\delta O\, \delta E_{p'}} = r_0^2 \gamma_n^2 \frac{p'}{p} \sum_{\alpha\beta} \sum_{fj, f'j'} (\delta_{\alpha\beta} - e_\alpha e_\beta)\, F_j(q)\, F_{j'}(q) \times$$

*See also Izyumov's review (1963a).

$$\times \frac{1}{2\pi} \int_{-\infty}^{\infty} dt\, e^{i\omega t} \langle e^{-i(q,\, r_{fj}(0))} e^{i(q,\, r_{f'j'}(t))} S^{\alpha}_{fj}(0) S^{\beta}_{f'j'}(t) \rangle. \qquad (41.2)$$

We shall neglect the spin–phonon interactions. Then,

$$\langle e^{-i(q,\, r_{fj}(0))} e^{i(q,\, r_{f'j'}(t))} S^{\alpha}_{fj}(0) S^{\beta}_{f'j'}(t) \rangle =$$
$$= \langle e^{-i(q,\, r_{fj}(0))} e^{i(q,\, r_{f'j'}(t))} \rangle \langle S^{\alpha}_{fj}(0) S^{\beta}_{f'j'}(t) \rangle. \qquad (41.3)$$

The first correlation function in Eq. (41.3) will be calculated in the lowest approximation, using decoupling of the Eq. (40.29) type. Using the formulas (40.20) and (40.30), we see that

$$\langle e^{-i(q,\, r_{fj}(0))} e^{i(q,\, r_{f'j'}(t))} \rangle \approx$$
$$\approx e^{i(q,\, f'-f)} e^{i(q,\, j'-j)} e^{-W_j(q) - W_{j'}(q)}. \qquad (41.4)$$

In the second correlation function in Eq. (41.3), we shall replace the operators S^{α}_{fj} ($\alpha = 1, 2, 3$) with the operators $S'^{\pm}_{fj}, S'^{z}_{fj}$ using the formulas of Eq. (5.8). We shall confine ourselves to systems with parallel (ferromagnetic) spin configurations:

$$\gamma^{\alpha}_{fj} = \gamma^{\alpha}, \quad A^{\alpha}_{fj} = A^{\alpha} \qquad (\alpha = 1,\ 2,\ 3). \qquad (41.5)$$

It can be easily shown that

$$\sum_{\alpha\beta} (\delta_{\alpha\beta} - e_{\alpha} e_{\beta}) \langle S^{\alpha}_{fj}(0) S^{\beta}_{f'j'}(t) \rangle =$$
$$= [1 - (e,\ \gamma)^2] \langle S^{z}_{fj}(0) S^{z}_{f'j'}(t) \rangle +$$
$$+ \left[\frac{1}{2} - |(e,\ A)|^2\right] \langle S^{+}_{fj}(0) S^{-}_{f'j'}(t) + S^{-}_{fj}(0) S^{+}_{f'j'}(t) \rangle + \ldots \qquad (41.6)$$

The notation used for the new operators on the right-hand side of Eq. (41.6) is the same as that for the original operators on the left-hand side of the same equation; this should cause no misunderstanding, because we shall not use the original operators again.

Equation (41.6) does not contain the terms including the correlation functions of the type $\langle S^{+}_{fj}(0) S^{+}_{f'j'}(t) \rangle$, $\langle S^{+}_{fj}(0) S^{z}_{f'j'}(t) \rangle$, etc. We can show that in the case of ferromagnets these terms make a small contribution compared with the functions which are left in Eq. (41.6) (see, for example, the solutions of the equations for the Green's functions in Sec. 39). We shall take into account that, on the strength of the conditions (5.9) for γ and A,*

$$\frac{1}{2}-|(e,\ A)|^2=\frac{1}{4}[1+(e,\ \gamma)^2]. \tag{41.7}$$

Consequently, Eq. (41.6) can be written approximately in the form

$$\sum_{\alpha\beta}(\delta_{\alpha\beta}-e_\alpha e_\beta)\langle S^\alpha_{fj}(0)\,S^\beta_{f'j'}(t)\rangle\approx$$
$$\approx[1-(e,\ \gamma)^2]\langle S^z_{fj}(0)\,S^z_{f'j'}(t)\rangle+$$
$$+\frac{1}{4}[1+(e,\ \gamma)^2]\langle S^+_{fj}(0)\,S^-_{f'j'}(t)+S^-_{fj}(0)\,S^+_{f'j'}(t)\rangle. \tag{41.8}$$

We note that the vector γ, which determines the direction of the quantization axis, is selected so that it coincides with the direction of the magnetization vector. Then,

$$\langle S^z_{fj}\rangle=S_j\sigma_{S_j},\quad \langle S^\pm_{fj}\rangle=0, \tag{41.9}$$

where σ_{S_j} is the relative magnetization of the j-th site in a unit magnetic cell.

Substituting into Eq. (41.2) the expressions (41.3) and (41.8), we obtain the following equation for the effective magnetic scattering cross section:

$$\frac{\delta^2\Sigma^{(m)}_{pp'}}{\delta O\,\delta E_{p'}}=r_0^2\gamma_n^2\frac{p'}{p}\sum_{fj,\,f'j'}F_j(q)\,F_{j'}(q)\,e^{i(q,\,f'-f)}e^{i(q,\,j'-j)}\times$$
$$\times e^{-W_j(q)-W_{j'}(q)}\frac{1}{2\pi}\int_{-\infty}^{\infty}dt\,e^{i\omega t}\Big\{[1-(e,\ \gamma)^2]\langle S^z_{fj}(0)\,S^z_{f'j'}(t)\rangle+$$
$$+\frac{1}{4}[1+(e,\ \gamma)^2]\langle S^+_{fj}(0)\,S^-_{f'j'}(t)+S^-_{fj}(0)\,S^+_{f'j'}(t)\rangle\Big\}$$
$$(\omega=E_{p'}-E_p). \tag{41.10}$$

---

* To prove this, we shall consider two vector identities which follow from Eq. (5.9):

$$[A^*\times A]=-\frac{i}{2}\gamma,\quad i[A^*\times[e\times A]]=[A^*\times[e\times[\gamma\times A]]]\quad (e^2=1).$$

Multiplying out the second identity, we obtain, using Eq. (5.9) and the first identity,

$$\frac{1}{2}e-A(A^*,\ e)=A^*(A,\ e)+\frac{1}{2}\gamma(\gamma,\ e).$$

The scalar multiplication of this equation by the vector e gives Eq. (41.7).

where e is the unit scattering vector of Eq. (40.23), and γ is the unit vector parallel to the magnetization vector. Next, having determined the correlation functions, for example, in terms of the Green's functions, we shall obtain an explicit expression for the differential effective cross section.

We shall represent approximately the correlation function of the z-components of the spin operators in the form of a product of averages, and we shall use Eq. (41.9):

$$\langle S^z_{fj}(0)\, S^z_{f'j'}(t)\rangle \approx \langle S^z_{fj}(0)\rangle \langle S^z_{f'j'}(0)\rangle = S_j S_{j'} \sigma_{s_j} \sigma_{s_{j'}}. \qquad (41.11)$$

We shall apply the Fourier transformation to the spin operators:

$$S^{\pm}_{fj} = \frac{1}{\sqrt{N'}} \sum_{\nu} e^{\pm i(f,\, \nu)} S^{\pm}_{\nu j}, \qquad (41.12)$$

where N' is the number of unit cells in the magnetic lattice. Since they are translationally invariant, functions of the $\langle S_{fj}(0)\, S_{f'j'}(t)\rangle$ type depend only on the differences $f - f'$. Consequently,

$$\langle S^+_{fj}(0)\, S^-_{f'j'}(t)\rangle = \frac{1}{N'} \sum_{\nu} e^{i(f-f',\, \nu)} \langle S^+_{\nu j}(0)\, S^-_{\nu j'}(t)\rangle. \qquad (41.13)$$

Substituting Eqs. (41.11) and (41.13) into Eq. (41.10) and using Eqs. (40.33) and (A2.17), we obtain

$$\frac{\delta^2 \Sigma^{(m)}_{pp'}}{\delta O\, \delta E_{p'}} = N' \frac{(2\pi)^3}{v} r_0^2 \gamma_n^2 \left| \sum_j S_j \sigma_{s_j} F_j(q)\, e^{-i(q,\, j)} e^{-W_j(q)} \right|^2 \times$$

$$\times [1 - (e,\, \gamma)^2] \sum_k \delta(q - 2\pi g'_k)\, \delta(E_{p'} - E_p) +$$

$$+ N' \frac{(2\pi)^3}{4v} r_0^2 \gamma_n^2 \frac{p'}{p} \sum_{jj'} F_j(q)\, F_{j'}(q)\, e^{i(q,\, j'-j)} e^{-W_j(q) - W_{j'}(q)} \times$$

$$\times [1 + (e,\, \gamma)^2] \frac{1}{N'} \sum_{\nu} \left\{ \sum_k \delta(q - \nu - 2\pi g'_k) \times \right.$$

$$\times \frac{1}{2\pi} \int_{-\infty}^{\infty} dt\, e^{i\omega t} \langle S^+_{\nu j}(0)\, S^-_{\nu j'}(t)\rangle + \sum_k \delta(q + \nu - 2\pi g'_k) \times$$

$$\left. \times \frac{1}{2\pi} \int_{-\infty}^{\infty} dt\, e^{i\omega t} \langle S^-_{\nu j}(0)\, S^+_{\nu j'}(t)\rangle \right\} \qquad (\omega = E_{p'} - E_p).$$

To express the above formula in a clearer form, we shall introduce the notation

$$\varphi^{(m)}(q) = \frac{(2\pi)^3}{v} r_0^2 \gamma_n^2 \left| \sum_j S_j \sigma_{S_j} F_j(q) e^{-i(q,j)} e^{-W_j(q)} \right|^2 e^{2W(q)}, \tag{41.14}$$

$$\psi_{jj'}^{(m)}(q) = \frac{(2\pi)^3}{4v} r_0^2 \gamma_n^2 F_j(q) F_{j'}(q) e^{i(q,j'-j)} e^{-W_j(q) - W_{j'}(q)} e^{2W(q)}, \tag{41.15}$$

$$I_{S_{\nu j}^+, S_{\nu j}^-}(\omega) = \frac{1}{2\pi} \int_{-\infty}^{\infty} dt\, e^{i\omega t} \langle S_{\nu j}^-(0) S_{\nu j'}^+(t) \rangle; \tag{41.16}$$

the average thermal factor W(q) of Eq. (40.31) is introduced for clarity. Using the notation of Eqs. (41.14)-(41.16), the expression for the differential effective cross section becomes

$$\begin{aligned} \frac{\delta^2 \Sigma_{pp'}^{(m)}}{\delta O\, \delta E_{p'}} &= N' \varphi^{(m)}(q) e^{-2W(q)} [1 - (e, \gamma)^2] \sum_k \delta(q - 2\pi g_k') \times \\ &\times \delta(E_{p'} - E_p) + \frac{p'}{p} N' e^{-2W(q)} [1 + (e, \gamma)^2] \sum_{jj'} \psi_{jj'}^{(m)}(q) \times \\ &\times \frac{1}{N'} \sum_\nu \Big\{ I_{S_{\nu j}^-, S_{\nu j}^+}(\omega) \sum_k \delta(q - \nu - 2\pi g_k') + \\ &\quad + I_{S_{\nu j}^+, S_{\nu j}^-}(\omega) \sum_k \delta(q + \nu - 2\pi g_k') \Big\} \\ &\quad (\omega = E_{p'} - E_p, \quad q = p - p'). \end{aligned} \tag{41.17}$$

The first term in Eq. (41.17) describes the elastic scattering of neutrons by the spin system (cf., Sec. 40). It gives sharp intensity maxima for the scattering angles satisfying the Bragg reflection condition of Eq. (40.49). The second term describes the inelastic scattering of neutrons, accompanied by the emission or absorption of one spin wave. The intensities of the elastic and inelastic scattering maxima depend on the angle through the form-factors, the thermal Debye–Waller factors, and the multipliers $[1 \pm (e, \gamma)^2]$.

We note that $I_{S_{\nu j}^+, S_{\nu j}^-}$ of Eq. (41.16) is, according to Eq. (26.2), nothing else but the spectral density for the correlation function

$$\langle S_{\nu j}^{-}(t')\, S_{\nu j'}^{+}(t)\rangle$$

or, according to Eq. (26.13), the spectral density for the Green's function

$$\langle\langle S_{\nu j'}^{+}(t) \,|\, S_{\nu j}^{-}(t')\rangle\rangle.$$

In the lowest approximation, when damping is neglected, the spectral functions (41.16) have the δ-type nature:

$$I(\omega) \sim \delta(\omega \pm E_r(\nu)), \tag{41.18}$$

where $E_r(\nu)$ is the spin-wave energy, and r is the number of the mode of the spin-wave spectrum. Since the sum with respect to ν in Eq. (41.17) is replaced by a δ-function of the momentum, we see that the energy distribution of the scattered neutrons should have δ-type maxima, whose positions are found from the equations

$$E_{p'} - E_p \pm E_r(p - p' - 2\pi g'_k) = 0.$$

Since the spin-wave energy is a periodic function of the wave vector with a period equal to the period of the reciprocal lattice, these equations can be rewritten as follows:

$$\varphi^{\pm}(p, p') = E_{p'} - E_p \pm E_r(p - p') = 0. \tag{41.19}$$

The equation $\varphi^{-}(p, p') = 0$ governs the conditions for the scattering of a neutron accompanied by the absorption of a spin wave, and the equation $\varphi^{+}(p, p') = 0$ gives the conditions for the scattering accompanied by the creation of a spin wave.

If we allow for the finite lifetime of spin waves, assuming, for example,

$$I(\omega) \sim \frac{\Gamma_r(\nu)}{[\omega - E_r(\nu)]^2 + \Gamma_r^2(\nu)} \qquad (\omega = E_{p'} - E_p), \tag{41.20}$$

the energy distribution of neutrons at the intensity maxima will be described by a Lorentz curve with a half-width $\Gamma_r(\nu)$ of the type

$$\frac{\Gamma_r(\nu)}{[E_{p'} - E_p - E_r(p - p')]^2 + \Gamma_r^2(\nu)} \cdot \tag{41.21}$$

Neglecting the spin-wave damping, we shall consider the scattering accompanied by the absorption of a spin wave. According to Eq. (41.19), such scattering is possible if p' satisfies the equation $\varphi^-(p, p') = 0$. We note that $E_r(\nu) \geq 0$ and that a maximum value E_r^{max} is reached at some ν. We can easily see then, that

$$\varphi^-(p, p') \begin{cases} > 0 \text{ when } p'^2 = p^2 + 2m_n E_r^{max}, \\ < 0 \text{ when } p'^2 = p^2. \end{cases}$$

Consequently, we can always find such a value of p', lying in the range

$$p^2 \leq p'^2 \leq p^2 + 2m_n E_r^{max}, \tag{41.22}$$

which satisfies the equation $\varphi^-(p, p') = 0$ for any scattering angle.

The scattering accompanied by the creation of a spin wave is possible if the equation $\varphi^+(p, p') = 0$ is satisfied. If $p^2 > 2m_n E_r^{max}$, we can always select such a value of p', lying in the range

$$p^2 - 2m_n E_r^{max} \leq p'^2 \leq p^2, \tag{41.23}$$

which satisfies the equation $\varphi^+(p, p') = 0$ for any scattering angle. If $p^2 < 2m_n E_r^{max}$, then, in general, the scattering of neutrons accompanied by the creation of a spin wave is not possible for an arbitrary scattering angle. At sufficiently low values of p, the scattering accompanied by the creation of a spin wave is altogether impossible.

The form of the spin-wave spectrum can be determined from the positions of the maxima of the inelastic magnetic scattering. In fact, since the value of p is given, and p' is found from the position of the intensity maximum, the spin-wave energy $E_r(\nu)$ ($\nu = p - p'$) can be determined from Eq. (41.19).

Sec. 42. Further Applications of the Green's Function Method

The applications of the Green's function method are not exhausted by the few elementary examples considered in the preceding sections, and we shall now touch on the applications of the method to some other problems in the theory of magnetism.

Ising Model. The Hamiltonian of a ferromagnet in the Ising model is written in the form

$$\mathcal{H} = -\mu H \sum_{f} S_f^z - \frac{1}{2} \sum_{f_1, f_2} I(f_1 - f_2) S_{f_1}^z S_{f_2}^z, \tag{42.1}$$

where S_f^z is the z-component of the spin operator at a site f. For this component, Doman and ter Haar (1962) deduced the following expression for spin $S = 1/2$:

$$\langle S_f^z \rangle = \frac{1}{2} - \left\langle \frac{2S_f^z}{e^{\frac{1}{\theta}\left[\mu H + \sum_{f'} I(f-f') S_{f'}^z\right]} - 1} \right\rangle. \tag{42.2}$$

If the average value of the function of S_f^z on the right-hand side of the above equation is replaced by the same function of the average values, $\langle S_f^z \rangle$, the molecular field equation is obtained. Tahir-Kheli, Doman, and ter Haar (1963) also considered the case of arbitrary spin. Callen (1963a) points out that the results of Tahir-Kheli et al. cannot be regarded as the solution of the Ising problem.

Antiferromagnets. P'u Fu-ch'o (1960a, b) investigated the ground state and the temperature and field dependences of the magnetization of an isotropic ferromagnet in the case $S = 1/2$. Chervonko (1963), Hewson and ter Haar (1963a) used somewhat different methods to investigate the statistical properties of an isotropic antiferromagnet with arbitrary spin. Lines (1964b) considered in detail the case $S \geq 1/2$. In all these investigations, the decoupling method was the same as that used in Secs. 29 and 32. This made it possible to obtain results valid over a wide range of temperatures. The sublattice magnetizations were found to be much less than the total magnetic moments even at zero temperature. The reason for this was the presence of zero-point vibrations in the spin system. Lines (1964a) used the Green's function method to calculate the Curie and Néel temperatures for layered structures.

The same method was employed by P'u Fu-ch'o (1961) to obtain general formulas for the resonance in uniaxial antiferromagnets.

Ferromagnets. Attempts to extend the results of the approximate solution of the equations for the Green's functions for spin $S = 1/2$ to arbitrary spin S were also made by Izyumov and Yakovlev (1960) and Kawasaki and Mori (1961); a detailed discussion of these results is given in Tahir-Kheli and ter Haar's paper (1962a). Callen (1963b) and Tahir-Kheli (1963) obtained approximate equations for the magnetization in the case of arbitrary S using the decoupling of equations for the Green's functions proposed in Callen's paper (see also Sec. 35). A comparison of the results of the first approximation for various cases was also made by Hewson and ter Haar (1963b). Using this first approximation, Tahir-Kheli and Callen (1964) calculated the correlation function for the z-components of the spin operators. Tahir-Kheli and Jarrett (1964) calculated the Curie temperature for cubic lattices allowing for the second-nearest neighbors. Potapkov (1963a) investigated the anisotropy of uniaxial ferromagnets allowing for the spin–orbit interaction; he also studied the influence of the magnetic anisotropy on the width of the ferromagnetic resonance line in uniaxial crystals [Potapkov (1962a, b; 1963b)]. Haubenreiser (1963) and Haas (1963) considered ferromagnets allowing for the spin–spin (dipole) interaction.

The introduction of damping as the imaginary part of the Green's function pole allowed Meng Hsien-chen (1963) to extend the well-known results of the classical theory of the angular dependence of the ferromagnetic resonance line width [Skrotskii and Kurbatov (1958)] and to establish a number of general relationships. Meng Hsien-chen and P'u Fu-ch'o (1961) and Meng Hsien-chen (1961) used the Green's function method to extend to arbitrary temperatures the results of Clogstone et al. (1956) and Callen and Pitteli (1960a, b) on the dependence of the line width on impurities. The behavior of the resonance curve near the Curie point was dealt with by Morkowski (1963), using the Green's function method.

A detailed discussion of the theory of magnetic resonance has been given by Tomita and Tanaka (1963a, b), who used an original method – based on Kubo's work (1962) – for decoupling the chains of equations.

Haas and Jarrett (1964) compared the available methods in the theory of ferromagnetism using the Heisenberg model of an isotropic ferromagnet.

Ferromagnets with Complex Structures. Izyumov (1963b) and Bar'yakhtar and Shishkin (1964) investigated the properties of ferromagnets consisting of two sublattices, including helical structures.

Mills et al. (1964) discussed a three-sublattice model of an isotropic ferrimagnet and calculated the magnetization of the sublattices.

Khachaturyan (1963) investigated the interpolation solutions for magnetic substances, similar to those considered in Sec. 32, without assuming that the average deviation of the spin at a site was independent of the lattice site number. The use of the group-theoretical methods made it possible to deduce various types of structure permitted by the lattice symmetry, and to investigate the stability of these structures at various temperatures.

Model with the s – d Exchange. The influence of the interaction between the conduction electrons and the electrons in the partly filled shells on the properties of transition metals was considered, within the framework of the s–d exchange model (see Sec. 5), and using the Green's function method, by Vonsovskii and Izyumov (1960), Potapkov and Tyablikov (1960), and Tahir-Kheli and ter Haar (1963). A detailed discussion of various applications was given in the review by Vonsovskii and Izyumov (1962b).

Scattering of Neutrons in Magnetic Materials. Kashcheev and Krivoglaz (1961) investigated, using the Green's functions, the influence of the spin–spin and spin–phonon interactions in ferromagnets on the energy distribution of scattered neutrons; similar investigations of antiferromagnets and ferrimagnets were carried out by Kashcheev (1962a, b, c). Bar'yakhtar and Maleev (1963) and Maleev et al. (1962) considered the scattering of neutrons in ferromagnets with the helical structure and took into account the inelastic magnetic scattering; Kashcheev (1964a, b), on the other hand, allowed for the spin–phonon interaction.

Appendices*

A1. Bravais and Reciprocal Lattices

We shall consider a simple lattice. We shall introduce three fundamental translation vectors a_1, a_2, a_3, which join any site in the lattice with three other sites, and we shall construct a parallelepiped from these vectors. The remaining sites of the lattice can be obtained by the parallel translation of this parallelepiped along the directions of the fundamental translation vectors, by amounts which are multiples of the lengths of the parallelepiped edges. The smallest parallelepiped whose translation can generate a lattice is called a primitive cell. Each primitive cell contains one lattice site. The selection of the basic cell of a crystal is not unambiguous, since there are many triplets of noncoplanar vectors from which cells of the same volume can be constructed. Usually, they are selected so that the translation periods have the smallest possible values. Such cells are called unit cells.

We shall select some site in a simple lattice as the origin of coordinates. Then the radius vectors of the sites in the lattice can be written in the form

$$R_l = l_1 a_1 + l_2 a_2 + l_3 a_3, \tag{A1.1}$$

where l_1, l_2, and l_3 are integers. The vectors a_r (r = 1,2,3) have the dimensions of length.

In simple lattices, all the sites are occupied by atoms of one kind. A complex lattice can, in general, be represented as an assembly of several simple lattices superimposed on one another.

*See Born and Kun Huang (1958) and Brillouin and Parodi (1959).

The sites in each of them can be occupied by atoms of a given type. Consequently, a unit cell of such a lattice will contain as many atoms as there are simple lattices of which the complex lattice is composed. These atoms form the basis of the lattice.

We shall select one of the simple lattices as the principal lattice and we shall number its cells in accordance with the radius vectors R_l of its sites; the number of atoms in a unit cell will be denoted by n. Then, the radius vector of any lattice site will be written as follows:

$$R_{lj} = R_l + R_j \quad (j = 1, 2, \ldots, n), \tag{A1.2}$$

where R_l is the radius vector of a site (or of a unit cell) in the principal lattice, R_j is the radius vector of an atom in a unit cell, measured relative to R_l. The index j is also the number of the simple lattice. The volume of a unit cell is equal to the mixed product of the fundamental translation vectors:

$$v_a = (a_1, [a_2 \times a_3]). \tag{A1.3}$$

If the lattice sites are occupied by atoms with nonzero magnetic moments, we shall speak of a magnetic lattice. The concepts of a unit cell and of the basis (for complex lattices) will be used also in magnetic lattices.

A unit cell of a magnetic lattice need not, in general, be identical with a unit cell of a crystal lattice, since, in a magnetic lattice the sites are regarded as identical if the spins at the sites are equal in magnitude and direction.

We shall consider an antiferromagnet in which magnetically active atoms are identical and located at the sites of a simple cubic lattice, separated by distance a fron one another. Let the spins at neighboring sites have antiparallel orientations (cf., Fig. 7a on p. 138. The (+) and (–) spin systems each form an fcc lattice with a period of $2a$. The antiferromagnet as a whole is an assembly of two fcc lattices superimposed on each other (NaCl-type structure). An example of a lattice in which the magnetic and crystal unit cells are identical is the antiferromagnetic lattice with the structure shown in Fig. 8a (p. 138).

We shall now introduce a reciprocal lattice. We shall use the following three fundamental translation vectors:

$$b^{(1)} = v_a^{-1}[a_2 \times a_3], \qquad b^{(2)} = v_a^{-1}[a_3 \times a_1], \qquad b^{(3)} = v_a^{-1}[a_1 \times a_2]. \tag{A1.4}$$

The vectors $b^{(r)}$ $(r = 1, 2, 3)$ are also noncoplanar and we can construct a lattice from them. Such a lattice is known as the reciprocal lattice, in contrast to the Bravais or space lattice, discussed earlier in the present section, which was constructed from the vectors a_r $(r = 1, 2, 3)$. The term "reciprocal" is used because the vectors $b^{(r)}$ have the dimensions of the reciprocal of length [in particular, if the vectors a_r are mutually perpendicular, the length of the vector $b^{(r)}$ is equal to the reciprocal of the length of the vector a_r].

The radius vectors of the sites in a reciprocal lattice are written in the form

$$g_k = k_1 b^{(1)} + k_2 b^{(2)} + k_3 b^{(3)}, \tag{A1.5}$$

where k_1, k_2, k_3 are integers.

The fundamental translation vectors of the Bravais and reciprocal lattices satisfy the relationships

$$(a_r, b^{(r')}) = \delta_{r,r'} \qquad (r, r' = 1, 2, 3). \tag{A1.6}$$

Again, the volume of a unit cell in a reciprocal lattice is equal to the reciprocal of the volume of the unit cell of the corresponding Bravais lattice:

$$v_b = (b^{(1)}, [b^{(2)} \times b^{(3)}]) = \frac{1}{(a_1, [a_2 \times a_3])} = \frac{1}{v_a}. \tag{A1.7}$$

The radius vector of each site in a reciprocal lattice, g_k, is perpendicular to the family of parallel planes passing through sites in the corresponding Bravais lattice. These planes will be called crystal planes. The vector equation of a plane perpendicular to g_k has the form

$$\frac{1}{|g_k|}(g_k, R_l) = c, \tag{A1.8}$$

where c is the distance from the origin of coordinates to this plane. Substituting in Eq. (A1.8) the expressions (A1.1) and (A1.5), we obtain

$$\frac{1}{|g_k|}(k_1 l_1 + k_2 l_2 + k_3 l_3) = c.$$

Since k_r, l_r (r = 1,2,3) are integers, the numerator in the left-hand part of the above equation must also be an integer; its smallest value is unity. Consequently, the distance between nearest crystal planes, perpendicular to g_k, is

$$\delta = |g_k|^{-1}. \tag{A1.9}$$

An arbitrary vector R in a Bravais lattice can always be written in the form

$$R = \xi_1 a_1 + \xi_2 a_2 + \xi_3 a_3. \tag{A1.10}$$

Taking the scalar product of this equation with $b^{(r)}$, and using Eq. (A1.6), we obtain

$$\xi_r = (R, b^{(r)}) \qquad (r = 1, 2, 3). \tag{A1.11}$$

Consequently, the expansion of an arbitrary vector in terms of the fundamental translation vectors has the form

$$R = (R, b^{(1)})\, a_1 + (R, b^{(2)})\, a_2 + (R, b^{(3)})\, a_3. \tag{A1.12}$$

We shall consider the expansion of the function F(R) as a Fourier series with a period equal to the lattice period. By definition,

$$F(R) = F(R + R_l), \tag{A1.13}$$

where R_l is an arbitrary vector of the form given by Eq. (A1.1). The function F(R) can be considered as a function of three variable, ξ_1, ξ_2, ξ_3, of Eq. (A1.11), andwe can expand it as a formal Fourier series in terms of each of the variables:

$$F(R) = F(\xi_1, \xi_2, \xi_3) =$$

$$= \sum_{k_1, k_2, k_3} \tilde{F}(k_1, k_2, k_3)\, e^{2\pi i (k_1\xi_1 + k_2\xi_2 + k_3\xi_3)}, \tag{A1.14}$$

where k_1, k_2, k_3 are integers, and where

$$\tilde{F}(k_1, k_2, k_3) = \int\int\int_0^1 F(\xi_1, \xi_2, \xi_3)\, e^{-2\pi i (k_1\xi_1 + k_2\xi_2 + k_3\xi_3)}\, d\xi_1 d\xi_2 d\xi_3. \tag{A1.15}$$

The expressions in the exponents of the exponential functions will be rewritten, using Eq. (A1.11), in the form

$$k_1\xi_1 + k_2\xi_2 + k_3\xi_3 = k_1(R, b^{(1)}) + k_2(R, b^{(2)}) + k_3(R, b^{(3)}) = \\ = (R, g_k),$$

where g_k is the reciprocal lattice vector of Eq. (A1.5). We can now write Eqs. (A1.14) and (A1.15) in the following way:

$$F(R) = \sum_{g_k} \tilde{F}(g_k)\, e^{2\pi i (R, g_k)}, \tag{A1.16}$$

where

$$\tilde{F}(g_k) = \frac{1}{v_a} \int_{(v_a)} F(R)\, e^{-2\pi i (R, g_k)}\, dR; \tag{A1.17}$$

the integral in Eq. (A1.17) is taken over the volume of a unit cell. The factor v_a^{-1} in front of the integral is due to the Jacobian of the transformation from the variables ξ_r to the variables R^α ($\alpha = 1, 2, 3$). In fact, in the Cartesian coordinates, Eq. (A1.11) becomes

$$\xi_r = R^x b^{(r)x} + R^y b^{(r)y} + R^z b^{(r)z} \qquad (r = 1, 2, 3).$$

Hence,

$$\frac{\partial(\xi_1, \xi_2, \xi_3)}{\partial(R^x, R^y, R^z)} = \begin{vmatrix} b^{(1)x} & b^{(1)y} & b^{(1)z} \\ b^{(2)x} & b^{(2)y} & b^{(2)z} \\ b^{(3)x} & b^{(3)y} & b^{(3)z} \end{vmatrix} = \\ = (b^{(1)}, [b^{(2)} \times b^{(3)}]) = v_a^{-1}.$$

A2. Wave Vectors and Zones

A typical problem in the theory of crystal lattices is the determination of the eigenvalues of equations of the type

$$E\varphi(R_{lj}) = \sum_{l'j'} I_{jj'}(R_l - R_{l'})\, \varphi(R_{l'j'}). \tag{A2.1}$$

where l is the number of a cell, j is the number of a site in the cell, and $I_{jj'}(R_l - R_{l'})$ is some function of the coordinates of a pair of sites; it is assumed that this function can undergo a Fourier transformation. Such problems occur in the determination of the spectrum of the normal vibrational modes in a crystal lattice or the spectrum of spin waves in a spin system.

We shall consider a simple lattice. In this case, Eq. (A2.1) simplifies to

$$E\varphi(R_l) = \sum_{l'} I(R_l - R_{l'})\varphi(R_{l'}). \tag{A2.2}$$

Let us assume that we are dealing with a lattice of finite dimensions. The number of sites in it will be denoted by N and the lattice volume by V. Equation (A2.2) must be supplemented by its boundary conditions. These conditions govern the form of the normal vibrational modes. However, if a sample is sufficiently large, the frequency distribution of the normal-mode vibrations ceases, in practice, to depend on the shape of the sample and on the form of the boundary conditions [Lederman (1944), Peierls (1954)]. Therefore, we shall assume the sample to be in the form of a parallelepiped constructed using the three vectors $N_r a_r$ (r = 1, 2, 3), where $N_1 N_2 N_3 = N$, $Nv_a = V$, and the true boundary conditions are replaced with Born's periodic boundary conditions [Born and Goeppert-Mayer (1938)], according to which the required quantity assumes identical values at corresponding points of opposite faces of a parallelepiped.* The periodic boundary conditions can be interpreted also as the conditions of the periodicity of the solution, with the periods equal to $N_1 a_1$, $N_2 a_2$, $N_3 a_3$:

$$\varphi(R_l) = \varphi(R_l + R_N). \tag{A2.3}$$

*In a crystal of finite dimensions there are surface vibrations, as well as volume vibrations. The replacement of the true boundary conditions with the periodic conditions does not affect greatly the volume vibrations, but it does alter considerably the surface vibrations and, in particular, it may destroy the latter. In calculations of the volume properties, this point is of no great importance, because the statistical weight of the surface vibrations is small compared with the statistical weight of the volume vibrations.

where R_l is the radius vector of a site, and R_N represents the vectors used to construct the volume V. The volume V is for this reason called the principal periodicity volume.

We shall seek the solution of Eq. (A2.2) in the form

$$\varphi(R_l) = \sum_{\nu} \tilde{\varphi}(\nu) e^{i(\nu, R_l)}, \tag{A2.4}$$

where ν is the wave vector. From this definition, it follows that the wave vectors are defined in the reciprocal lattice space. Substituting Eq. (A2.4) into Eq. (A2.2), we obtain

$$E\tilde{\varphi}(\nu) = E(\nu)\tilde{\varphi}(\nu), \quad E(\nu) = \sum_{l} I(R_l) e^{-i(\nu, R_l)}. \tag{A2.5}$$

The eigenvalue $E(\nu)$ (A2.5) of Eq. (A2.2) is a periodic function of the wave vector with a period equal to the reciprocal lattice period, multiplied by 2π:

$$E(\nu + 2\pi g_k) = E(\nu), \tag{A2.6}$$

where g_k is the reciprocal lattice vector of Eq. (A1.5).

We shall substitute into both parts of Eq. (A2.6) the expressions for $E(\nu)$ from Eq. (A2.5), g_k from Eq. (A1.5), and R_l from Eq. (A1.1). Using Eq. (A1.6), we obtain

$$(\nu + 2\pi g_k, R_l) = (\nu, R_l) + 2\pi (k_1 l_1 + k_2 l_2 + k_3 l_3) = \\ = (\nu, R_l) + 2\pi m,$$

where m is an integer. Hence, we obtain the equality given by Eq. (A2.6):

$$\sum_{l} I(R_l) e^{-i(\nu + 2\pi g_k, R_l)} = \sum_{l} I(R_l) e^{-i(\nu, R_l) - 2\pi i m} = \\ = \sum_{l} I(R_l) e^{-i(\nu, R_l)}.$$

Due to the periodicity referred to earlier, the wave-vector domain can be divided into subdomains, in each of which there is a single-valued relationship between $E(\nu)$ and ν. As the first such subdomain, we can use, for example, the first unit cell of a reciprocal lattice.

The selection of a unit cell in a reciprocal lattice as the principal wave-vector subdomain is not the only possible way of approaching the problem. In investigations of definite problems, it is frequently more convenient to use the first Brillouin zone as the principal subdomain, because it has a higher symmetry than a unit cell in the reciprocal lattice, while the volumes of the zone and the unit cell are equal. The remaining subdomains of $E(\nu)$ are defined as the Brillouin zones of higher orders.

The Brillouin zones are constructed as follows. We join the origin of coordinates O_b in a reciprocal lattice by rectilinear segments with the remaining sites in the reciprocal lattice. We then draw planes through the middle points of these segments at right angles to them. A polyhedron of the smallest volume with its center at O_b is the first Brillouin zone. The equality of its volume and the volume of a unit cell in a reciprocal lattice follows from the fact that the points ν uniformly fill the reciprocal lattice space (see below) and that any external point may be obtained from an internal point in the polyhedron by displacing the latter by a distance equal to the reciprocal lattice vector.

In other words, the first Brillouin zone contains all the points ν which can be reached by continuous translation from the point O_b without intersecting any of the planes bounding the zone. The second zone contains all the points ν which can be reached from O_b crossing only one of these planes; the third zone contains the points reached by crossing two different planes, etc.

We shall now determine the permissible values of the vectors ν in the first subdomain. The boundary conditions (A2.3) must be satisfied by each particular solution of Eq. (A2.2):

$$\varphi(R_l) = \varphi(R_l + R_N),$$
$$\varphi(R_l) = \text{const}\, e^{i(\nu, R_l)}.$$

Hence we see that the vectors ν should satisfy the conditions

$$(\nu, R_N) = 2\pi n, \qquad \text{(A2.7)}$$

where n is an integer, and R_N is one of the vectors used to construct the principal periodicity domain.

We shall bear in mind that, by analogy with Eq. (A1.12), the vector ν can be expanded in terms of the fundamental translation vectors of a reciprocal lattice:

$$\nu = \nu_1 b^{(1)} + \nu_2 b^{(2)} + \nu_3 b^{(3)},$$
$$\nu_r = (\nu, a_r) \quad (r = 1, 2, 3), \tag{A2.8}$$

and that the vector R_N in Eq. (A2.3) can be used in one of the following forms:

$$N_1 a_1, \quad N_2 a_2, \quad N_3 a_3. \tag{A2.9}$$

Substituting Eqs. (A2.8) and (A2.9) into Eq. (A2.7), we finally obtain

$$\nu_r = \frac{2\pi}{N_r} n_r \quad (r = 1, 2, 3) \tag{A2.10}$$

and then

$$\nu = 2\pi \left(\frac{n_1}{N_1} b^{(1)} + \frac{n_2}{N_2} b^{(2)} + \frac{n_3}{N_3} b^{(3)} \right), \tag{A2.11}$$

where n_r are integers which take the values

$$n_r = 0, 1, 2, \ldots, N_r - 1 \quad (r = 1, 2, 3). \tag{A2.12}$$

Thus, the wave vectors are represented in the form of linear combinations of the fundamental translation vectors of the reciprocal lattice. The number of different wave vectors ν is $N = N_1 N_2 N_3$; they fill uniformly the first unit cell in the reciprocal lattice.

In making calculations, it is frequently necessary to evaluate sums of the form

$$F = \frac{1}{N} \sum_{\nu} F(\nu), \tag{A2.13}$$

where $F(\nu)$ is some function of ν; summing is carried out with respect to all vectors ν contained in a unit cell of the reciprocal lattice (in the first Brillouin zone). In statistical mechanics problems, the number of atoms N in the volume occupied by a crystal lattice is of the order of Avogadros's number. Therefore, the vectors ν fill the first zone almost continuously (the distance between two neighboring points will be of the order of $2\pi |b^{(r)}|/N_r$) and the sum with respect to ν, given by Eq. (A2.13), may be replaced with an integral.

We shall divide a unit cell in a reciprocal lattice into volumes $\Delta v(\nu)$, constructed near the points ν using three vectors

$$\Delta\nu_r = \frac{2\pi}{N_r} b^{(r)} \Delta n_r \qquad (r = 1,\ 2,\ 3).$$

We shall use $\Delta n(\nu)$ to denote the number of points ν contained in a volume $\Delta v(\nu)$. Since

$$\Delta v(\nu) = \frac{(2\pi)^3}{N_1 N_2 N_3}\left(b^{(1)}, \left[b^{(2)} \times b^{(3)}\right]\right)\Delta n_1 \Delta n_2 \Delta n_3 =$$

$$= \frac{(2\pi)^3}{N v_a} \Delta n_1 \Delta n_2 \Delta n_3 = \frac{(2\pi)^3}{N v_a} \Delta n(\nu),$$

therefore

$$\Delta n(\nu) = \Delta n_1 \Delta n_2 \Delta n_3 = \frac{N v_a}{(2\pi)^3} \Delta v(\nu).$$

We shall replace the sum with respect to ν in Eq. (A2.13) by a sum over all $\Delta v(\nu)$, assuming $F(\nu)$ to be constant within each elementary volume $\Delta v(\nu)$. Then,

$$F = \frac{1}{N} \sum_{\Delta v(\nu)} F(\nu)\, \Delta n(\nu) = \frac{v_a}{(2\pi)^3} \sum_{\Delta v(\nu)} F(\nu)\, \Delta v(\nu).$$

The above expression is simply a sum of integrals and, therefore, in the limit we have

$$\frac{1}{N} \sum_{\nu} F(\nu) \to \frac{v_a}{(2\pi)^3} \int_{(v_b)} F(\nu) dv(\nu), \tag{A2.14}$$

where the integral is taken over the volume of a unit cell in the reciprocal lattice (or over the volume of the first Brillouin zone).

In various applications, one also meets with sums over unit cells of Bravais lattices of the form

$$\varphi(\nu) = \frac{1}{N} \sum_{l} e^{i(\nu, R_l)}, \tag{A2.15}$$

where ν is the wave vector of Eq. (A2.11), R_l is the radius vector of sites in a Bravais lattice, given by Eq. (A1.1).

If $\nu = 2\pi g_k$, where g_k is one of the reciprocal lattice vectors, then

$$(\nu, R_l) = 2\pi (g_k, R_l) = 2\pi \sum_{r=1}^{3} k_r l_r = 2\pi m,$$

where m is an integer, and the sum of Eq. (A2.15) is equal to unity.

If $\nu \neq 2\pi g_k$, then

$$\frac{1}{N} \sum_{l} e^{i(\nu, R_l)} = \prod_{r=1}^{3} \frac{1}{N_r} \sum_{l_r=0}^{N_r-1} e^{2\pi i \frac{n_r l_r}{N_r}} =$$

$$= \prod_{r=1}^{3} \frac{1}{N_r} \frac{1 - e^{2\pi i l_r}}{1 - e^{2\pi i \frac{l_r}{N_r}}} = 0.$$

Combining these two cases, we can write the following expression for a quasi-continuous index ν:

$$\frac{1}{N} \sum_{l} e^{i(\nu, R_l)} = \sum_{k} \Delta (\nu - 2\pi g_k), \tag{A2.16}$$

where $\Delta(x) = 1$ if $x = 0$, and $\Delta(x) = 0$ if $x \neq 0$.

We shall now determine the form of the sum of Eq. (A2.15) in the limiting case when the index ν varies continuously. We shall denote it by $\overline{\varphi}(\nu)$.

Let $F(\nu)$ be some smoothly varying function. Then, according to Eq. (A2.16),

$$\sum_{\nu} F(\nu) \varphi(\nu) = \sum_{k} F(2\pi g_k).$$

Going over to a continuous distribution of the values of ν, we shall replace $\varphi(\nu)$ on the left-hand side of the above equation with $\overline{\varphi}(\nu)$, and the sum with respect to ν with an integral [cf., Eqs. (A2.13) and (A2.14)]:

$$N \frac{v_a}{(2\pi)^3} \int_{(v_b)} F(\nu) \overline{\varphi}(\nu)\, dv(\nu) = \sum_{k} F(2\pi g_k).$$

Hence, it follows that the function $\overline{\varphi}(\nu)$ may be written in the form

$$\frac{1}{N}\sum_{l} e^{i(\nu,\, R_l)} = \frac{(2\pi)^3}{Nv_a}\sum_{k}\delta(\nu - 2\pi g_k), \tag{A2.17}$$

where δ (x) is the Dirac delta-function.

In other words, on going over from a discrete to a continuous distribution of the wave vectors, we replace the Kronecker delta with the Dirac delta-function:

$$\Delta(\nu) \to \frac{(2\pi)^3}{Nv_a}\delta(\nu) = \frac{(2\pi)^3}{V}\delta(\nu). \tag{A2.18}$$

In calculating repeated sums over a lattice, it is always possible to use Eq. (A2.16), while the use of Eq. (A2.17) requires a certain amount of care, because it cannot always be used for each sum separately. For example,

$$\Big|\sum_{l} e^{i(\nu,\, R_l)}\Big|^2 = \sum_{l,\, l'} e^{i(\nu,\, R_l - R_{l'})} = N\sum_{l''} e^{i(\nu,\, R_{l''})} =$$

$$= N\frac{(2\pi)^3}{v_a}\sum_{k}\delta(\nu - 2\pi g_k). \tag{A2.19}$$

A3. Bragg Reflection Conditions

A wave is reflected from a crystal lattice if the latter has a system of crystal planes satisfying the Bragg condition

$$2\delta\cos\varphi = n\lambda, \tag{A3.1}$$

where δ is the distance between neighboring planes, φ is the angle of incidence, λ is the wavelength, and n is an integer. We shall express Eq. (A3.1) in a different form.

The condition (A3.1) is satisfied if the wave vectors of the incident (ν) and scattered (ν') waves satisfy the conditions

$$\nu' = \nu + 2\pi g_k, \qquad |\nu'| = |\nu| \tag{A3.2}$$

(g_k is the reciprocal lattice vector). In fact, it follows from Eq. (A3.2) that

$$2\pi g_k^2 = 2(\nu',\, g_k) = -2(\nu,\, g_k).$$

Using φ to denote the angle between the vectors g_k and ν' (or g_k and ν), we can write

$$2\,|\nu'|\cos\varphi = 2\pi g_k. \qquad \text{(A3.3)}$$

We shall bear in mind that the vector g_k is always normal to the family of parallel planes passing through the sites of a Bravais lattice, and that the distance between two neighboring planes is $\delta = 1/|g_k|$ [cf., remarks made about Eq. (A1.8)]. Since the wavelength is related to the wave vector by $\lambda = 2\pi\nu^{-1}$, we can easily see that Eq. (A3.3) reduces to the Bragg condition of Eq. (A3.1).

Literature Cited*

Abrahams, E. (1954), Phys. Rev., 98:387 [Sec. 1].

Abrahams, S.C. (1963), J. Phys. Chem. Solids, 24:589 [Sec. 1].

Abrikosov, A.A., Gor'kov, L.P., and Dzyaloshinskii, I.E. (1962), Quantum Field Theory Methods in Statistical Physics, Fizmatgiz, Moscow [Sec. 25].

Akhiezer, A.I., Bar'yakhtar, V.G., and Kaganov, M.I. (1960a), Usp. Fiz. Nauk, 71:533 [Sec. 19].

Akhiezer, A.I., Bar'yakhtar, V.G., and Kaganov, M.I. (1960b), Usp. Fiz. Nauk, 72:3 [Sec. 19].

Akhiezer, A.I., and Pomeranchuk, I.Ya. (1950), Some Problems in the Theory of the Nucleus, Gostekhizdat [Sec. 40].

Akulov, N.S. (1939), Ferromagnetism, GITTL, Moscow-Leningrad [Secs. 1,2,19].

Alekseev, A.I. (1961), Usp. Fiz. Nauk, 73:41 [Sec. 25].

Alexander, S. (1962), Phys. Rev., 127:420 [Sec. 7].

Al'tshuler, S.A., and Kozyrev, B.M. (1961), Electron Paramagnetic Resonance, Fizmatgiz, Moscow [Sec. 7].

Amatuni, A.Ts. (1956), Fiz. Metal. i Metalloved., 3:411 [Sec. 19].

Amatuni, A.Ts. (1957), Fiz. Metal. i Metalloved., 4:17 [Sec. 19].

Amatuni, A.Ts. (1958), Fiz. Metal. i Metalloved., 6:395 [Sec. 19].

Anderson, P.W. (1950a), Phys. Rev., 79:350 [Sec. 1].

Anderson, P.W. (1950b), Phys. Rev., 79:705 [Secs. 16,19].

Anderson, P.W. (1952), Phys. Rev., 84:694 [Sec. 19].

Anderson, P.W. (1959), Phys. Rev., 115:2 [Sec. 1].

Antiferromagnetism (1956) (Collection, ed. by S.V. Vonsovskii) [Russian translation], IL, Moscow [Secs. 1,16,23].

* The square brackets indicate the sections or appendices in which the reference is made.

Arai, T. (1962), Phys. Rev., 126:471 [Secs. 6,7].
Bacon, G.E. (1957), Diffraction of Neutrons [Russian translation], IL, Moscow [Sec. 40].
Bacon, G.E. (1963), Usp. Fiz. Nauk, 81(2):335 [Sec. 40].
Baker, G.A., and Gammel, J.L. (1961), J. Math. Anal. and Applications, 2:21 [Sec. 24].
Baker, G.A., Gammel, J.L., and Wills, J.G. (1961), J. Math. Anal. and Applications, 2:405 [Sec. 24].
Baker, G.A., Rushbrooke, G.S., and Gilbert, H.E. (1964), Phys. Rev., 135:A1272 [Sec. 24].
Bar'yakhtar, V.G., and Maleev, S.V. (1963), Fiz. Tverd. Tela, 6:1175 [Sec. 42].
Bar'yakhtar, V.G., and Peletminskii, S.V. (1960), Zh. Éksperim. i Teor. Fiz., 39:651 [Sec. 1].
Bar'yakhtar, V.G., Savchenko, M.A., and Shishkin, L.A. (1964), Fiz. Tverd. Tela, 6:1435 [Sec. 19].
Bar'yakhtar, V.G., and Shishkin, L.A. (1964), Fiz. Metal. i Metalloved., 17:664 [Sec. 42].
Baym, G., and Sessler, A. (1963), Phys. Rev., 131:2345 [Sec. 34].
Belov, K.P. (1951), Elastic, Thermal, and Electrical Phenomena in Ferromagnetic Metals, GITTL, Moscow-Leningrad [Secs. 1,2].
Belov, K.P. (1959), Magnetic Transitions, Fizmatgiz, Moscow [Secs. 1,2].
Belov, K.P., Levitin, R.Z., and Nikitin, S.A. (1964), Usp. Fiz. Nauk, 82:449 [Sec. 1].
Belov, K.P., and Zaitseva, M.A. (1958), Usp. Fiz. Nauk, 66:141 [Sec. 19].
Belov, N.V., Neporova, N.N., and Smirnova, T.S. (1956), Tr. Inst. Kristallogr. Akad. Nauk SSSR, 2:33 [Sec. 7].
Berdyshev, A.A., and Vonsovskii, S.V. (1954), Izv. Akad. Nauk SSSR, Ser. Fiz., 18:328 [Sec. 1].
Bertaut, F. (1960), Compt. Rend. Acad. Sci. Paris, 251:1733 [Sec. 7].
Bertaut, F. (1961a), J. Phys. Chem. Solids, 21:256 [Sec. 7].
Bertaut, F. (1961b), J. Phys. Chem. Solids, 21:295 [Sec. 7].
Bethe, H.A., and Salpeter, E.E. (1960), Quantum Mechanics of One- and Two-Electron Atoms [Russian translation] Fizmatgiz, Moscow [Sec. 7].

Bloch, C., and de Dominicis, C. (1958), Nucl. Phys., 7:459 [Secs. 10,24].
Bloch, F. (1929), Z. Physik, 57:545 [Sec. 1].
Bloch, F. (1930), Z. Physik, 61:206 [Secs. 12,15,33].
Bloch, F. (1932), Z. Physik, 74:295 [Secs. 12,15,33].
Bloch, F., and Gentile, G. (1931), Z. Physik, 70:395 [Sec. 7].
Blokhintsev, D.I. (1963), Fundamentals of Quantum Mechanics, Vysshaya Shkola, Moscow [Secs. 3,4].
Blombergen, N. (1950), Phys. Rev., 78:572 [Sec. 38].
Bogolyubov, N.N. (1946), Problems of Dynamical Theory in Statistical Physics, Gostekhizdat, Moscow [Sec. 25].
Bogolyubov, N.N. (1947), Izv. Akad. Nauk SSSR, Ser. Fiz., 9:77 [Sec. 10].
Bogolyubov, N.N. (1949), Lectures on Quantum Statistics [in Ukrainian], Radyanska Shkola, Kiev [Secs. 1,3,4,6,7,8,9, 12,13].
Bogolyubov, N.N. (1956), Private communication [Sec. 20].
Bogolyubov, N.N. (1958), Zh. Éksperim. i Teor. Fiz., 34:58 [Sec. 10].
Bogolyubov, N.N. (1961), Quasi-Averages in Statistical Mechanics Problems, Rotaprint Report OIYaI, D-781, Dubna [Secs. 11,26,28,40].
Bogolyubov, N.N., and Gurov, K.P. (1947), Zh. Éksperim. i Teor. Fiz., 17:614 [Sec. 9].
Bogolyubov, N.N., and Parasyuk, O.S. (1956), Dokl. Akad. Nauk SSSR, 109:717 [Sec. 26].
Bogolyubov, N.N., Tolmachev, V.V., and Shirkov, D.V. (1958), New Method in the Theory of Superconductivity, Izd. Akad. Nauk SSSR, Moscow [Sec. 10].
Bogolyubov, N.N., and Tyablikov, S.V. (1949a), Zh. Éksperim. i Teor. Fiz., 19:251 [Secs. 4,6,7].
Bogolyubov, N.N., and Tyablikov, S.V. (1949b), Vestn. Mosk. Univ., No. 3:35 [Sec. 6].
Bogolyubov, N.N., and Tyablikov, S.V. (1949c), Zh. Éksperim. i Teor. Fiz., 19:256 [Secs. 12,13,17].
Bogolyubov, N.N., and Tyablikov, S.V. (1957), Izv. Akad. Nauk SSSR, Ser. Fiz., 21:849 [Sec. 12].
Bogolyubov, N.N., and Tyablikov, S.V. (1959), Dokl. Akad. Nauk SSSR, 126:53 [Secs. 26,32].
Bohm, D., and Pines, D. (1951), Phys. Rev., 82:625 [Sec. 1].

Bohm, D., and Pines, D. (1953), Phys. Rev., 92:609 [Sec. 1].
Bolsterli, M. (1960), Phys. Rev. Letters, 4:82 [Sec. 26].
Bonch-Bruevich, V. L. (1956), Zh. Éksperim. i Teor. Fiz., 31:522 [Sec. 26].
Bonch-Bruevich, V. L. (1959), Dokl. Akad. Nauk SSSR, 129:529 [Secs. 25, 31].
Bonch-Bruevich, V. L. (1962), Dokl. Akad. Nauk SSSR, 147:1049 [Sec. 30].
Bonch-Bruevich, V. L., and Kogan, Sh. M. (1960), Ann. Phys. N.Y., 9:125 [Sec. 26].
Bonch-Bruevich, V. L., and Tyablikov, S. V. (1961), Green's Function Method in Statistical Mechanics, Fizmatgiz, Moscow [Sec. 25].
Born, M., and Goeppert-Mayer, M. (1938), Solid State Physics [Russian translation], ONTI, Moscow-Leningrad [A2].
Born, M., and Green, H. S. (1947), Proc. Roy. Soc. (London), A191 : 168 [Sec. 9].
Born, M., and Kun Huang (1958), Dynamical Theory of Crystal Lattices, Oxford University Press, Inc., New York, 1954 [Russian translation], IL, Moscow [A1].
Borovik-Romanov, A. S. (1962), Antiferromagnetism, Science Progress, Physico-Mathematical Sciences, No. 4, Izd. Akad. Nauk SSSR, Moscow [Secs. 1, 16, 19, 23].
Bozorth, R. M. (1956), Ferromagnetism, D. Van Nostrand Co., Inc., Princeton, New Jersey, 1951 [Russian translation], IL, Moscow [Sec. 1].
Brillouin, L., and Parodi, M. (1959), Propagation of Waves in Periodic Structures [Russian translation], IL, Moscow [A1].
Brockhouse, B. N., and Watanabe, H. (1962), Symposium on Inelastic Scattering of Neutrons in Solids and Liquids, Chalk River, Ontario, Sept. 10-14 [Sec. 36].
Bross, H. (1964a), Phys. Stat. Sol., 4:645 [Sec. 37].
Bross, H. (1964b), Phys. Stat. Sol., 4:661 [Sec. 37].
Brown, H. A. (1956), Phys. Rev., 104:624 [Sec. 24].
Brown, H. A., and Luttinger, J. M. (1955), Phys. Rev., 100:685 [Secs. 24, 35].
Brout, R., and Haken, H. (1960), Bull. Am. Phys. Soc., 5:148 [Sec. 32].
Cällen, G. (1952), Helv. Phys. Acta, 25:417 [Sec. 26].
Callen, H. B. (1963a), Phys. Letters, 4:161 [Sec. 42].

Callen, H. B. (1963b), Phys. Rev., 130:890 [Secs. 35, 42].
Callen, H. B., and Welton, T. A. (1951), Phys. Rev., 83:34 [Secs. 26, 31].
Callen, H. B., and Pitteli, E. (1960a), Phys. Rev., 119:5 [Sec. 42].
Callen, H. B., and Pitteli, E. (1960b), Phys. Rev., 119:1523 [Sec. 42].
Charap, S. H., and Boyd, E. L. (1964), Phys. Rev., 133:A811 [Sec. 15].
Chervonko, E. (1963), Rotaprint Report OIYaI, P-1215, Dubna [Sec. 42].
Clogstone, A. M., Suhl, H., Anderson, P. W., and Walker, L. R. (1956), J. Phys. Chem. Solids, 1:129 [Sec. 42].
Condon, E. U., and Shortley, G. H. (1949), The Theory of Atomic Spectra, Cambridge University Press, New York, 1935 [Russian translation], IL, Moscow [Sec. 1].
Cooke, A. H., Edmonds, D. T., Finn, C. B. P., and Wolf, W. P. (1959), Proc. Phys. Soc. (London), 74(6): 791 [Sec. 1].
Cooper, B. R., and Elliott, R. J. (1963), Phys. Rev., 131:1043 [Sec. 19].
Cooper, B. R., Elliott, R. J., Nettel, S. J., and Suhl, H. (1962), Phys. Rev., 127:57 [Secs. 18, 19].
Davis, H. L. (1962), J. Phys. Chem. Solids, 23:1348 [Sec. 16].
Dembinski, S. T. (1964), Physica, 30:1217 [Sec. 5].
Dimmock, J. O., and Wheeler, R. G. (1962), Phys. Rev., 127:391 [Sec. 7].
Dirac, P. A. M. (1929), Proc. Roy. Soc. (London), A123:714 [Sec. 6].
Dirac, P. A. M. (1960), The Principles of Quantum Mechanics [Russian translation], Fizmatgiz, Moscow [Secs. 6, 26].
Doman, B. G. S., and ter Haar, D. (1962), Phys. Letters, 2:15 [Sec. 42].
Domb, C., and Sykes, M. F. (1962), Phys. Rev., 128:168 [Sec. 24].
Dyson, F. (1956a), Phys. Rev., 102:1217 [Secs. 5, 14, 15, 24, 37].
Dyson, F. (1956b), Phys. Rev., 102:1230 [Secs. 5, 14, 15, 37].
Dzyaloshinskii, I. E. (1957), Zh. Éksperim. i Teor. Fiz., 32:1547 [Sec. 7].
Dzyaloshinskii, I. E. (1958), J. Phys. Chem. Solids, 4:241 [Sec. 7].
Dzyaloshinskii, I. E. (1964a), Zh. Éksperim. i Teor. Fiz., 46:1420 [Sec. 1].

Dzyaloshinskii, I.E. (1964b), Zh. Éksperim. i Teor. Fiz., 47:336 [Sec. 1].
Dzyaloshinskii, I.E. (1964c), Zh. Éksperim. i Teor. Fiz., 47:992 [Sec. 1].
Elliott, J. (1961), Phys. Rev., 124:346 [Sec. 1].
Elliott, J. (1962), J. Phys. Soc. Japan, 17, Suppl. Bl :'1 [Sec. 1].
Englert, F. (1960), Phys. Rev. Letters, 5:102 [Sec. 32].
Ferromagnetic Resonance (1952) (Collection, ed. by S. V. Vonsovskii) [Russian translation], IL, Moscow [Secs. 1, 23].
Ferromagnetic Resonance (1961) (Collection, ed. by S. V. Vonsovskii; in the series: Current Problems in Physics), Fizmatgiz, Moscow [Secs. 1,2,19,23,38].
Fock, V.A. (1930), Z. Physik, 61:126 [Secs. 12, 20].
Fock, V.A. (1932), Z. Physik, 75:622 [Sec. 3].
Fock, V.A. (1957), Papers on Quantum Field Theory [Russian translation], Leningrad State University Press [Sec. 3].
Freeman, A.J., Nesbet, R.K., and Watson, R.E. (1962), Phys. Rev., 125:1978 [Sec. 6].
Freeman, A.J., and Watson, R.E. (1960), Phys. Rev., 120:1439 [Sec. 6].
Frenkel', Ya.I. (1928), Z. Physik, 49:31 [Secs. 1,6].
Fukuchi, M. (1961a), Progr. Theoret. Phys. (Kyoto), 25:939 [Sec. 1].
Fukuchi, M. (1961b), Progr. Theoret. Phys. (Kyoto), 25:956 [Sec. 1].
Gammal, J., Marshall, W., and Morgan, L. (1963), Proc. Roy. Soc. (London), A275:257 [Sec. 24].
Gibbs, J.W. (1946), Fundamental Principles of Statistical Mechanics [Russian translation], Gostekhizdat, Moscow-Leningrad [Sec. 9].
Ginzburg, V.L., and Fain, V.M. (1960), Zh. Éksperim. i Teor. Fiz., 39:1323 [Sec. 32].
Goldstone, J. (1957), Proc. Roy. Soc. (London), A239:267 [Sec. 24].
Goodenough, J.B. (1960), Phys. Rev., 120:67 [Sec. 6].
Gor'kov, L.P. (1958), Zh. Éksperim. i Teor. Fiz., 34:735 [Sec. 26].
Gorter, E.W. (1955), Usp. Fiz. Nauk, 57(2,3) [Sec. 1].
Gossard, A.C., Jaccarino, W. and Remeika, J.P. (1961), Phys. Rev. Letters, 7:122 [Sec. 15].

Gurevich, A.G. (1962), Ferrites at Microwave Frequencies, [Sec. 1, 2, 19]. Consultants Bureau, New York.
Gurov, K.P. (1946), Dissertation for Candidate's Degree, presented at the Physics Faculty of the Moscow State University, Moscow [Sec. 9].
Gurov, K.P. (1947), Vestn. Mosk. Univ., No. 1:135 [Sec. 9].
Gusev, A.A. (1955), Zh.Éksperim. i Teor. Fiz., 29:181 [Sec. 19].
Gusev, A.A. (1959), Kristallografiya, 4:695 [Secs. 22,23].
Gusev, A.A. (1960), Kristallografiya, 5:420 [Sec. 23].
Gusev, A.A., and Pakhomov, A.S. (1961), Izv.Akad.Nauk SSSR, Ser. Fiz., No. 11:1327 [Secs. 19, 23].
Haas, C.W. (1963), Phys.Rev., 132:228 [Sec. 42].
Haas, C.W., and Jarrett, H.S. (1964), Phys.Rev., 135:A1089 [Sec. 42].
Halpern, O., and Johnson, M.H. (1939), Phys.Rev., 55:898 [Sec. 40].
Hatherly, M., Hirakawa, K., Lawde, R., Mallett, J.F., and Stringfellow, M.W. (1963), Abstracts of papers presented at the Symposium on Ferromagnetism and Ferroelectricity held in Leningrad, May 30-June 5, 1963, Izd. Akad. NaukSSSR [Sec. 36].
Haubenreiser, W. (1963), Phys. Letters, 6:43 [Sec. 42].
Heber, G. (1954), Z. Naturforsch., 9a:91 [Sec. 32].
Heisenberg, W. (1928), Z. Physik, 49:619 [Secs. 1, 6].
Heisenberg, W. (1930), Metallwirtschaft, 9:843 [Sec. 7].
Heitler, W., and London, H. (1927), Z. Physik, 44:455 [Sec. 6].
Herpin, A. (1962), J. Phys. Radium, 23:453 [Sec. 1].
Herring, C., and Kittel, C. (1951), Phys.Rev., 81:869 [Sec. 1].
Herring, C., and Kittel, C. (1952a), Phys. Rev., 85:1003 [Sec. 1].
Herring, C., and Kittel, C. (1952b), Phys.Rev., 87:60 [Sec. 1].
Hewson, A.C., and ter Haar, D. (1963a), Preprint No. 121/63, Clarendon Lab., Oxford [Secs. 35, 42].
Hewson, A.C., and ter Haar, D. (1963b), Preprint, Clarendon Lab., Oxford [Sec. 42].
Hewson, A.C., and ter Haar, D. (1963c), Phys. Letters, 6:136 [Sec. 37].

Hill, T. L. (1960), Statistical Mechanics, McGraw-Hill Book Company, New York, 1956 [Russian translation], IL, Moscow [Sec. 8].
Holstein, T., and Primakoff, H. (1940), Phys. Rev., 58:1098 [Secs. 5, 7].
Hugenholtz, N.M. (1957a), Physica, 23:481 [Sec. 24].
Hugenholtz, N.M. (1957b), Physica, 23:533 [Sec. 24].
Hulten, L. (1936), Proc. Roy. Acad. Sci. Amsterdam, 39:190 [Secs. 12,16,17].
Husimi, K. (1940), Proc. Phys.-Math. Soc. Japan, 22:264 [Sec. 9].
Indenbom, V. L. (1960), Kristallografiya, 5:513 [Sec. 7].
Irkhin, Yu. P., and Turov, E.A. (1957), Fiz. Metal. i Metalloved., 4:9 [Sec. 1].
Izyumov, Yu.A. (1959), Dokl. Akad. Nauk SSSR, 125:1227 [Secs. 5, 14].
Izyumov, Yu.A. (1963a), Usp. Fiz. Nauk, 80:41 [Secs. 36, 40, 41, 42].
Izyumov, Yu.A. (1963b), Fiz. Tverd. Tela, 5:717 [Sec. 42].
Izyumov, Yu.A., and Yakovlev, E.N. (1960), Fiz. Metal. i Metalloved., 9:667 [Sec. 41].
Kaplan, T.A. (1958), Phys. Rev., 109:782 [Sec. 19].
Kaplan, T.A. (1959), Phys. Rev., 116:888 [Sec. 7].
Kaplan, T.A. (1960), Phys. Rev., 119:1460 [Sec. 15].
Kaplan, T. A. (1961), Phys. Rev., 124:329 [Secs. 1, 18, 19].
Kaplan, T.A., and Lyons, D.H. (1960), Phys. Rev., 120:1580 [Sec. 7].
Kaplan, T.A., and Lyons, D.H. (1963), Phys. Rev., 129:2073 [Sec. 1].
Kashcheev, V.N. (1962a), Fiz. Tverd. Tela, 4:759 [Sec. 42].
Kashcheev, V.N. (1962b), Fiz. Tverd. Tela, 4:1432 [Sec. 42].
Kashcheev, V.N. (1962c), Fiz. Tverd. Tela, 5:909 [Sec. 42].
Kashcheev, V.N. (1964a), Acta Phys. Polon., 25:337 [Sec. 42].
Kashcheev, V.N. (1964b), Acta Phys. Polon., 25:349 [Sec. 42].
Kashcheev, V.N., and Krivoglaz, M.A. (1961), Fiz. Tverd. Tela, 3:1541 [Sec. 42].
Kasuya, T. (1956), Progr. Theoret. Phys. (Kyoto), 16:45 [Sec. 1].
Kasuya, T. (1958), Progr. Theoret. Phys. (Kyoto), 20:980 [Sec. 1].
Kawasaki, K., and Mori, H. (1961), Progr. Theoret. Phys. (Kyoto), 25:1045 [Sec. 42].

Keffer, F., and Oguchi, T. (1959), Phys. Rev., 115:1428 [Sec.1].
Khachaturyan, A.G. (1963), Fiz. Tverd. Tela, 5:2178 [Sec. 42].
Kittel, C. (1947), Phys. Rev., 71:270 [Sec. 39].
Kittel, C. (1948), Phys. Rev., 73:155 [Sec. 39].
Klein, A., and Prange, R. (1958), Phys. Rev., 112:994 [Sec. 29].
Kogan, Sh. M. (1959), Dokl. Akad. Nauk SSSR, 126:546 [Sec. 26].
Kondorskii, E.I., and Pakhomov, A.S. (1953), Dokl. Akad. Nauk SSSR, 93:431 [Sec. 19].
Kondorskii, E.I., Pakhomov, A.S., and Siklos, T. (1956), Dokl. Akad. Nauk SSSR, 109:931 [Sec. 19].
Kramers, H.A. (1934), Physica, 1:182 [Sec. 1].
Kubo, R. (1953), Rev. Mod. Phys., 25:344 [Sec. 19].
Kubo, R. (1957), J. Phys. Soc. Japan, 12:570 [Secs. 28, 31, 38].
Kubo, R. (1962), J. Phys. Soc. Japan, 17:1100 [Sec. 42].
Kubo, R., and Tomita, K. (1954), J. Phys. Soc. Japan, 9:888 [Secs. 31, 38].
Kubo, R., Yokota, M., and Nakajima, S. (1957), J. Phys. Soc. Japan, 12:1203 [Sec. 31].
Kvasnikov, I.A. (1957), Dokl. Akad. Nauk SSSR, 113:544 [Sec. 23].
Landau, L.D. (1927), Z. Physik, 45:430 [Sec. 8].
Landau, L.D. (1934), Sow. Phys., 4:675 [Sec. 16].
Landau, L.D. (1958), Zh. Éksperim. i Teor. Fiz., 34:262 [Sec. 26].
Landau, L.D., and Lifshits, E.M. (1935a), Sow. Phys., 8:153 [Sec. 1].
Landau, L.D., and Lifshits, E.M. (1935b), Sow. Phys., 8:175 [Sec. 38].
Landau, L.D., and Lifshits, E.M. (1959), Electrodynamics of Continuous Media, Fizmatgiz, Moscow [Secs. 1, 2, 21, 24].
Landau, L.D., and Lifshits, E.M. (1963), Quantum Mechanics, Pt. I, Fizmatgiz, Moscow-Leningrad [Secs. 1, 3, 4].
Lavrent'ev, M.A., and Shabat, B.V. (1958), Methods in the Theory of Functions of the Complex Variable, Gostekhizdat, Moscow [Sec. 27].
Lax, M. (1955), Phys. Rev., 97:629 [Sec. 33].
Lederman, W. (1944), Proc. Roy. Soc. (London), A182:362 [A2].
Lehmann, H. (1954), Nuovo Cimento, 11:342 [Sec. 26].
Li Yin-Yuan (1950), Phys. Rev., 80:457 [Sec. 16].
Lines, M.E. (1963), Proc. Roy. Soc. (London), A271:105 [Sec. 19].

Lines, M.E. (1964a), Phys.Rev., 133:A841 [Sec. 42].

Lines, M.E. (1964b), Phys.Rev., 135:A1337 [Sec. 42].

Löwdin, P.O. (1962), Rev.Mod.Phys., 34:80 [Sec. 6].

Luttinger, J.M. (1951), Phys.Rev., 81:1015 [Sec. 16].

Luttinger, J.M. (1961a), Lectures presented in the Physics Faculty of the Moscow State University in 1961 [Sec. 10].

Luttinger, J.M. (1961b), Phys.Rev., 121:942 [Sec. 30].

Lyons, D.H. (1963), Phys.Rev., 132:122 [Sec. 19].

McCollum, D.C., Jr., and Callaway, J. (1962), Phys.Rev. Letters, 9:376 [Sec. 15].

Magnetic Structure of Ferromagnets (1959) (Collection, ed. by S.V. Vonsovskii) [Russian translation], IL, Moscow [Sec. 1].

Maleev, S.V. (1957), Zh.Éksperim. i Teor.Fiz., 33:1010 [Sec. 5].

Maleev, S.V. (1961), Zh.Éksperim. i Teor. Fiz., 41:1675 [Sec. 30].

Maleev, S.V., Bar'yakhtar, V.G., and Suris, R.A. (1962), Fiz. Tverd.Tela, 4:3461 [Sec. 42].

Marshall, W. (1955a), Proc.Roy.Soc.(London), A232:48 [Sec. 19].

Marshall, W. (1955b), Proc.Roy.Soc.(London), A232:69 [Sec. 19].

Martin, P.C., and Schwinger, J. (1958), Bull.Am.Phys.Soc., 3:202 [Sec. 26].

Matsubara, T. (1954), J.Fac.Sci.Hokkaido Univ., Ser. II, 4:292 [Sec. 1].

Matsubara, T. (1955), Progr.Theoret.Phys.(Kyoto), 14:351 [Sec. 25].

Meng Hsien-chen (1961), Izv.Akad.Nauk SSSR, Ser.Fiz., No. 11:1353 [Sec. 42].

Meng Hsien-chen (1963), Fiz.Tverd.Tela, 5:1988 [Sec. 42].

Meng Hsien-chen and P'u Fu-ch'o (1961), Acta Phys. Sinica, 17:214 [Sec. 42].

Meyer, H., and Harris, A.B. (1962), J.Appl.Phys., 31:49S [Sec. 19].

Mills, R.E., Kenan, R.P., and Milford, F.J. (1964), Phys. Letters, 12:173 [Sec. 42].

Mitchell, A.H. (1957), Phys.Rev., 105:1439 [Sec. 1].

Miwa, H. (1963), Progr.Theoret.Phys.(Kyoto), 29:477 [Sec.19].

Mori, H. (1956), J.Phys.Soc.Japan, 11:1029 [Sec. 31].

Moria, T. (1960), Phys. Rev., 120:91 [Sec. 7].
Morkowski, J. (1963), Acta Physica Polon., 23:469 [Sec. 42].
Mott, N. F. (1935), Proc. Phys. Soc. (London), 47:571 [Sec. 1].
Nagamiya, T. (1962), J. Appl. Phys., 33:1029 [Sec. 19].
Nagamiya, T., Nagato, K., and Kitano, Y. (1962a), Progr. Theoret. Phys. (Kyoto), 27:1253 [Secs. 1, 19].
Nagamiya, T., Nagato, K., and Kitano, Y. (1962b), J. Phys. Soc. Japan, 17, Suppl. Bl:10 [Secs. 1, 19].
Nagamiya, T., Yosida, K., and Kubo, R. (1955), Advan. Phys., 4(13) [Sec. 1, 16].
Naish, V. E. (1963), Izv. Akad. Nauk SSSR, Ser. Fiz., 27:1496 [Sec. 1].
Nakamura, T. (1952), Progr. Theoret. Phys. (Kyoto), 7:539 [Sec. 19].
Néel, L. (1932), Ann. Phys. (Paris), 17:61 [Secs. 1, 16, 23].
Néel, L. (1936), Ann, Phys. (Paris), 5:232 [Secs. 1, 16, 23].
Néel, L. (1948), Ann, Phys. (Paris), 3:137 [Secs. 1, 16, 23].
Oguchi, T. (1957), Progr. Theoret. Phys. (Kyoto), 17:659 [Sec. 7].
Oguchi, T. (1960), Phys. Rev., 117:117 [Sec. 14].
Oguchi, T. (1961), Progr. Theoret. Phys. (Kyoto), 25:721 [Sec. 5].
Opechowski, W. (1937), Physica, 4:181 [Sec. 24].
Opechowski, W. (1960), Physica, 25:476 [Sec. 37].
Pakhomov, A. S., and Smol'kov, N. A. (1962), Ferrites, Science Progress, Physico-Mathematical Sciences, No. 4, Izd. Akad. Nauk SSSR, Moscow [Secs. 1, 19, 23].
Pal, L. (1954), Acta Phys. Hung., 3:287 [Sec. 19].
Pauthenet, R. (1952), Ann. Phys. (Paris), 7:710 [Sec. 36].
Peierls, R. E. (1938), Phys. Rev., 54:918 [Sec. 20].
Peierls, R. E. (1954), Proc. Nat. Inst. Sci. India, A20:121 [A2].
Physics of Ferromagnetic Domains (1951) (Collection, ed. by S. V. Vonsovskii) [Russian translation], IL, Moscow [Sec. 1].
Pike, E. R. (1964), Proc. Phys. Soc. (London), 84:83 [Sec. 26].
Pines, D. (1953), Phys. Rev., 92:626 [Sec. 1].
Pines, D., and Bohm, D. (1952), Phys. Rev., 85:338 [Sec. 1].
Potapkov, N. A. (1957), Dokl. Akad. Nauk SSSR, 117:965 [Sec. 19].
Potapkov, N. A. (1958), Dokl. Akad. Nauk SSSR, 118:269 [Sec. 19].
Potapkov, N. A. (1962a), Dokl. Akad. Nauk SSSR, 144:297 [Secs. 7, 42].
Potapkov, N. A. (1962b), Fiz. Tverd. Tela, 4:1803 [Sec. 42].
Potapkov, N. A. (1963a), Dokl. Akad. Nauk SSSR, 151:543 [Sec. 42].

Potapkov, N.A. (1963b), Izv.Akad.Nauk SSSR, Ser. Fiz., 28:495 [Sec. 42].
Potapkov, N.A., and Tyablikov, S.V. (1960), Fiz. Tverd. Tela, 2:2733 [Secs. 1, 42].
Potter, H.H. (1934), Proc.Roy.Soc. (London), A146:362 [Sec.36].
Powell, F.C. (1930), Proc.Roy.Soc. (London), A130:176 [Sec.7].
Praveczki, E. (1961), Fiz.Metal. i Metalloved., 12:296 [Sec.5].
Praveczki, E. (1963), Phys. Letters, 6:147 [Secs. 35, 37].
P'u Fu-ch'o (1960a), Dokl.Akad.Nauk SSSR, 130:1244 [Sec. 42].
P'u Fu-ch'o (1960b), Dokl.Akad.Nauk SSSR, 131:546 [Sec. 42].
P'u Fu-ch'o (1961), Fiz. Tverd. Tela, 3:476 [Sec. 42].
Puzei, I.M. (1957), Izv.Akad.Nauk SSSR, Ser. Fiz., 21:1088 [Sec.19].
Puzei, I.M. (1963), Izv.Akad.Nauk SSSR, Ser. Fiz., 27:1469 [Sec.19].
Rode, V.E., and Vedyaev, A.V. (1963), Zh.Éksperim. i Teor. Fiz., 45:415 [Sec. 1].
Rozing, B.L. (1892), Zh.Russ.Fiz.-Khim.Obshchest. (Chast' Fizich.), 24:105 [Sec. 1].
Rozing, B.L. (1896), Zh.Russ.Fiz.-Khim.Obshchest. (Chast' Fizich.), 28:59 [Sec. 1].
Rozing, B.L. (1910), Zh.Russ.Fiz.-Khim.Obshchest. (Chast' Fizich.), 42:71 [Sec. 1].
Rudoi, Yu.G. (1963a), Fiz. Tverd. Tela, 5:534 [Sec. 19].
Rudoi, Yu.G. (1963b), High-Temperature Expansions in the Theory of Ferromagnetism, Rotaprint Report of MIAN SSSR [Sec. 24].
Rushbrooke, G.S., and Wood, P.J. (1958), Mol. Phys., 1:257 [Sec. 24].
Ryzhik, I.M., and Gradshtein, I.S. (1951), Tables of Integrals, Sums, Series, and Products, Gostekhizdat, Moscow-Leningrad [Sec. 33].
Saénz, A.W. (1962), Phys.Rev., 125:1940 [Sec. 15].
Savchenko, M.A., and Bar'yakhtar, V.G. (1963), Fiz. Tverd. Tela, 5:2747 [Sec. 19].
Schafroth, M.R. (1954), Proc. Phys. Soc. (London), A67:33 [Sec. 32].
Scott, G.G. (1962), Rev. Mod. Phys., 34:102 [Sec. 1].
Seidov, Yu.M. (1963), Private communication [Sec. 7].
Shimizu, M. (1952), Progr. Theoret. Phys. (Kyoto), 8:416 [Sec.1].
Shimizu, M. (1960), J. Phys. Soc. Japan, 15:376 [Sec. 1].

Siklos, T. (1957), Acta Phys.Hung., 7:141 [Sec. 19].
Skrotskii, G.V., and Kurbatov, L.V. (1958), Zh.Éksperim. i Teor.Fiz., 35:216 [Sec. 42].
Slater, J.C. (1936), Phys.Rev., 49:537, 931 [Sec. 1].
Slater, J.C. (1953), Rev.Mod. Phys., 25:199 [Sec. 6].
Smart, J.S. (1952), Phys.Rev., 86:968 [Sec. 16].
Smart, J.S. (1953), Phys.Rev., 90:55 [Sec. 16].
Smit, J., and Wijn, H.P.J. (1962), Ferrites, John Wiley and Sons, Inc., New York, 1959 [Russian translation], IL, Moscow [Secs. 1, 19, 23].
Stinchcombe, R.B., Horwitz, G., Englert, F., and Brout, R. (1962), Thermodynamic Behavior of the Heisenberg Ferromagnet, Preprint, Cornell University [Sec. 24].
Stoner, E.C. (1936), Proc.Roy.Soc.(London), A154:656 [Sec. 1].
Stoner, E.C. (1938), Proc.Roy.Soc.(London), A165:372 [Sec. 1].
Stoner, E.C. (1939), Proc.Roy.Soc.(London), A169:339 [Sec. 1].
Stoner, E.C. (1948), Rept.Progr.Phys., 11:43 [Sec. 1].
Stuart, R., and Marshall, W. (1960), Phys.Rev., 120:353 [Sec. 6].
Szaniecki, J. (1962), Acta Physica Polon., 21:481 [Sec. 37].
Tahir-Kheli, R.A. (1963), Phys.Rev., 132:689 [Sec. 42].
Tahir-Kheli, R.A., and Callen, H.B. (1964), Phys. Rev., 135: A679 [Sec. 42].
Tahir-Kheli, R.A., Doman, B.G.S., and ter Haar, D. (1963), Phys. Letters, 4:5 [Sec. 42].
Tahir-Kheli, R.A., and Jarrett, H.S. (1964), Phys.Rev., 135 A1096 [Sec. 42].
Tahir-Kheli, R.A., and ter Haar, D. (1962a), Phys.Rev., 127: 88 [Secs. 35, 42].
Tahir-Kheli, R.A., and ter Haar, D. (1962b), Phys.Rev., 127: 95 [Sec. 37].
Tahir-Kheli, R.A., and ter Haar, D. (1963), Phys.Rev., 130: 108 [Secs. 1, 42].
Tavger, B.A., and Zaitsev, V.M. (1956), Zh.Éksperim. i Teor. Fiz., 30:564 [Sec. 7].
ter Haar, D., and Lines, M.E. (1962), Phil. Trans. Roy. Soc. London, A255:1 [Sec. 19].
ter Haar, D., and Parry, W.E. (1962), Phys.Letters, 1 : 145 [Sec. 26].
Tessman, J. (1952), Phys.Rev., 88:1132 [Sec. 19].

Theory of the Ferromagnetism of Metals and Alloys (1963) (Collection, ed. by S. V. Vonsovskii) [Russian translation], IL, Moscow [Sec. 1].
Tomita, K., and Tanaka, M. (1963a), Progr. Theoret. Phys. (Kyoto), 29:528 [Sec. 42].
Tomita, K., and Tanaka, M. (1963b), Progr. Theoret. Phys. (Kyoto), 29:651 [Sec. 42].
Turov, E.A. (1953), Zh. Éksperim. i Teor. Fiz., 25:352 [Sec. 1].
Turov, E.A. (1954), Dokl. Akad. Nauk SSSR, 98:945 [Sec. 1].
Turov, E.A. (1955a), Izv. Akad. Nauk SSSR, Ser. Fiz., 19:462 [Sec. 1].
Turov, E.A. (1955b), Izv. Akad. Nauk SSSR, Ser. Fiz., 19:474 [Sec. 1].
Turov, E.A. (1957), Fiz. Metal. i Metalloved., 4:183 [Sec. 1].
Turov, E.A. (1958), Fiz. Metal. i Metalloved., 6:203 [Sec. 1].
Turov, E.A. (1963), Physical Properties of Magnetically Ordered Crystals, Izd. Akad. Nauk SSSR, Moscow [Secs. 1, 2, 7]. English translation published by Academic Press, Inc., New York, 1965.
Turov, E.A., and Irkhin, Yu. P. (1958), Izv. Akad. Nauk SSSR, Ser. Fix., 22:1168 [Secs. 17, 19].
Turov, E.A., and Shavrov, V.G. (1958), Tr. Inst. Fiz. Metal., No. 20:101, Sverdlovsk [Sec. 1].
Turov, E.A., and Vonsovskii, S.V. (1953), Zh. Éksperim. i Teor. Fiz., 24:501 [Sec. 1].
Tyablikov, S.V. (1947), Dissertation for Candidate's Degree, Physics Faculty, Moscow State University [Sec. 13].
Tyablikov, S.V. (1950), Zh. Eksperim. i Teor. Fiz., 20:661 [Sec. 19].
Tyablikov, S.V. (1956a), Fiz. Metal. i Metalloved., 2:193 [Secs. 16, 17].
Tyablikov, S.V. (1956b), Fiz. Metal. i Metalloved., 3:3 [Sec. 19].
Tyablikov, S.V. (1959a), Fiz. Metal. i Metalloved., 8:152 [Sec. 19].
Tyablikov, S.V. (1959b), Ukr. Mat. Zh., 11:287 [Secs. 32, 33].
Tyablikov, S.V. (1960a), Fiz. Tverd. Tela, 2:361 [Secs. 38, 39].
Tyablikov, S.V. (1960b), Fiz. Tverd. Tela, 2:2009 [Sec. 38].
Tyablikov, S.V. (1963a), Dokl. Akad. Nauk SSSR, 149:573 [Sec. 37].
Tyablikov, S.V. (1963b), Fiz. Metal. i Metalloved., 15:641 [Sec. 37].

Tyablikov, S. V. (1963c), Fiz. Metal. i Metalloved., 15:801 [Sec. 37].
Tyablikov, S. V. (1963d), Fiz. Metal. i Metalloved., 16:321 [Sec. 35].
Tyablikov, S. V. (1964), Fiz. Metal. i Metalloved., 17:283 [Sec. 37].
Tyablikov, S. V., and Amatuni, A. Ts. (1956), Dokl. Akad. Nauk SSSR, 108:69 [Sec. 19].
Tyablikov, S. V., and Bonch-Bruevich, V. L. (1962), Perturbation Theory for Double-Time Green's Temperature-Dependent Functions, Rotaprint Report of MIAN SSSR, T-7 [Sec. 29].
Tyablikov, S. V., and Gusev, A. A. (1956), Fiz. Metal. i Metalloved., 2:385 [Sec. 19].
Tyablikov, S. V., and Moskalenko, V. A. (1964), Dokl. Akad. Nauk SSSR, 158:839 [Sec. 10].
Tyablikov, S. V., and P'u Fu-ch'o (1961), Fiz. Tverd. Tela, 3:142 [Secs. 25, 31].
Tyablikov, S. V., and Siklos, T. (1959), Acta Phys. Hung., 10:259 [Sec. 19].
Tyablikov, S. V., and Yakovlev, E. N. (1962), Dokl. Akad. Nauk SSSR, 144:303 [Sec. 37].
Tyablikov, S. V., and Yakovlev, E. N. (1963), Fiz. Tverd. Tela, 5:137 [Sec. 37].
van den Handel, J., Gijsman, H. M., and Poulis, H. J. (1952), Physica, 18:862 [Sec. 19].
Van Hove, L. (1954a), Phys. Rev., 95:249 [Sec. 40].
Van Hove, L. (1954b), Phys. Rev., 95:1374 [Sec. 40].
Van Hove, L. (1955), Physica, 21:901 [Sec. 24].
Van Hove, L. (1956), Physica, 22:343 [Sec. 24].
Van Kranendonk, J., and Van Vleck, J. H. (1958), Rev. Mod. Phys., 30:1 [Secs. 15, 19].
Van Vleck, J. H. (1937), Phys. Rev., 52:1178 [Sec. 7].
Van Vleck, J. H. (1941), J. Chem. Phys., 9:85 [Secs. 16, 23].
Van Vleck, J. H. (1947), Ann. Inst. Henri Poincaré, 10:57 [Sec. 7].
Van Vleck, J. H. (1951), J. Phys. Radium, 12:262 [Secs. 1, 16, 23].
Villain, T. (1959), J. Phys. Chem. Solids, 11:303 [Secs. 1, 7].
Vlasov, K. B., and Ishmukhametov, B. Kh. (1954), Zh. Éksperim. i Teor. Fiz., 27:75 [Sec. 23].

von Neumann, J. (1927a), Göttingen Nachr., No. 3:245 [Sec. 8].
von Neumann, J. (1927b), Göttingen Nachr., No. 3:273 [Sec. 8].
von Neumann, J. (1932), Mathematische Grundlagen der Quantenmechanik, J. Springer, Berlin [Sec. 8].
Vonsovskii, S.V. (1940a), J. Phys. (USSR), 3:83 [Sec. 7].
Vonsovskii, S.V. (1940b), Zh. Éksperim. i Teor. Fiz., 10:762 [Sec. 19].
Vonsovskii, S.V. (1946), Zh. Éksperim. i Teor. Fiz., 16:981 [Sec. 1].
Vonsovskii, S.V. (1952), Modern Science of Magnetism, Gostekhizdat, Moscow-Leningrad [Sec. 1, 2].
Vonsovskii, S.V., and Izyumov, Yu.A. (1960), Fiz. Metal. i Metalloved., 10:321 [Secs. 1, 42].
Vonsovskii, S.V., and Izyumov, Yu.A. (1962a), Usp. Fiz. Nauk, 77:379 [Sec. 1].
Vonsovskii, S.V., and Izyumov, Yu.A. (1962b), Usp. Fiz. Nauk, 78:1 [Secs. 1, 42].
Vonsovskii, S.V., and Kobelev, L.Ya. (1961), Fiz. Metal. i Metalloved., 11:820 [Sec. 1].
Vonsovskii, S.V., and Seidov, Yu.M. (1954), Izv. Akad. Nauk SSSR, Ser. Fiz., 18:319 [Sec. 19].
Vonsovskii, S.V., and Seidov, Yu.M. (1956), Dokl. Akad. Nauk SSSR, 107:37 [Sec. 1].
Vonsovskii, S.V., and Shur, Ya.S. (1948), Ferromagnetism, Gostekhizdat, Moscow-Leningrad [Secs. 1, 2, 21].
Vonsovskii, S.V., and Turov, E.A. (1953), Zh. Éksperim. i Teor. Fiz., 24:419 [Secs. 1, 7].
Wallace, D.C. (1962), Phys. Rev., 128:1614 [Sec. 15].
Weiss, P. (1907), J. Phys. Radium, 6:661 [Sec. 1].
Wohlfarth, E.P. (1949), Nature, 163:57 [Sec. 6].
Wojtowicz, P., and Joseph, R.I. (1964), Phys. Rev., 135:A1314 [Sec. 24].
Wortis, M. (1963), Ph.D. Thesis, Harvard University, Cambridge, Massachusetts [Secs. 34, 37].
Yafet, Y., and Kittel, C. (1952), Phys. Rev., 87:290 [Secs. 19, 23].
Yakovlev, E.N. (1957), Dokl. Akad. Nauk SSSR, 115:699 [Sec. 19].
Yakovlev, E.N. (1958), Fiz. Metal. i Metalloved., 6:976 [Sec. 19].
Yamashita, J. (1954), J. Phys. Soc. Japan, 9:339 [Sec. 1].

Yoshimori, A. (1959), J. Phys. Soc. Japan, 14:807 [Secs. 1, 7].
Yosida, K. (1957), Phys. Rev., 106:893 [Sec. 1].
Yosida, K., and Miwa, H. (1961a), J. Appl. Phys., 32:8S [Sec. 1].
Yosida, K., and Miwa, H. (1961b), Progr. Theoret. Phys. (Kyoto), 26:693 [Sec. 1].
Zener, C. (1951), Phys. Rev., 81:440 [Sec. 6].
Ziman, J. M. (1952a), Proc. Phys. Soc. (London), 65:540 [Sec. 19].
Ziman, J. M. (1952b), Proc. Phys. Soc. (London), 65:548 [Sec. 19].
Zubarev, D. N. (1960), Usp. Fiz. Nauk, 71:71 [Secs. 25, 31].
Zubarev, D. N. (1961a), Dokl. Akad. Nauk SSSR, 140:92 [Sec. 31].
Zubarev, D. N. (1961b), Private communication [Sec. 34].

Index

A

B

transition metals
electron configurations 3-4
electronic specific heats 20
typical ferromagnets 2